Cities Research Series

Series Editor

Paul Burton, Gold Coast campus, Cities Research Institute, Griffith University, Southport, QLD, Australia

This book series brings together researchers, planning professionals and policy makers in the area of cities and urban development and publishes recent advances in the field. It addresses contemporary urban issues to understand and meet urban challenges and make (future) cities more sustainable and better places to live. The series covers, but is not limited to the following topics:

- Transport policy and behaviour
- Architecture, architectural science and construction engineering
- Urban planning, urban design and housing
- Infrastructure planning and management
- Complex systems and cities
- Urban and regional governance
- Smart and digital technologies

More information about this series at https://link.springer.com/bookseries/16474

Dominic Ek Leong Ong · Marco Barla ·
Jason Wen-Chieh Cheng · Chung Siung Choo ·
Minmin Sun · Mohammud Irfaan Peerun

Sustainable Pipe Jacking Technology in the Urban Environment

Recent Advances and Innovations

≙ Springer

Dominic Ek Leong Ong
School of Engineering and Built
Environment & Cities Research Institute
Griffith University
Nathan, Queensland, Australia

Jason Wen-Chieh Cheng
School of Civil Engineering
Xi'an University of Architecture
and Technology
Xi'an, China

Minmin Sun
W&M Consultants
Singapore, Singapore

Marco Barla
Department of Structural, Geotechnical
and Building Engineering
Politecnico di Torino
Turin, Italy

Chung Siung Choo
Faculty of Engineering, Computing
and Science
Swinburne University of Technology
Sarawak Campus
Kuching, Sarawak, Malaysia

Mohammud Irfaan Peerun
School of Engineering and Built
Environment & Cities Research Institute
Griffith University
Nathan, Queensland, Australia

ISSN 2662-4842   ISSN 2662-4850 (electronic)
Cities Research Series
ISBN 978-981-16-9374-8   ISBN 978-981-16-9372-4 (eBook)
https://doi.org/10.1007/978-981-16-9372-4

© The Editor(s) (if applicable) and The Author(s), under exclusive license to Springer Nature Singapore Pte Ltd. 2022
This work is subject to copyright. All rights are solely and exclusively licensed by the Publisher, whether the whole or part of the material is concerned, specifically the rights of translation, reprinting, reuse of illustrations, recitation, broadcasting, reproduction on microfilms or in any other physical way, and transmission or information storage and retrieval, electronic adaptation, computer software, or by similar or dissimilar methodology now known or hereafter developed.
The use of general descriptive names, registered names, trademarks, service marks, etc. in this publication does not imply, even in the absence of a specific statement, that such names are exempt from the relevant protective laws and regulations and therefore free for general use.
The publisher, the authors and the editors are safe to assume that the advice and information in this book are believed to be true and accurate at the date of publication. Neither the publisher nor the authors or the editors give a warranty, expressed or implied, with respect to the material contained herein or for any errors or omissions that may have been made. The publisher remains neutral with regard to jurisdictional claims in published maps and institutional affiliations.

This Springer imprint is published by the registered company Springer Nature Singapore Pte Ltd.
The registered company address is: 152 Beach Road, #21-01/04 Gateway East, Singapore 189721, Singapore

To our family members and friends for their eternal trust, support and love

Foreword by Paul Burton

I am delighted to introduce the third contribution to the Cities Research Series supported and published by Springer Nature. Reflecting the commitment of the Cities Research Institute at Griffith University to rigorous scholarship allied to practical relevance, this collaborative work brings together leading-edge research with an appreciation of the challenges faced by contemporary engineers. As an urban planner from beyond the engineering profession, my initial reaction as Series Editor, on being presented with the proposal for this book, was to think it is somewhat specialized for our intended audience. But a moment's reflection made me realize that its focus on the literal foundations of our towns and cities made it of fundamental importance. While planners like me might identify the areas we deem suitable for certain types of development and my architect colleagues focus on the form and function of what is built in these places, it is the engineers who ensure they are structurally sound. This book presents state-of-the-art research on all aspects of pipe jacking and microtunneling operations, including new applications of artificial intelligence and rigorous articulations of complex soil–pipe interaction in the challenging geologies found in many existing urban areas.

I hope that, like me, readers will appreciate the contemporary application of Karl Terzaghi's conception of good engineering as a marriage of art and science, where imaginative possibilities are subjected to rigorous empirical field testing. This way of thinking that respects disciplinary boundaries but does not let them stand in the way of trying to solve complex urban problems is, I hope, the hallmark of this series, and I am very pleased that this latest book upholds and applies this principle.

If you would like to propose your own contribution to the series, please do not hesitate in contacting me.

Professor Paul Burton, Ph.D
Series Editor, Cities Research Institute
Griffith University
Nathan, Queensland, Australia
p.burton@griffith.edu.au

Foreword by Raymond Sterling

It is a great pleasure to be asked to write the foreword for this new book on advances in pipe jacking technology. It is a welcome addition to the existing literature on this specialized but versatile form of underground construction.

With the collaboration of six different authors from five different countries on three continents, the book brings an international perspective to the discussion of theory and practice of microtunneling and pipe jacking. Also, the disciplinary backgrounds of the authors combine with their research and practice specialties to highlight a number of specific current and future areas of advancement.

This book builds on earlier books by Craig (1984), Thomson (1993), and Stein (2005) describing pipe jacking principles and practice. It also provides up-to-date information on selected research advances applied to pipe jacking and microtunneling problems—using the specific background and experience of the authors to show how research and geotechnical understanding can be applied to solve construction challenges. Such challenges include appropriate site investigation practices and laboratory testing, deep shaft construction and using artificial intelligence techniques to improve jacking force predictions and performance issues such as cutterhead clogging. The important area of estimating required pipe jacking thrusts receives significant attention, and the book includes a variety of case study discussions that highlight specific problems that may be encountered in pipe jacking projects. Two chapters of the book are devoted to rectangular pipe jacking which is finding increasing application for common utility tunnels and pedestrian connections in dense urban areas. The effective space use in the rectangular cross section allows more efficient use of urban underground space—particularly in the most important vertical direction. The final chapter looks at various aspects in the progress in assessing and improving the contributions of tunneling and pipe jacking to sustainable methods of infrastructure provision.

Overall, this book consolidates and explains various recent research findings and practice improvements in the field and illustrates the potential for pipe jacking to overcome challenging applications and difficult site conditions. The compilation of knowledge provided here can assist in the faster spread of innovations and improved performance and reliability in pipe jacking and underground infrastructure provision.

Professor Emeritus Raymond Sterling,
Ph.D., P.E.
Louisiana Tech University
Ruston, USA

Foreword by Keh-Jian Shou

Pipe jacking has been identified as being used in the USA and Europe back in the late nineteenth century. The early form of pipe jacking required person access to the face in order to carry out the excavation work and to load and remove the spoil. Later on, in the twentieth century, this technique was improved to have a rotary cutting head in conjunction with the jacking of a steel pipe. Nowadays, this technology has been further developed for the challenges of longer driving distance, larger diameter, irregular cross section, etc.

As the demands of sustainability, including environment friendliness and reduction in CO_2 emission, the trenchless technologies now provide a better option for the installation of underground pipelines. This trend not only promotes the popularity of trenchless technology (No-Dig), but also advances its development. Pipe jacking, as one of the most popular trenchless technologies, has been very demanding for the past few decades. And it is expected that, with further developments, pipe jacking will still be popular for underground pipeline installation for quite some time.

This book, with emphasis on sustainable pipe jacking technologies in the urban environment, includes the fundamental theories of the behavior of soils and rocks to the broad concept of trenchless technology. Recent advances and innovations in different aspects are presented and discussed in detail. Ten case studies from five different countries, namely Australia, China, Italy, Malaysia, and Singapore, were included in this book for the readers, especially for the practitioners, to have better understanding of the advances of this trenchless technology. As this book has made a benchmark in the development of trenchless technology, I would like to recommend this book to you and honor the authors for their efforts on this challenging work.

Distinguished Professor Keh-Jian Shou, Ph.D
Vice Chairman, International Society
for Trenchless Technology (ISTT);
Elected Vice President, International
Society of Soil Mechanics and
Geotechnical Engineering (ISSMGE)
Department of Civil Engineering
National Chung-Hsing University
Taichung, Taiwan

Preface

The book "Sustainable Pipe Jacking Technology in the Urban Environment—Recent Advances and Innovations" connects fundamental theories of the behavior of soils and rocks to the broad concept of trenchless technology, but with special emphasis on sustainable pipe jacking technologies in the urban environment. It is written in such a way that it complements standard textbooks covering basic principles of trenchless technology, soil mechanics, and geotechnical engineering. This book takes its readers a step further and immerses them in the world of pipe jacking and microtunneling operations and applications practiced in five countries, namely Australia, China, Italy, Malaysia, and Singapore based on all of the authors' research and practical experiences in these countries. Altogether, there are ten detailed case studies found in the various chapters of this book to benefit readers on how impactful research outcomes are translated to geotechnical engineering practice.

The book does not only encapsulates the finer points in design and construction aspects related to microtunneling or tunneling by pipe jacking methods, but also discusses the challenges faced by engineers and the eventual solutions that have been successfully conceived and applied to tame some of the world's more complicated geology that we have the opportunities to work on or in, such as:

- Very soft estuarine and marine sediments
- Anisotropic permeability of organic soils
- Highly weathered soft rocks
- Erratic highly fractured rocks
- Cavity-laden karst with continuous water-bearing channels

Reader benefits derived from this book include:

- Use of layman's language without discounting the engineering technicalities
- Practical case studies to reinforce readers' understanding
- Detailed discussions and validation of laboratory and field testing results against established journals, guidelines or codes of practice
- Up-to-date information on innovative irregular-shaped pipe jacking and irregular-shaped shield technologies

- Dedicated evaluation of AI-based techniques in predicting jacking forces and mTBM health monitoring
- Extensive list of references should the readers wish to deep dive into any relevant topics
- Recommendations and glimpses on the future direction of pipe jacking.

After reading and understanding the contents of this book, readers are expected to be able to:

- Identify the benefits of trenchless technologies in an urban environment
- Interpret nonlinear soil strength parameters based on established results from advanced laboratory tests
- Articulate complex soil–pipe interaction in challenging geology
- Distinguish the various types of retaining wall suitable for construction of deep shafts based on geology
- Assess the benefits and innovations in the design of irregular-shaped pipe jacking and irregular-shaped shield technologies
- Reflect on the prowess and advancement of AI-based mTBM health monitoring techniques
- Devise concepts to decarbonize infrastructure with reference to government policies
- Appraise the use of tunnel spoil in the construction industry.

In closing, we would like to share an immortalized quote of Karl Terzaghi, '*A well-documented case history should be given as much weight as ten ingenious theories and the results of laboratory investigations should not receive too much attention unless the validity of the conclusions has been demonstrated by adequate field observations on full-sized structures.*' More than 70 years later, Terzaghi's words hold true in the ongoing efforts to shed light on new challenges in pipe jacking. This book documents valuable case studies of current state-of-the-art and recent technological advances in microtunneling or tunneling by pipe jacking method as well as noncircular tunneling methods, which are fast becoming a popular and alternative concept since their first installations in Tokyo and Shanghai in 1970 and 1995, respectively. The case studies presented here have been investigated through the use of fundamental analytical solutions, improved laboratory tests, field in situ tests, sensitivity analyses, and back-analyses validated by calibrated numerical simulations, as well as reliable field measurements to provide quantifiable and mechanical solutions to new challenges in pipe jacking. It is the hope of all the authors that this book becomes a demonstration of evidence-based and data-driven solutions to construction challenges and serves as a reliable reference to future research and applications in the ever-growing field of pipe jacking.

Brisbane, Australia — Dominic Ek Leong Ong
Turin, Italy — Marco Barla
Xi'an, China — Jason Wen-Chieh Cheng
Kuching, Malaysia — Chung Siung Choo
Singapore, Singapore — Minmin Sun
Brisbane, Australia — Mohammud Irfaan Peerun

Acknowledgements

The authors would like to acknowledge the tremendous efforts rendered by Siaw Chian Jong and Dr. Fredrik Phangkawira who have stood by the team so steadfastly and readily to contribute their assistance in proof-reading and formating all the materials presented in this book. They are also indebted to Hock Seng Lee Berhad, Nishimatsu Construction Co. Ltd., Jurutera Jasa (Sarawak) Sdn. Bhd., Jurutera Perunding Geoteknik Sdn. Bhd., Swinburne University of Technology Sarawak Campus, Red Dot Sdn. Bhd., BGE Sdn. Bhd., the Sewerage Services Department of the Sarawak State Government (Malaysia) as well as individuals like Lee Yoon Tai, Dr. Wen Hui Ting, Yoshiaki Chikushi, Simon Lau, and Yih Seng Sim for their long-lasting relationships, guidance, generosity in knowledge sharing as well as the support rendered in grants and in-kind contributions. Some contents of the book have also been benefited by the review work performed during the doctoral studies of Dr. Marco Camusso at Politecnico di Torino. Guidance by Prof. James C. Ni and the beneficial studies of Dr. Xue-Dong Bai and Dr. Ge Li are also respectfully acknowledged.

Sincere gratitude is also extended to Binquan Yu and Dr. Guoyang Fu for their valuable contributions especially to Chaps. 5 and 6. Special emphasis should be given to Binquan Yu for contributing the basic information, without which these two chapters would not have been possible. The support of Shandong Wan Guang Construction Engineering Co. Ltd. (China) for contributing the latest information to the case study in Chap. 6 is also acknowledged and appreciated. Special thanks are also directed to Yangzhou Guangxin Heavy Machinery Co. Ltd. (China) and Kim Hua Cheng (Singapore) for their support.

The authors would also like to acknowledge the contributions from VAC Group, Translational Research Institute (Kamil Sokolowski, Preclinical Imaging Officer), Advanced Design and Prototyping Technologies Institute of Griffith University, Queensland Research and Innovation Services cloud (QRIScloud), and Bentley Software Systems (Kuthur Sriram, Global Manager). Without everyone's contributions, this book would not have been able to take off as planned.

Contents

1 **Introduction: Pipe Jacking in the Urban Environment** 1
   1.1 Background 1
   1.2 A Brief History 2
   1.3 Focus of the Book 2
   1.4 Chapters of the Book 3
   References 6

2 **Investigation Techniques: Pipe Jacking in Complex Geology** 7
   2.1 Introduction 7
   2.2 Data to be Obtained 8
   2.3 Geophysical Investigations 8
   2.4 Site Investigation 10
      2.4.1 In Situ Boreholes and In Situ Geotechnical Tests 10
      2.4.2 Using Drilling Parameters 11
      2.4.3 Geostructural Survey for Rock Masses 13
      2.4.4 Case Study: Pressuremeter Testing of Highly
              Fractured Phyllite 17
   2.5 Geotechnical Laboratory Testing 31
      2.5.1 Interface Tests 32
      2.5.2 Case Study: Direct Shear Testing of Reconstituted
              Tunneling Spoil from Highly Fractured Rocks 33
   2.6 The Geological and Geotechnical Model 38
   References 39

3 **Complex Soil–Pipe Interaction: Challenges in Geological
   Characterization and Construction** 43
   3.1 Background 43
   3.2 Pipe Jacking Forces 45
      3.2.1 Experimental Results Relating to Frictional Stress 47
      3.2.2 Calculation Methodology for Frictional Stress 49
   3.3 The Influence of Construction Operational Parameters 60
      3.3.1 Influence of Stoppages 62

		3.3.2	Influence of Lubrication	63
		3.3.3	Influence of Overcut Size	66
		3.3.4	Influence of Misalignment	67
		3.3.5	Influence of Steering	68
	3.4	Case Study 1: Pipe Jacking in the Altamura Limestone		70
		3.4.1	Characterization of the Altamura Limestone	70
	3.5	Case Study 2: Pipe Jacking in the Highly Weathered 'Soft Rocks' of the Tuang Formation		73
		3.5.1	Developing Geotechnical Strength Parameters Using Direct Shear Testing of Reconstituted Tunneling Rock Spoil	75
		3.5.2	Estimating Geotechnical Strength Parameters Using In Situ Pressuremeter Test	91
	3.6	Conclusions		97
	References			98
4	**Deep Shaft Excavation: Design, Construction, and Their Challenges**			**103**
	4.1	Introduction		103
	4.2	Types of Shafts		103
		4.2.1	Circular Concrete Caisson	103
		4.2.2	Steel Sheet Pile	104
		4.2.3	Diaphragm Wall	105
		4.2.4	Sinking Circular Concrete Caisson	106
		4.2.5	Cased or Oscillation-Drilled Shaft	106
	4.3	Design Philosophy and Selection Guide for Shafts		107
		4.3.1	Bottom-Up or Top-Down Construction	108
		4.3.2	Jacking and Receiving Shafts	109
		4.3.3	Shaft Excavation in Soils	109
		4.3.4	Shaft Excavation in Rocks	112
		4.3.5	Strut-Less or Strutted Shaft Design	117
		4.3.6	Nearby Infrastructure Risk Assessment Documentation and Analyses	119
	4.4	Soil–Structure Interactions		123
		4.4.1	Analytical Method	123
		4.4.2	Empirical or Closed-Form Method	124
		4.4.3	Field Observational Method	124
		4.4.4	Numerical Method and Finite Element Method (FEM)	124
		4.4.5	Artificial Intelligence	125
		4.4.6	Consistency of Methods	125
	4.5	Typical Construction Challenges		126
		4.5.1	Soil Conditions	126
		4.5.2	Seepage Pressure	127
		4.5.3	Ground Water Intrusion	129
		4.5.4	Space Constraints	131

		4.5.5	Utility Detection in Urban Areas	131
	4.6	Case Studies		133
		4.6.1	Case Study 1: Shaft X in Organic Soils	133
		4.6.2	Case Study 2: Shaft Y in Karst	139
	4.7	Conclusions		140
	References			142
5	**Irregular-Shaped Pipe Jacking Technique**			147
	5.1	Introduction		147
	5.2	Types of Irregular-Shaped Pipe Jacking Machines		148
		5.2.1	Rectangular Jacking Machines with Combined Multi-cutterheads	149
		5.2.2	Eccentric Multi-axis Pipe Jacking Machines	156
		5.2.3	Square Planetary Gear-Type Jacking Machine	166
	5.3	Selection of Machine		172
		5.3.1	Introduction	172
		5.3.2	Soil and Site Conditions	172
		5.3.3	Construction Method and Skill	175
		5.3.4	Project Cost and Construction Schedule	176
		5.3.5	Environmentally Friendly and Sustainable Development	178
	5.4	Case Studies		178
		5.4.1	Rectangular Pipe Jacking Project with Four Parallel Box Sections in Foshan, Guangdong, China	179
		5.4.2	Rectangular Pipe Jacking Project in Complex Soil Conditions in Baotou, Inner Mongolia, China	186
	References			197
6	**Irregular-Shaped Shield**			199
	6.1	Introduction		199
	6.2	6.0 m × 3.3 m Rectangular Shield		200
		6.2.1	Rectangular Lining Rings and Seals	200
		6.2.2	Configuration of the Segment Erection Shield	202
		6.2.3	Transportation of Lining Segments Inside the Tunnel	206
		6.2.4	Erector System for Lining Segments	207
		6.2.5	Other Accessories	212
		6.2.6	Erection Process of Lining Segments	213
	6.3	Case Study—An Irregular-Shaped Shield Project in Shandong, China		224
		6.3.1	Project Background	224
		6.3.2	U-shaped Lining Segments and Connections	224
		6.3.3	Procedure of the Shield Construction	227
		6.3.4	Introduction of Hybrid Method Using Irregular-Shaped Pipe Jacking and Irregular-Shaped Shield Methods	234
		6.3.5	Site Photos	234

		6.3.6	Conclusions	235
	6.4	Summary		237
	References			237

7 Pipe Jacking Performance: Mechanistic Behavior and Maintenance Challenges—An Artificial Intelligence-Based Approach 239

	7.1	Introduction		239
	7.2	Mechanistic Behavior of a Tunnel Boring Machine		240
		7.2.1	Operational Parameters	240
		7.2.2	Soil Characteristics	243
	7.3	Maintenance Challenges		247
		7.3.1	Overview of Artificial Intelligence Techniques	247
		7.3.2	Implementation	252
		7.3.3	Tunneling in Alluvial Soils	253
		7.3.4	Tunneling in Loess Soil	253
		7.3.5	*Case Study 1: Identification of Geological Conditions*	257
		7.3.6	*Case Study 2: Clogging Detection*	262
	References			274

8 Decarbonizing Tunnel Design and Construction 277

	8.1	Why 'Decarbonize' Infrastructure?		277
		8.1.1	Global Sustainability Goals	278
		8.1.2	Circular Economy Goals	278
		8.1.3	Life Cycle Assessment (LCA)	279
	8.2	Decarbonizing Tunneling Processes		279
		8.2.1	Open-Cut Versus Trenchless Technologies	279
		8.2.2	Advances in Tunnel Design and Construction	282
	8.3	Decarbonizing Through Spoil Reuse		301
		8.3.1	Bedding Material for Pipe Installation	301
		8.3.2	Backfill Material for Trenches	302
		8.3.3	Road Pavement	303
	8.4	Decarbonizing in Practice: From Compliance to Best Practice		304
		8.4.1	State Government Requirements	304
		8.4.2	Case Study: The Cross River Rail Project	306
		8.4.3	Local Government Requirements	308
		8.4.4	Rating Scheme Incentives	308
	8.5	Conclusions		309
	References			309

Contributors

The authors would like to thank the following colleagues who have contributed some contents, photos, advice, and time in making this book a reality. Without their friendships, motivation, and support, this book would not have seen the light of day. Most sincere thanks and gratitude from all our hearts!

Chapters 2 and 3: Dr. Alessandra Insana and Dr. Fredrik Phangkawira
Chapter 4: Brandon Kueh
Chapters 5 and 6: Binquan Yu and Dr. Guoyang Fu
Chapter 7: Siaw Chian Jong
Chapter 8: Prof. Cheryl Desha and Siaw Chian Jong

Chapter 1
Introduction: Pipe Jacking in the Urban Environment

1.1 Background

Urbanization has increased the demand for underground infrastructure to be built in order to minimize disruptions to road users, hence the popularization of trenchless technology. Pipe jacking is just one of the many trenchless technology methods, others being the displacement (e.g., pipe ramming), drilling (e.g., horizontal directional drilling), and conventional large tunneling methods. Conventionally, microtunneling comes as a subset of the pipe jacking method, but in recent decades, larger tunneling shields have been incorporated into pipe jacking works as well.

Pipe jacking is described as the process of installing segmental pipes through the ground by means of applying jacking pressures to propel the microtunnel or tunnel boring machine forward in order to form a continuous string of pipes. A distinguishing fact about pipe jacking as compared to other methods of tunneling is that the lining (i.e., precast elements) of the eventual tunnel is jacked through the ground from the initial launch point rather than being built piece-by-piece just behind the excavation face of a typical tunnel boring machine (TBM) (Sterling 2020). Therefore, pipe jacking via the microtunneling or tunneling method is an often-used terminology to effectively describe the popular combined technique, which is widely adopted around the world today.

With Japan leading the industrialization of pipe jacking in the 1980s, pipe jacking operations have become heavily mechanized, often incorporating laser guidance systems, remote controls, and real-time instrumentation. These improvements to quality assurance and control created a boom in the global popularity of the pipe jacking technique. Since then, pipe jacking has been popularly used in the installation of sewer lines (sanitary and storm water), especially in densely populated metropolitan areas. Most documented pipe jacking drives have been reported in soils, such as sands and clays. The pipe jacking method is becoming more commonly performed in rocks due to increasing demands for deep sewer systems for a centralized wastewater treatment process. This is evidenced by the assessment of 29 microtunneling drives in rocks, as documented by Hunt and Del Nero (2009).

1.2 A Brief History

Thomson (1993) reported that pipe jacking had been used as early as 1896 to install a concrete culvert under the Northern Pacific Railroad in the USA. Augustus Griffin was credited with popularizing the technique in California when he developed techniques to jack cast iron pipe culverts under rail tracks. Concrete pipes were then used in the late 1920s and ranged from 750 to 2400 mm. Pipe jacking did not evolve much since then, until well into the 1960s. By this time, pipe jacking was possible in person-entry sized diameters used for crossing installations.

In the 1970s, a Japanese company, Komatsu appeared to have developed the world's first microtunneling machines, which for the first time, promoted nonperson-entry smaller diameter machines, which signify greater use of automation. The first pressurized slurry microtunneling boring machine (mTBM) was launched in the market around 1979. Since then, modern mTBMs were principally developed in Japan, Germany, and the UK. The first microtunneling project in the USA was recorded in 1984. Readers who are interested in the greater history of pipe jacking are encouraged to read Thomson (1993), Drennon (1979), and Craig (1984).

However, in recent years innovations and demands have led to the development of noncircular, nonconventional tunneling shields via pipe jacking method, which are covered in Chaps. 5 and 6 of this book. With irregular-shaped (e.g., rectangular or arch door) tunnels coming into the market recently, the use of pipe jacked infrastructure will now include pedestrian crossings or mall connectors, apart from conventional uses for utility, water, and sewage transfers.

1.3 Focus of the Book

This book is the brainchild of the lead author, who subsequently invited his peers, ex Ph.D. students and current Ph.D. students to collaborate and contribute to writing a geotechnically advanced book that encompasses the authors' and contributors' research and practical experiences gained principally on microtunneling or tunneling via pipe jacking works in challenging and problematic ground conditions. As such, this book briefly covers the conventional norms of pipe jacking but focuses in detail the recent innovations and great challenges faced in executing pipe jacking works as THE sustainable trenchless solution to the urbanization process experienced worldwide.

The book is written in such a way that it does not flow like a conventional beginners' textbook, nor is it a collection of jargon-heavy journal papers but somewhat in between. Concepts are lightly laid out at first before leading the readers to deep-dive into unconventional complexity in performing pipe jacking works in perhaps some of the region's most challenging geology involving very soft clays, organic soils, karst, and highly weathered soft rocks with practical case studies compiled from Europe, East & Southeast Asia, and Down Under (Australia).

1.3 Focus of the Book

The book is indeed a labor of love for all the authors, who, like everyone else, witnessed how the COVID-19 virus ravaged and changed the world forever. This book will forever be remembered as one that is 'born' in the pandemic era in the hope that it will be as impactful as the virus, only in a good way!

1.4 Chapters of the Book

This book starts off with Chap. 1 as an introduction to pipe jacking in the urban environment and dedicates the chapter to briefly explain how pipe jacking is differentiated from other trenchless methods. A brief history on the evolution of the pipe jacking method is also presented. Content summaries of all other chapters are reported here to give readers a broad appreciation of the theme and flow of the book.

Chapter 2 is written to help readers appreciate how a geotechnical investigation program, laboratory tests, field in situ tests, and nondestructive tests are chosen and interpreted to achieve the key design requirements for pipe jacking force assessment and shaft design. The concepts presented here would require readers to have some experience in basic laboratory or field tests in conventional Site or Soil Investigation (S.I.) works. A section on advanced testing using (i) unconventional field pressuremeter test (PMT) for highly weathered in situ rock and (ii) laboratory direct shear test (DST) for lower-bound characterization of reconstituted rock spoil is presented to keep readers up-to-date on the possibility of supplementing standard testing methods with other established techniques in order to obtain greater accuracy or to validate the eventual test results. For example, particle image velocimetry (PIV) technique has been successfully applied to develop fundamental understanding in particulate behavior and strength development using DST. Key observations are then interpreted based on supplementary techniques involving the application of the power law function and the generalized tangential approach to establish reliable geomaterial strength parameters via the elastic-perfectly plastic Mohr–Coulomb (MC) model but considering nonlinear soil behavior. Readers could then use this established method to improve their own design parameters if necessary.

Chapter 3 assesses the complex soil–pipe interaction in challenging geology based on empirical and analytical methods, supplemented by appropriate material property selection and advanced numerical modeling. In order to develop a more reliable prediction of pipe jacking forces, factors such as geological settings, stability of tunnel excavation, and effect of arching are taken into consideration. The impacts of real-life challenges such as jacking stoppages, lubrication, sizes of overcuts, and tunnel misalignments are discussed based on reliable, measured field data collated from construction sites in different countries. Two case studies are also presented; one involving a fractured rock mass in Italy and the other involves the notorious highly weathered soft rocks of Malaysia, a country located on the Equator thus its challenging highly weathered geological setting. Both case studies shed light on how careful interpretation of in situ and laboratory testing techniques yield favorable test results that help engineers improve their design accuracy in predicting the often elusive jacking forces.

Chapter 4 dwells into the construction and challenges faced in constructing deep shafts. The features and advantages of various types of shafts are discussed with reference to the surrounding geological settings. Broad design philosophy of shaft excavation in challenging ground conditions is emphasized based on sound soil or rock mechanics principles and practical geotechnical engineering applications. Detrimental effects of ground water lowering and the associated geotechnical risks assessment triggered by adjacent shaft excavation works are also broadly discussed for greater appreciation. Such a cause-and-effect situation often requires a higher level of analytical skillsets and practical field experience to devise effective geotechnical solutions. The specialized field of soil–structure interactions focusing on performance-based design is normally linked to such cause-and-effect scenarios, and the use of a finite element code and/or field observational method is often regarded as indispensable tools or techniques. Two case studies are presented and discussed in detail. The first case study involves a deep shaft excavation in organic soil overlying the highly weathered soft rock that unforeseeably triggers a substantial ground water drawdown through a section of fractured geology, which is eventually mitigated by installing a recharge well system. The second case study also involves excavation for a deep shaft, but in a karst environment that inevitably triggers substantial ground water drawdown and is subsequently mitigated via Tube-a-Manchette (TAM) compaction grouting technique.

Chapter 5 covers the irregular-shaped jacking technology, in which three types of irregular-shaped pipe jacking technology are discussed, including (i) rectangular pipe jacking with combined multi-cutterheads, (ii) eccentric multi-axis pipe jacking, and (iii) square planetary gear-type pipe jacking. Proper selection of jacking machine that suits the project requirements is essential to ensure successful completion of the works. Several selection factors that should be considered, such as soil condition, construction method, project cost and schedule, and sustainability, are discussed in detail to help readers understand the various aspects that are involved in the process of selecting the most suitable jacking machine for a specific project. Finally, two case studies are presented to discuss the feasibility of applying irregular-shaped pipe jacking technology to existing projects in the industry. The first case study involves the construction of four parallel box tunnels with 500 mm distance in between each tunnel segment using rectangular pipe jacking, which is constrained by the presence of existing structures in the vicinity of the project location, such as concrete drains, Mass Rapid Transit (MRT) tunnels, and in-service utilities. The second case study discusses the design and construction of an underground passage using rectangular pipe jacking in complex soil condition consisting sandy pebbles, where the situation was mitigated by means of grouting as a form of ground improvement method.

Chapter 6 focuses on the innovative irregular-shaped (noncircular) shield tunnel construction method via pipe jacking technology. The shield construction method is introduced with a specific focus on the newly proposed erection system for rectangular shield. Some innovations that have been developed for this technology include

1.4 Chapters of the Book

the maximization of the space available for erecting lining segments, grouting system in the shield tail, and the use of square columns for erecting segments. To illustrate the potential of irregular-shaped shield construction method, a case study is presented. The underground utility project that is discussed in the case study is the first successful application of rectangular shield technology. The design and procedure of shield construction are discussed in detail to highlight the advantages of adopting irregular-shaped pipe jacking method as compared to conventional (circular) pipe jacking practice.

Chapter 7 discusses the mechanistic behavior and maintenance challenges affecting jacking performance via artificial intelligence (AI)-based approaches. Understanding the mechanistic behavior of a mTBM is important because it can help to improve the efficiency of tunnel excavations and project costs. Some operational parameters of mTBM and typical soil characteristics are introduced to help readers understand the factors that may influence the mechanistic behavior of the mTBM. Subsequently, challenges involved in the maintenance of the mTBM, particularly the variation in geological conditions and development of clay-clogging that could lead to costly tunnel construction, are discussed, utilizing AI-based techniques. Specifically, three popular anomaly detection (AD) techniques are presented in the case studies to demonstrate the potential of supplementing existing pipe jacking design methods using AI techniques.

Chapter 8 explores the concept of 'decarbonizing infrastructure' in the pipe jacking or tunnel industry by providing an overview on the achievable global sustainability and circular economy goals as well as the importance of life cycle assessment. A comparative study is presented to demonstrate the contrast between conventional open-cut method and the trenchless method in terms of costs, expensed carbon footprint, and energy consumption. Pipe jacking is fast becoming the preferred option compared to open trenching due to the former's cost-effectiveness, efficiency, and being a more sustainable solution over the long run. Advances in the field and laboratory research (see Chaps. 2 and 3) as well as the introduction of digital and immersive technologies, e.g., PIV, software development including discrete element modeling (DEM), micro-CT scan, and 3D printing, have contributed to greater understanding of fundamental particle behavior and strength properties of granular geomaterials via friction development, leading to better prediction of jacking forces. In addition, AI techniques have emerged to be a potential solution in solving complex engineering problems, as they are capable of processing nonlinear relationships between different variables in a model. Thus, AI techniques can be used to supplement the existing pipe jacking design methods in the industry. In recent years, state and local government policies as well as national standards have evolved to welcome the integration of recycled materials (e.g., tunnel spoil, mining by-products) in the design of road pavement to minimize the disposal loads from choking our landfills. A case study on the possible reuse of tunneling spoil as subbase for road pavement and as pipe bedding is discussed with respect to available specifications set by local authorities. A rating scheme to incentivize innovation when tendering for projects is recommended to encourage the take-up rate of recycled materials. To be effective, state and local governments must take the lead, while industry must be receptive to embrace new technology and invest in research for the betterment of the future environment and community.

References

Craig RN (1984) Pipe jacking: a state-of-the-art review. Construction Industry Research & Information Assoc, London

Drennon CB (1979) Pipe jacking: state of the art. J Constr Div 105(C03):217–223. https://doi.org/10.1061/JCCEAZ.0000834

Hunt SW, Del Nero DE (2009) Rock microtunneling—an industry review. In: International No-Dig 2009, Toronto, Ontario, Canada, 29 Mar–3 Apr 2009

Sterling RL (2020) Developments and research directions in pipe jacking and microtunneling. Undergr Space 5:1–19. https://doi.org/10.1016/j.undsp.2018.09.001

Thomson J (1993) Pipejacking and microtunnelling. Chapman & Hall, Glasgow, UK, p 273

Chapter 2
Investigation Techniques: Pipe Jacking in Complex Geology

2.1 Introduction

For the design of any buried infrastructure such as pipelines, it is crucial for investigations to be conducted. Such investigations include subsurface exploration, in situ field testing, and laboratory tests. These findings will back up the geological setting and other underground data gathered. During the planning stage, it is important to have these data available for interpretation. That way, the data can be useful and reliable for the final design.

Prior to a pipe jacking project, geotechnical investigations include the gathering of geological, hydrogeological, and geotechnical data, as well as the detection of natural or artificial hidden obstacles that could cause problems during the construction process. Because of the small diameter of the pipeline structures and the variable depths at which they must be installed, the digging operation is extremely sensitive to heterogeneities in the subsurface profile, ranging from the natural (boulders or blocks) to the artificial (old foundations, existing structures) heterogeneities. Microtunneling, unlike trenching, necessitates thorough investigations for several reasons:

- The findings drawn from the investigations will affect plans and decisions for pipe jacking. Once the microtunnel boring machine (mTBM) has entered the ground, it is very difficult to reverse these decisions.
- Overlooking any significant variations in the geology along the pipeline alignment could cause significant overestimation (or underestimation) of the advancement rate or 'choking' of the mTBM. This could necessitate costly recovery of the mTBM through the excavation of an additional intermediate rescue shaft.
- The adjacent built environment close to the drive could sustain potential damage if soft soils were overlooked during site investigation.

Borehole sampling and laboratory testing of recovered samples, field testing, nondestructive testing, and geophysical techniques are all methods for getting thorough information on soil conditions.

The benefits are primarily related to project geometry optimization (course plan, longitudinal profile, number, location, and dimensions of shafts), mTBM selection and setup (jacking system, spoil removal system, lubrication, need for intermediate jacking stations, and thrust wall dimensions), and project cost and construction time estimation.

Geotechnical investigations must be carried out at the earliest possible stage of the project because this is when the most money may be saved, especially when it comes to project geometry optimization. It should be highlighted, for example, that shaft construction is a significant cost of the microtunneling project, ranging from 20% to 40%. This cost may rapidly increase if the shafts must be made watertight. This emphasizes the significance of geotechnical research in ensuring proper design (number, position, dimensions, and reinforcements).

2.2 Data to be Obtained

It is important to create a map of geological, hydrogeological, and geotechnical risks that are as close to reality as possible. That way, the construction project may be better integrated, with the risk level outcomes measured or even understood. The fundamental goal is to understand the geological configuration of the site, which results in a longitudinal geological cross-section. This includes identifying and characterizing geological features that are detrimental to boring equipment, such as:

- Mechanically heterogeneous constituents, such as man-made backfill, moraines, burrstone clay, slope scree, irregularly altered rocks, and flint embedded in a softer matrix
- Coarse soils, such as alluvial
- Plastic clays, which can cause swelling or sticking in the presence of water
- Soft soils, which causes steering and pipeline alignment problems
- Interface between different layers, which is difficult to overcome with the mTBM
- The presence of natural cavities.

Water table level and its fluctuations are referred to as hydrogeological circumstances that can affect shaft sizing and machine advancement rate. In order to establish a method for the execution and waterproofing of the shafts, information on the soil permeability is also required. The horizontal flow speed and chemical composition of the ground water are two crucial hydrogeological characteristics to understand since they influence lubricating effects and the type of pipe that can be jacked.

2.3 Geophysical Investigations

Electromagnetic, radar, magnetic, electrical, gravimetric, and seismic surveys are the six primary categories of geophysical methods. These methods have several

2.3 Geophysical Investigations

Fig. 2.1 Example of ground probing radar for utility detection and mapping (IDS Georadar n.d.)

advantages, including providing a 2D image of soil distribution along the proposed alignment (continuous profile), optimizing the layout of exploratory borings, and allowing lateral extrapolation of boring data after the geophysical profiles have been tested on these borings. They also highlight localized heterogeneities that boreholes are unlikely to encounter.

The most widely used approach is ground probing radar (GPR). This involves sending an electromagnetic radiation pulse into the ground via a directional antenna and measuring the reflected radiation owing to soil discontinuities (Fig. 2.1). Rather than detecting discontinuities or unknown items, it is more interested in detecting and locating objects of known nature but uncertain position (e.g., networks and horizontal and vertical interfaces). However, due to the available depth of study (about 5 m) and the nature of the soil, this method has certain practical limits. In reality, the GPR is ineffective in silty soils, clayey soils, or soils below the water table (due to the water absorption of this type of electromagnetic radiation).

Other methods (such as radio magnetotelluric (RMT), electrostatic quadripole, and electromagnetic prospection) are better suited for describing the distribution and nature of soils than GPR, but they are less sensitive in locating obstacles and horizontal interfaces, or when used in densely urban areas.

Seismic methods take advantage of the differences in soil and rock properties that affect the speed with which vibrations travel through them. They can work with a variety of vibration sources, including high-frequency resonance, reflection, and refraction. Seismic methods can be useful for detecting stratum changes. Typical applications include profiling bedrock or subsurface features such as caverns or big boulders. Because the strata profile revealed by seismic waves does not necessarily correspond to drill data, the results of these geophysical investigations must be properly connected with borehole data.

The surface wave method (SWM) was used in a recent study on seismic investigation methods in the realm of trenchless technology. This is a cost-effective and time-saving surveying approach, whose data can be used to examine the soil's mechanical qualities as well as to deduce the geological profile. The method uses surface waves propagation in heterogeneous media to infer the shear wave velocity profile (which is linked to the shear stiffness of the soil strata at small strain), and the resulting soil classification in terms of drillability and excavation tool performance. A quick SWM test approach was devised (Foti et al. 2002) to acquire the stiffness profile of the first 3 m of soil with appropriate precision, employing a high-frequency range, and accounting for the impacts of a particularly stiff top layer (i.e., the pavement system commonly found in an urban environment).

Finally, microgravimetry, a localized method for detecting the presence of cavities regardless of the type of host ground, must be noted. This method is only effective for cavities with a shallow depth (a tunnel with a diameter of 2 m remains invisible if its roof is less than 5 m), and it is very expensive because it necessitates the use of measuring stations that are very close to one another, especially when detecting small cavities.

In all circumstances, one must first attempt to build a projected geological section based on the available material, evaluate its uncertainty, and then choose the most appropriate geophysical method(s).

For a microtunneling project at shallow depths in an urban area where the geological structure is well understood, GPR or electrical and electromagnetic prospections (in the case of clayey or wet soil) are recommended. Meanwhile, for a project in a terrain with a less obstructed surface, an RMT or seismic method may be more appropriate to clarify the geology of the soil, followed by GPR or electromagnetic methods for the detection of isolated obstacles.

2.4 Site Investigation

2.4.1 In Situ Boreholes and In Situ Geotechnical Tests

Exploratory digging is required to accurately identify the geological section of the drive, assess the water table level and soil permeability, collect disturbed or undisturbed samples for laboratory geotechnical analyses, and maybe conduct in situ geotechnical tests. Boreholes are bored following the findings of geophysical studies, as they aid in their interpretation, just as geophysics aids in the extrapolation of their findings.

The number of boreholes required is highly dependent on the site's geological complexity and the level of prior knowledge. Boreholes should be drilled at each shaft point and along the project section at a distance of 30–50 m. The latter should be taken down by roughly 2 m or 2 pipe diameters below the invert, whichever is greater, while a continuous sampling from one diameter above and below the intended pipe is recommended.

2.4 Site Investigation

The principal types of in situ tests that can be used are as follows:

- The standard penetration test (SPT) delivers revised samples as well as an SPT 'N' index of penetration resistance (SPT 'N' values are the number of blows required to drive a standard sampler 45 cm, using a standard size hammer falling a standard distance). Many soil parameters, such as modulus of elasticity, cohesiveness in fine-grained soils, and friction angle in granular soils, have been roughly associated with blow count.
- The cone penetration test (CPT) or static cone penetrometer comprises a cone at the end of a line of rods that is pushed into the ground at a steady rate, generating a continuous resistance log that can be connected with stratigraphy. If necessary, measurements are taken of the resistance to penetration of the outside surface (lateral friction) of the rod and the cone (cone stress), whose ratio, known as the skin friction ratio, is experimentally associated with the soil's nature.
- The pressuremeter (PMT) comprises an inflatable galvanized rubber probe which is lowered into a borehole to the required depth and expanded using pressurized gas or water to apply pressure to the borehole wall. The material's pressure and distortion are measured as it expands. The test, which takes around 15 min, aids in the identification of the ground (if it has been drilled with an auger) as well as providing deformability (ASTM D4719 2020) and resistance values (see Gibson and Anderson 1961 for undrained shear strength in cohesive soils, Hughes et al. 1977 for friction angle in sands, Yu and Houlsby 2015 for cohesion and friction angle based on cavity expansion theory). Pre-bored pressuremeters are commonly utilized, while newer techniques including self-boring pressuremeters and rock pressuremeters are gaining popularity. The former may drill a hole while simultaneously lowering the probe, reducing the amount of ground disturbance at the test depth. For use in hard soils and rocks, the latter is built with a higher pressure capacity (up to 20 MPa) than standard pressuremeters (up to 5–6 MPa) (Haberfield and Johnson 1990; Dafni 2013; Isik et al. 2008; Sharo and Liang 2012; Phangkawira et al. 2016, 2017a, b).

Table 2.1 gives indications of the adequacy of various methods of geotechnical investigation for five types of frequently encountered soils.

2.4.2 Using Drilling Parameters

Information can be obtained by recording boring parameters during drilling operations, assisted by conventional logging of borehole stratigraphy, which allows the distribution of specific energy against depth to be determined. The specific energy is the amount of energy required to drill a unit volume of soil and can be used to classify the soil encountered after proper calibration. It should be noted that specific energy does not allow for the evaluation of the soil's geotechnical properties, thus the findings must be interpreted using other well-known data collected through further testing. The following is how the specific energy E_s is defined:

Table 2.1 Adequacy of different geotechnical investigation methods (FSTT 2004)

Ground	Mechanical shovel	Core sampling	Destructive drilling	Auger	SPT	CPT	Pressuremeter	Notes
Urban backfills	**	*	**	0	0	0	*	Depending on the nature of backfill material
Fine soils (silt, clay)	*	**	*	**	**	**	**	
Sands and gravels	**	0	*	**	*	*	**	Depending on the maximum diameter of cobbles
Soft rocks	0	**	**	*	*	0	0	Depending on the depth
Rocks	00	**	**	00	0	00	00	Depending on the degree of weathering

Key ** very well suited, * quite suited, 0—not suited, 00—unsuitable

$$E_s = \frac{F}{A} + 2 \cdot \pi \cdot \frac{N \cdot T}{u \cdot A} \quad (2.1)$$

where F is the thrust force applied on the drilling bit, N is the number of revolutions per second, T is the torque, A is the cross-sectional area of the borehole, and u is the rate of penetration. The first ratio of Eq. 2.1 represents the mean pressure exerted by the bit at the bottom of the borehole and is equal to the specific energy if rotation is not applied.

When tests are carried out with a low and constant thrust force as well as a constant rotation velocity (i.e., the torque is practically constant during the tests), the monitored specific energy is simply a function of the rate of penetration, which in turn is only a function of the soil's drillability. This means that by continuously recording the boring parameters of Eq. 2.1, it is feasible to explain the fluctuation of specific energy versus depth and, as a result, develop a criterion for evaluating the consistency of the ground.

Figure 2.2 illustrates the interpreted vertical profile of the ground near two drilling tests as an example (DAC4 and D20). The data are from an alluvial soil deposit with random distribution cementation.

The results demonstrated that, beyond the first meters of man-made backfill, the profile is primarily defined by the existence of a geotechnical unit with a degree of

2.4 Site Investigation

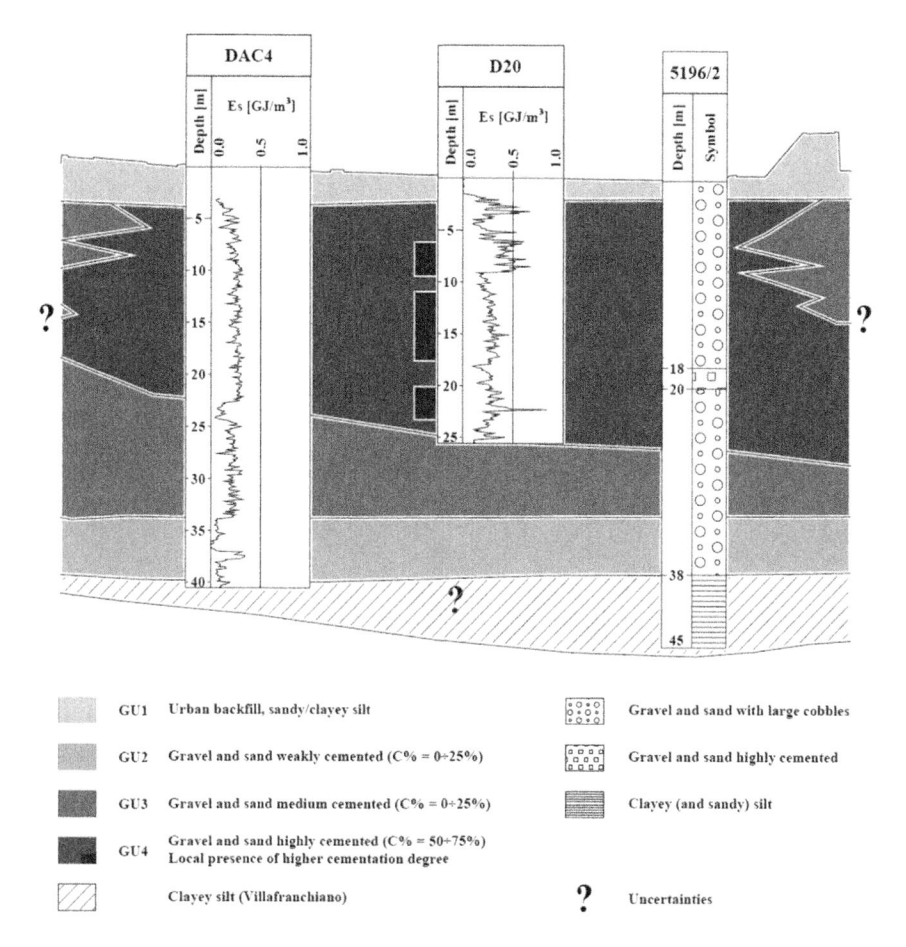

Fig. 2.2 Interpreted vertical profile of the Turin subsoil in the Porta Susa railway station area based on drilling tests and borehole sampling carried out during the construction of the Metro Line 1 (after Camusso 2008)

compaction/cementation, which extends to a depth of 20–25 m. Beyond this depth, the specific energy decreases, resulting in a weaker geotechnical unit. This information is critical during the design phase of a microtunneling installation since it offers details on the projected jacking forces and tunnel stability conditions.

2.4.3 Geostructural Survey for Rock Masses

In order to get meaningful insights and predictions into complicated challenges like microtunneling operations, accurate and detailed knowledge of the in situ geomechanical features is required. Among the activities that must be completed during

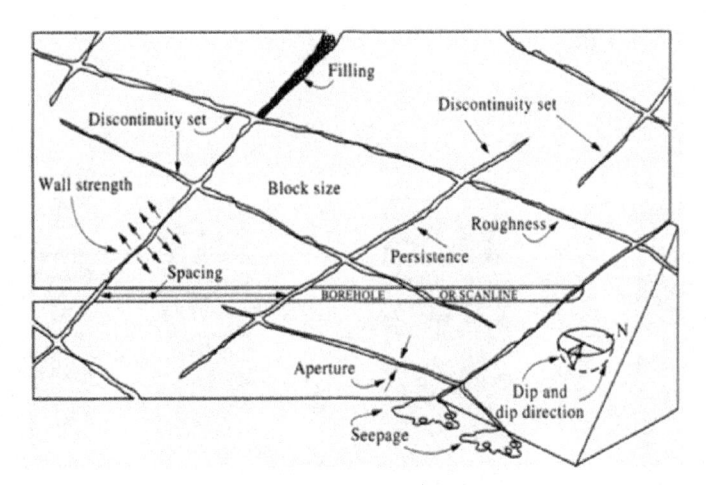

Fig. 2.3 Schematic representation of the main geometrical properties of discontinuities (after Hudson 1989)

the design stage, the geostructural survey is critical for accurately anticipating jacking forces and preventing unexpected stoppages caused for example, by localized rock.block collapse. The survey can be carried out in accordance with outcrops or the walls of a jacking shaft, and it necessitates the identification of joint sets, i.e., clusters of discontinuities that occur at specific orientations, in terms of (see Fig. 2.3):

- Orientation: The dip and dip direction uniquely describe the discontinuity orientation since the discontinuity is considered to be planar.
- Spacing: The frequency, or the number of discontinuities per unit distance, is the reciprocal of spacing, which is the distance between adjacent discontinuity junctions.
- Persistence: The extent of the discontinuity in its own plane.
- Roughness: Despite the assumption of planar discontinuities for the evaluation of orientation and persistence, the discontinuity surface can be rough. Roughness can be defined by reference to standard charts or mathematically.
- Aperture: This is the distance between the adjacent rock surfaces measured perpendicularly to the discontinuity and can be a constant, linearly varying, or completely variable value.
- Filling and presence of water.

At this stage, it is also helpful to have an estimate of both the mean block size and the block size distribution, in view of excavation and support activities.

An example of the result of the identification of discontinuity sets is depicted in Fig. 2.4, where both poles and major planes identified are represented. The poles are the points associated with the line perpendicular to the plane. Once many poles have been plotted on the hemispherical projection, the normals can be clustered to detect the basic rock structure. At least, two rock exposures at various orientations would be necessary to gain an assessment of the three-dimensional nature of the rock mass.

2.4 Site Investigation

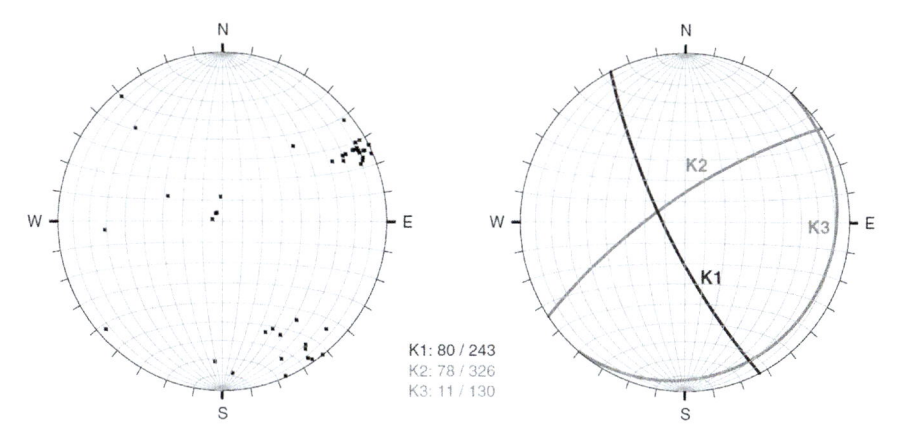

Fig. 2.4 Example of poles (left) and major planes (right) plots obtained from weighed analysis (Barla et al. 2006)

Borehole core obtained during drilling, as well as scans utilizing borehole television cameras or indirect methods, can be employed if the borehole core is not accessible (geophysical techniques). Although it offers minimal information on the lateral extent of the crossing discontinuities, the former is the most popular method for examining frequency along the drill direction.

This information is useful in determining whether a continuum or discontinuum numerical model is better suited to describe the subject under investigation (see Fig. 2.5). By assuming deterministic or statistical values (mean or mean with standard deviation, respectively), a discontinuum model can explicitly account for the structural traits identified.

The case of the construction of a new duct 200 m long and 5–9 m deep under a main road using trenchless technology for the renewal of the city sewer system in the city of Martina Franca in the South East of Italy was described by Barla et al. (2006). The construction of the new duct involved the use of Lovat MTS 1000 mTBM (excavation diameter = 825 mm) to jack a DN 600 mm vitrified clay pipe. The location is in a

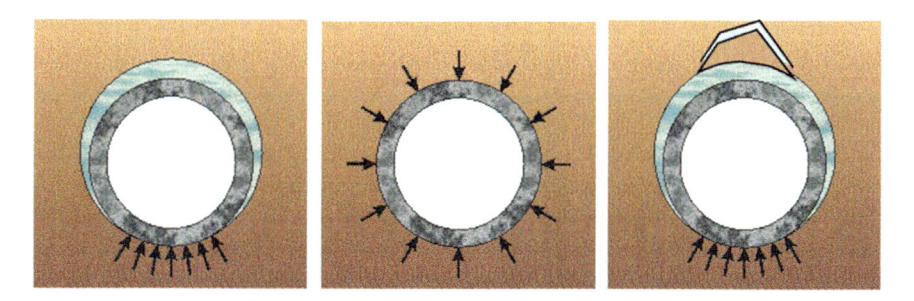

Fig. 2.5 Illustration showing the microtunnel condition in a stable ground (overcut open), in case of collapsed soil (overcut closed) and in a discontinuum media (local collapse) (Barla et al. 2006)

limestone hill known as Calcare di Altamura (Altamura limestone). One launching shaft was excavated, and two reception shafts were planned for the line's distant ends. This allowed the tunnel to be excavated in two phases, with the launching shaft being used twice.

During the construction of the first section, the machine's advance pace was extremely slow, and the jacking forces reached 1.8–2.0 MN, rendering it immobile. The microtunnel had to be retrieved by digging a new shaft, and the remaining extent of the section had to be constructed via cut and cover method. Concerns over the ability to excavate in rock led to the commissioning of some laboratory tests in various laboratories. The same machine was used to excavate the second section using the same jacking shaft. The cutting head became caught after only 3 m of excavation and was unable to continue. Direct observation during cut and cover operations to rescue the Lovat MTS 1000 mTBM after the stoppage, revealed a heavily fractured rock mass at that chainage.

For the first half of the drive in both cases, a reasonable agreement between computed jacking forces and monitored data was attained. The rate of rising of the friction forces was substantially faster than expected for the remainder of the drive, both in Sects. 2.1 and 2.2. Instabilities at the excavated tunnel barrier generated this behavior. As a result, jacking forces must be calculated in order to account for the weight of unstable rock blocks.

By using discontinuum numerical modeling to account for the discontinuum nature of the rock mass, a satisfactory description of ground behavior was produced. The distinct element approach was employed to accomplish this. To simulate the geometry and structure of the rock mass, two deterministic models were run, with mean values of inclination and spacing of joint sets and a trace length of 100%, and two statistical models, with average and standard deviation values for inclination, spacing, and trace length of joint sets randomly generated by the code. A pair of models with a mean GSI index of 58 and another couple of models with a GSI of 38 were created. The stress ratio K_0 was always equal to 0.3. Based on the geotechnical characterization, intact rock deformability and strength criteria were assigned to rock blocks and joints. To intact rock and rock joints, an elastic-perfectly plastic constitutive model was applied, with a failure criterion determined by the Mohr–Coulomb envelope.

Both the deterministic and statistical models responded to excavation in the same way: with GSI 58, only a few blocks were mobilized in the first, while the tunnel remained stable in the second. As a result, friction forces calculated using the rock load were still unable to justify the field performance. When the quality of the granite mass deteriorated, more blocks might fall as a result of stress release. Because of the consequence of failure at the tunnel surround, it was obvious that if the rock mass were more fragmented, the forces required to jack the pipe would increase correspondingly. These conditions were likely to have occurred in the example study given when the rock mass quality deteriorated in both Sects. 2.1 and 2.2.

Block instability gave rise to an increase of friction forces both at the top and the bottom of the pipe. At the same time, sidewall instability led to misalignments of the pipe train during jacking. If these effects were considered in the computation

2.4 Site Investigation

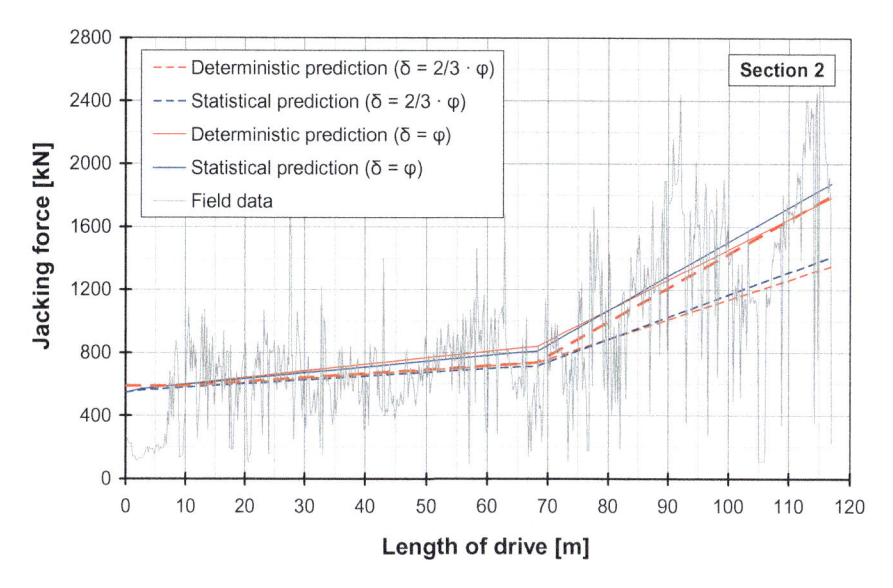

Fig. 2.6 Computed and measured jacking forces for Sect. 2.2, excavated by Herrenknecht AVN800A (discontinuum analyses, Barla et al. 2006)

of jacking forces, then a better comparison with field data may be obtained. Results obtained with GSI = 58 were used for computing jacking forces for the first part of the launch, while GSI = 38 were used to compute forces for the remaining part (Fig. 2.6).

The prediction of jacking forces was in good agreement with the monitoring for Sect. 2.2. Thus, a proper description of the ground behavior is required representing the discontinuum nature of the rock mass by discontinuum numerical modeling to capture the unstable blocks that developed at the tunnel crown and sidewalls. A more fractured rock mass implies higher jacking forces due to the more diffused failure at the tunnel boundary, with dramatic consequences on the works if a poor investigation is undertaken.

2.4.4 Case Study: Pressuremeter Testing of Highly Fractured Phyllite

The pressuremeter is an in situ test equipment that can be used to test the strength and stiffness of soil. Its main components include a probe that can expand via pressurized gas or water. The expansion of the probe can be controlled manually using a hand pump or automated via a control unit. Pre-bored pressuremeters are currently the most commonly available pressuremeter apparatus in the market.

In recent years, more advanced variants of pressuremeter technology have started to enter the industry such as self-boring pressuremeters and rock pressuremeters. Self-boring pressuremeters are equipped with drilling bits that can simultaneously drill a hole while lowering the probe to the desired test depth. This further minimizes the surrounding ground disturbance at the test depth, making it easier to obtain acceptable results. Rock pressuremeter is designed with a higher pressure capacity (up to 20 MPa) than normal pressuremeters (up to 5–6 MPa). This allows for pressuremeter tests to be performed in hard soils and rocks. However, it is worth noting that these pressuremeter technologies are still relatively new in the industry. As a result, they may not be as readily available as the common pressuremeter tests.

Pressuremeter test (PMT) is done by lowering the probe into a test borehole, to the desired test depth. Once positioned, the probe is expanded to apply pressure to the surrounding borehole wall. During the expansion, pressure and deformation of the material are measured. These measurements can be interpreted to characterize the properties of the material. PMT can generally be performed relatively quickly (within 15 min). Nowadays, PMT is commonly used for various applications, such as foundation design (Clarke and Smith 1992), settlement prediction (Briaud 1992), and bearing capacity checks for piles (Yu 2000).

Before performing PMT, calibration exercises are needed to ensure that the measurements during the actual test have accounted for pressure losses and volume losses. Pressure losses originate from the probe wall stiffness. Calibration against pressure loss can be done by inflating the probe under atmospheric pressure at the ground surface. The rate of expansion can be either controlled with a pressure increment of 10 kPa or a volume increment of 5% of the uninflated probe volume. At each increment, a maintained period of 1 min. is required. The increment shall continue until the maximum probe volume is reached. Meanwhile, volume loss originates from the compressibility of the testing equipment (such as the probe and liquid). Calibration for volume loss is done by pressurizing the probe within a steel casing within increments of 100 kPa (for maximum expansion pressure of 2.5 MPa) or 500 kPa (for maximum expansion pressure of 5 MPa). Each pressure increment should be applied within 20 s, and a maintained period of 1 min shall be applied once the probe is in contact with the steel tube. Following these 2 exercises, calibration curves are then produced to compute the pressure and volume losses. Using these data, the measured data during the actual test can then be corrected.

2.4 Site Investigation

During the actual PMT, a pre-bored hole is required in order to lower the pressuremeter probe into the required test section. It is generally advisable to maintain the pre-bored hole diameter between 1.03 and 1.2 times the pressuremeter probe diameter (ASTM D4719 2020). The material surrounding the pre-bored hole at the test section should experience minimal disturbance, especially during the pre-boring process. Furthermore, the PMT should be done immediately as soon as the test section is formed. These are essential steps to ensure that the test results are reliable.

While the application of PMT in soils has been established for a long time (Gibson and Anderson 1961; Houlsby and Carter 1993; Hughes et al. 1977), PMT in rocks is currently still limited. This is because pressuremeter tests are historically designed with a pressure capacity of up to 5 MPa. This limits the ability of normal PMTs to measure sufficient plastic deformation in rocks, which are needed for interpretation of strength parameters. It is worth noting that pressuremeters with higher pressure capacity (up to 20 MPa) have been developed recently. Despite this, recent studies have only used PMTs to derive stiffness parameters in intact rocks (Birid 2015; Cao et al. 2013; Dafni 2013; Mair and Wood 2013).

2.4.4.1 Testing procedure

Generally, the steps to perform PMT can be summarized as follows:

i .Pre-bore a test borehole toward the designated test depth
ii .Clean debris at the designated test depth
iii .Adjust all pressuremeter readings to zero at atmospheric pressure before lowering the probe
iv .Lower the pressuremeter probe to the required test depth
v .Expand the pressuremeter probe to begin the test
vi .Upon completion of test, deflate the probe to its original volume, V_0 and remove the probe from the test borehole.

Generally, the pressuremeter probe can be expanded by controlling either the pressure (Procedure A) or volume (Procedure B). Table 2.2 summarizes the criteria of performing both procedures, with an ideal pre-bored pressuremeter curve presented in Fig. 2.7.

The pressuremeter curve consists of three stages, representing three phenomena during the expansion of the pressuremeter probe. In Stage 1, the probe expands to be in contact with the borehole cavity walls. Once in contact, the probe pressure increases and restores the in situ stress condition at the test section. This is normal, as the pre-boring process causes stress relief on the soils surrounding the borehole cavity. Once the in situ stress condition is restored, further probe expansion initiates the elastic (Stage 2), and subsequently plastic (Stage 3) behavior of the tested soil.

It is worth noting that the presence of Stage 1 in the pre-bored pressuremeter curve is predominantly due to the effects of pre-boring. It is often argued that the actual response of the material toward pressuremeter probe expansion starts from

Table 2.2 Summary of criteria for performing pressuremeter test (PMT) using Procedure A (pressure-controlled) or Procedure B (volume-controlled)

Procedure	A (Pressure-controlled)	B (Volume-controlled)
Pressure or volume increment	25, 50, 100, or 200 kPa per increment	0.05–$0.1V_0$
No. of increments (or readings) per test	7–10	10–20
Reading accuracy	Within 5% of max. applied pressure	Within 0.2% of V_0
Maintained period per increment	30 s–1 min	
Termination criteria	Probe expansion at one load increment exceeds $0.25V_0$	Max. expansion limit of pressuremeter probe

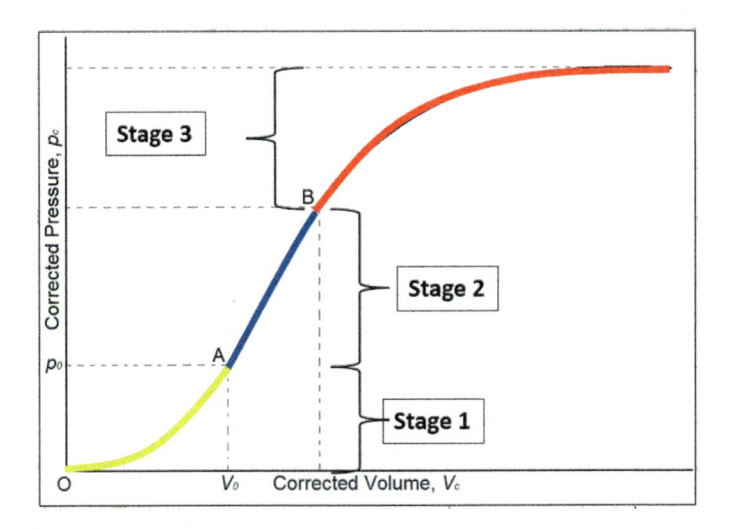

Fig. 2.7 Ideal pressure–volume curve for pre-bored pressuremeter curve

Stage 2 (Briaud 1992). This argument is further supported when a typical curve from self-boring PMT is compared against that of pre-bored pressuremeter, as shown in Fig. 2.8.

It is worth noting that self-boring PMT does not require any pre-boring. This is because a self-boring pressuremeter probe can simultaneously drill the ground while being lowered to the required test depth. Given the absence of pre-boring works in self-boring PMT, it was evident that the measurements presented in Stage 1 are not present in the results from self-boring PMT.

If the PMT is done correctly, a pressuremeter curve similar to Fig. 2.7 should be expected. From this result, pressuremeter modulus E_p can be interpreted using the equation below (ASTM D4719 2020).

2.4 Site Investigation

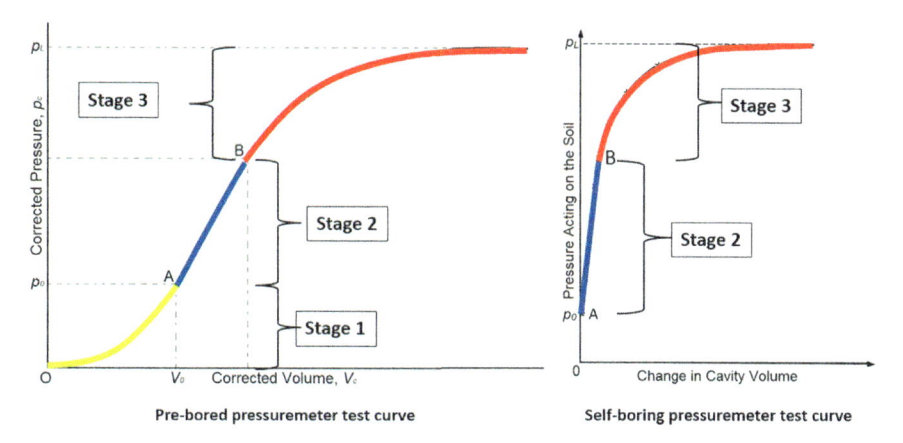

Fig. 2.8 Comparison between typical pre-bored pressuremeter test curve vs. self-boring pressuremeter test curve

$$E_p = 2(1 + v)(V_0 + V_m)\frac{\Delta P}{\Delta V} \qquad (2.2)$$

where V is Poisson's ratio, V_0 is the volume of the uninflated probe, $\frac{\Delta P}{\Delta V}$ is the gradient of the elastic deformation (Stage 2 in Fig. 2.7), and V_m is the corrected volume reading at the midpoint of volume increase in Stage 2.

External factors, such as the borehole size, or the tested material strength or stiffness may affect the shape of the pressuremeter curves. If the borehole size is much smaller/bigger than the probe, the shape of the pressuremeter curve may change. Furthermore, if the tested material is much stiffer and stronger than the maximum pressure of the pressuremeter test, the measured data may not show sufficient plastic behavior. These are illustrated in Fig. 2.9.

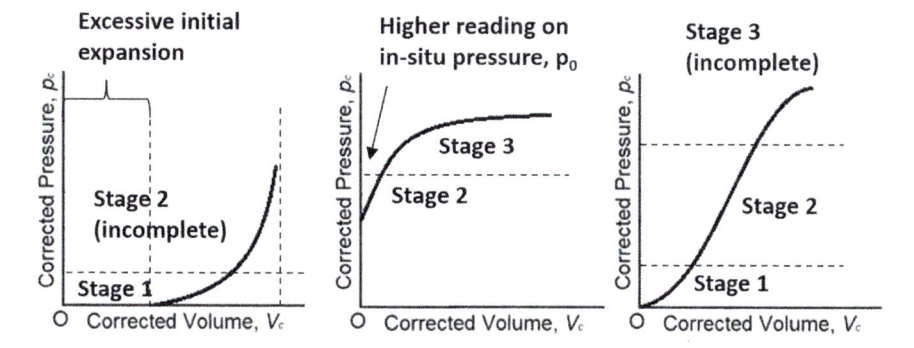

Fig. 2.9 (from left to right) Pressuremeter curve when the pre-bored hole is too large, pressuremeter curve when the pre-bored hole is too small, pressuremeter curve when the tested material is stiffer and stronger than the probe maximum pressure

It is worth noting that PMT is ideally done under undrained conditions. Generally, it is important to ensure that the test is done within 15–20 min. Performing the test much longer than this may result in the collapse of the pre-bored borehole cavity, which may damage the pressuremeter probe. An example of such a case for a test in highly weathered rock is presented, hereinafter.

2.4.4.2 Pressuremeter Testing in Highly Weathered Phyllite

A PMT (which shall be referred to as PMT-A hereinafter) was carried out within a highly fractured phyllite rock mass in Kuching City, Malaysia. The encountered subsoil condition at the test location is shown in Fig. 2.10.

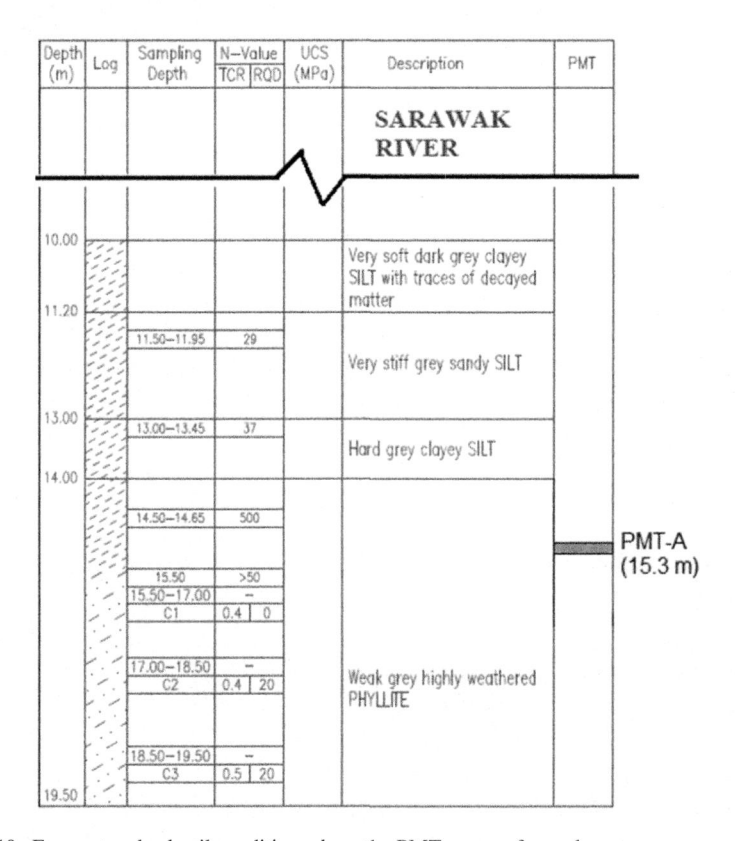

Fig. 2.10 Encountered subsoil condition where the PMT was performed

2.4 Site Investigation

Fig. 2.11 Monitoring board for the Menard pressuremeter test

The test was performed using Menard pre-bored pressuremeter. Figure 2.11 shows the monitoring board of the pre-bored pressuremeter. Measurements of pressure and deformation were manually read through the monitoring board. Figure 2.12 shows the exploded view of the inflatable pressuremeter probe.

During the test, the central cell membrane would expand when pressure is applied onto the surrounding rock mass. The guard cell sleeve helps to limit the vertical expansion of the central cell membrane. That way, the central cell membrane expansion can be maintained in the radial direction only.

The initial attempt of this test was abandoned due to a rupture in the central cell, detected when no change in the volumeter was visible with any increase in applied pressure. Replacement of the central cell resulted in a delay of approximately a day after the test section had been pre-bored. This was an unsatisfactory scenario as the PMT can be likened to an undrained unconsolidated triaxial test. Testing must be performed at the soonest instance after pre-boring of the test section. The test should be conducted between 15 and 20 min. Figure 2.13 shows some of the damage sustained from this prolonged test section.

Upon completion of the repairs, the second attempt at testing PMT-A was executed. Testing was carried out over 90 min, by which time it was speculated that the cavity wall had collapsed. This was later confirmed during the extraction of

Fig. 2.12 Exploded view of the pressuremeter probe

the probe. Figure 2.13 shows the extracted probe after testing of PMT-A as compared to a newly prepared probe. The guard cell along with the top collar ring had slid off during extraction, causing the guard cell sleeve to be partially folded inwards. Preparations for the next test required the replacement of the guard cell sleeve, as well as the damaged central cell membrane. Approximately 1 m of the gas and water line had to be severed due to damages imposed during the extraction of the probe. Fortunately, the central steel tube of the probe was undamaged.

Testing was only carried out to a maximum pressure of 60 bars. This limit was set as a safety precaution. However, this limit hindered the completeness of the PMT testing process at some test sections. In order to utilize the expression devised by Marsland and Randolph (1977), the full realization of the plastic region is required.

2.4.4.3 Interpretation of Test Results Using Cavity Expansion Theory

The measured data from PMT are commonly interpreted to characterize the pressuremeter modulus (ASTM D4719 2020). Gibson and Anderson (1961) developed one of the earliest methods to interpret undrained shear strength, c_u from pressuremeter tests in cohesive soils. Hughes et al. (1977) introduced an approach to

Fig. 2.13 (clockwise from top-left) pressuremeter probe alongside elastic membrane for central cell, pressuremeter probe after extraction from test section, ruptures in gas and water line after extraction of the probe, cutting open mesh-reinforced guard sleeve after damage from probe extraction

interpret friction angle from pressuremeter tests in sands while considering the dilatant behavior of sand. Yu and Houlsby (2015) used a semi-numerical approach to interpret pressuremeter curves using cavity expansion theory.

The cavity expansion theory used by Yu and Houlsby (2015) describes a material with linear elastic behavior (see Eq. 2.3). Meanwhile, the plastic behavior of the material is expressed using an infinite expansion equation, which is expressed in Eqs. 2.4–2.6.

$$\frac{a_0 - a}{a_0} = \frac{p - p_0}{2G} \text{ for } p < py \tag{2.3}$$

$$\frac{a-a_0}{a_0} = \ln\left\{\left[\frac{R^{-\gamma}}{(1-\delta)^{\frac{(\beta+1)}{\beta}} - \left(\frac{\gamma}{\eta}\right)\Lambda(R,\xi)}\right]^{\frac{\beta}{(\beta+1)}}\right\} \text{ for } p < p_y \quad (2.4)$$

$$R_\sigma = \frac{(1+\alpha)[Y+(\alpha-1)p]}{2\alpha[Y+(\alpha-1)p_0]} \quad (2.5)$$

$$\Lambda(R_\sigma,\xi) = \sum_{n=0}^{\infty} A_n^1 \text{ where, } A_n^1 = \frac{\xi^n}{n!}\ln R \text{ if } n = \gamma,$$

$$\text{or } A_n^1 = \frac{\xi^n}{n!(n-\gamma)}\left[R^{n-\gamma}-1\right] \text{ otherwise} \quad (2.6)$$

where a is the cavity radius, a_0 is the cavity radius corresponding to pressure, p, and G is the shear modulus. Material constants such as α, Y, γ, β, δ, ξ, η are functions of cohesion, c', friction angle φ', in situ pressure p_0, and Poisson's ratio, v. Using Eqs. 2.3–2.6, a theoretical pressuremeter curve can be generated. The theoretical curve is governed by the input parameters, such as c', φ', p_0, and G. Interpretation of strength and stiffness is done by fitting the theoretical curves against the actual pressuremeter curve. Yu and Houlsby's (2015) approach allows for characterization of soil strength in terms of Mohr–Coulomb (MC) strength parameters (cohesion, c' and friction angle, φ'). This makes it a suitable approach for characterizing soils that exhibit both cohesive and frictional properties.

Meanwhile, there are relatively limited studies on the use of PMT in rocks. Haberfield and Johnson (1990) performed laboratory-based PMT in synthetic mudstone and used the finite element method to interpret the results. The interpretation involved using Mohr–Coulomb strength parameters to govern the initiation of radial cracks during pressuremeter testing. With time and advancement of technology, there have been growing numbers of studies on PMT interpretation in rocks (Dafni 2013; Isik et al. 2008; Sharo and Liang 2012).

Recent studies done by Phangkawira et al. (2016, 2017a; b) have shown that cavity expansion theory can be used to interpret the results of PMTs in highly fractured and weathered phyllite. In the study, the semi-numerical approach developed by Yu and Houslby (2015) was used to characterize the stiffness and strength of the highly fractured and weathered phyllite rock mass. The stiffness and strength of the rock mass were expressed in terms of Young's modulus and Mohr–Coulomb strength parameters, respectively. The characterized parameters have been shown to be applicable for the assessment of jacking forces.

The results of PMTs in the highly fractured and weathered phyllite were interpreted using Yu and Houlsby's (2015) semi-numerical approach. The approach was based on the fundamental theory of cylindrical cavity expansion in a material with cohesive and frictional properties. The material was assumed to have linear elastic deformation and nonlinear plastic deformation.

The semi-numerical approach interpreted pressuremeter curves by generating a theoretical pressure-expansion curve using a combination of a linear function and infinite series expansion equations (Yu 2000). The theoretical pressure-expansion

2.4 Site Investigation

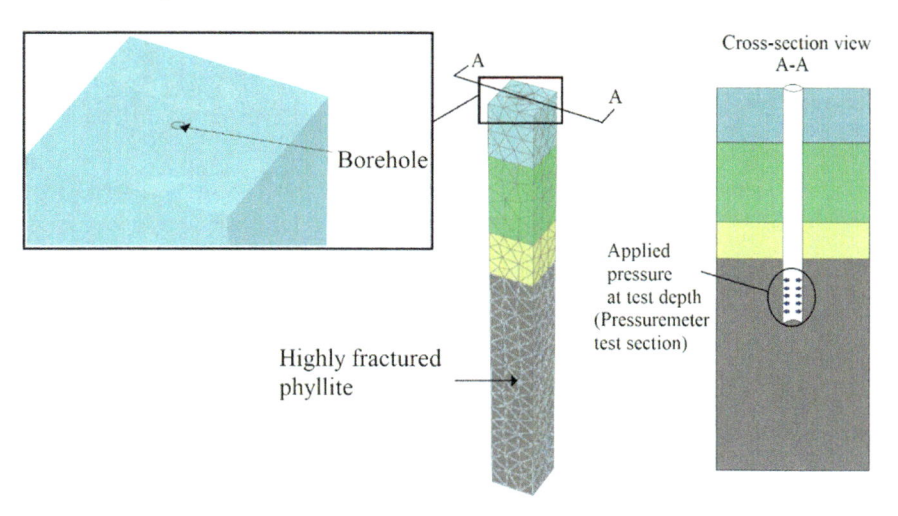

Fig. 2.14 Model geometry for the numerical pressuremeter test

curve is governed by several input parameters, such as in situ stress, p_0, Young's modulus, E, Poisson's ratio, v, cohesion, c', and friction angle, φ'. Interpretation of pressuremeter curves was done by optimizing the input parameters to generate a best-fit between the theoretical pressure-expansion curves and the actual pressuremeter curves. The optimized input parameters were then considered to be representative of the equivalent rock mass properties.

2.4.4.4 Validation of Strength Parameters Using Finite Element Analysis

The optimized parameters were validated by simulating the PMT via finite element analysis. This involved generating a numerical model of geology based on the soils and rock mass encountered in the borehole where the PMTs were performed. The optimized parameters were used to simulate the rock mass behavior in the finite element analysis. Afterward, a cylindrical cavity was introduced from the modeled ground surface to the respective pressuremeter test depth. The expansion of pressuremeter probe was modeled by applying radial pressure against the cylindrical cavity surface. The resulting normal stresses and hoop strains in the modeled rock mass were plotted and compared against the actual pressuremeter curves and the theoretical pressure-expansion curves. Figure 2.14 shows the numerical model geometry of the PMT.

2.4.4.5 Sensitivity Assessment of Developed Strength Parameters

The sensitivity of the input strength parameters was investigated in order to justify the use of the equivalent MC strength parameters. This was done using the finite

Table 2.3 Benchmark input parameters for sensitivity study (obtained via hardening soil model)

Tested material	Phyllite
Unit weight, γ (kN/m^3)	22
E_{50}^{ref} (kPa)	107×10^3
$E_{\text{oed}}^{\text{ref}}$ (kPa)	64×10^3
$E_{\text{ur}}^{\text{ref}}$ (kPa)	214×10^3
Stress-dependent power, m	0.5
Equivalent cohesion, c' (MPa)	3.0
Equivalent friction angle, φ' (°)	53

element model that was developed earlier, as shown in Fig. 2.14. This sensitivity analysis further confirmed that the numerical pressuremeter curve presented earlier can only be achieved with a specific set of strength values (i.e., unique combination of equivalent c' and φ').

To perform the sensitivity study, one example of pressuremeter curve was used as a benchmark. The pressuremeter curve was obtained from the test performed within the highly fractured and weathered phyllite. Using the earlier characterization method, benchmark input parameters for this sensitivity study will be established. These parameters are summarized in Table 2.3.

The benchmark input parameters are used as a starting point for the sensitivity study since this combination of parameters is capable of generating a numerical pressuremeter curve that fits the measured pressuremeter curve. The sensitivity study involves investigating the sensitivity of cohesion c' and friction angle φ'. This is done by changing one parameter at a time to evaluate how this parameter affects the generated numerical pressuremeter curve. Comparison between the actual pressuremeter curve and the generated numerical pressuremeter curves will be presented, hereinafter.

Effect of Equivalent Cohesion, c' on the Numerical Pressuremeter Curves

In this parametric study, several sets of numerical pressuremeter curves were generated with various values of cohesion, c', while φ' remains unchanged. Figure 2.15 shows the sensitivity of c' toward the generated numerical pressuremeter curves.

As shown in Fig. 2.15, c' impacts the behavior of the pressuremeter curves at higher strain levels. This was notable when the hoop strain, ε_θ was greater than 4%, which corresponded to the plastic deformation of the numerical pressuremeter curve. The tangential gradient of the pressuremeter curve at $\varepsilon_\theta > 4\%$ became steeper with increasing c' values. This is expected, as increasing c' value means higher material strength. Meanwhile, the changes of c' have no significant effects on the numerical pressuremeter curve at lower strain level. This was demonstrated by the relatively unchanged pressuremeter curve at $\varepsilon_\theta < 4\%$ despite the changes in c'. Therefore, this observation indicates that the equivalent c' mainly governs the material strength at higher strain levels, specifically where plastic deformation is expected to be achieved.

2.4 Site Investigation

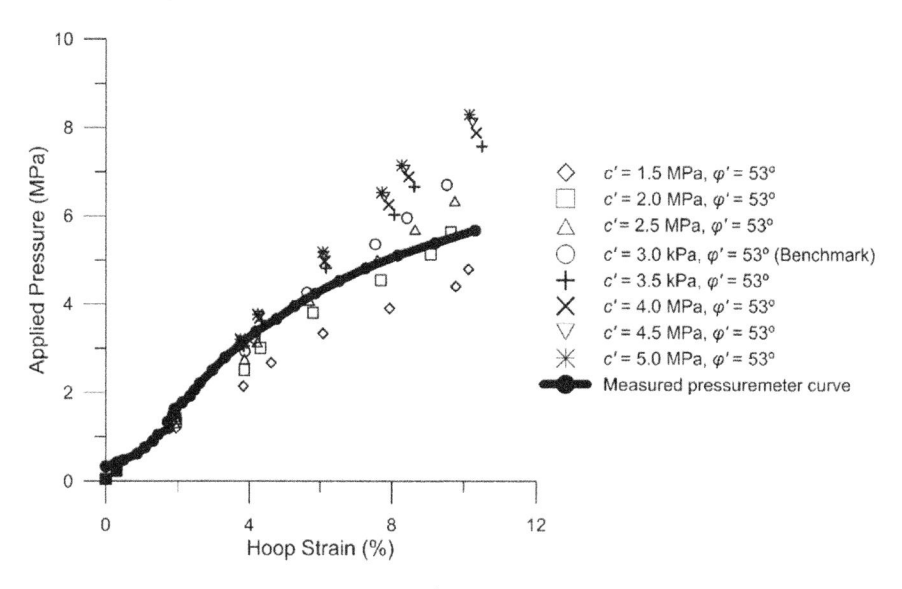

Fig. 2.15 Sensitivity of equivalent cohesion, c' toward the numerical pressuremeter curve

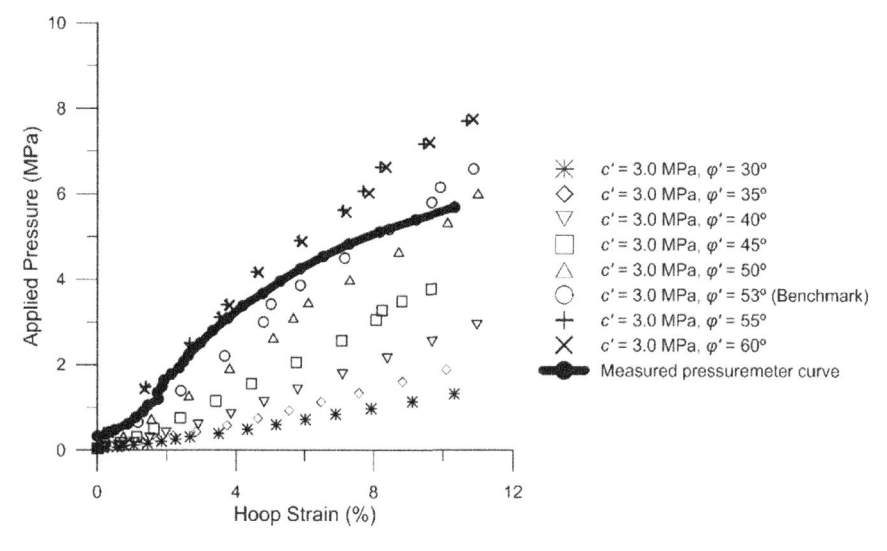

Fig. 2.16 Sensitivity of equivalent friction angle, φ' to the numerical pressuremeter curve

Effect of Equivalent Friction Angle, φ' on the Numerical Pressuremeter Curves

The sensitivity of φ' was studied by generating numerical pressuremeter curves with different variations of φ' values, while c' remains unchanged. Figure 2.16 shows the sensitivity of φ' toward the generated numerical pressuremeter curves.

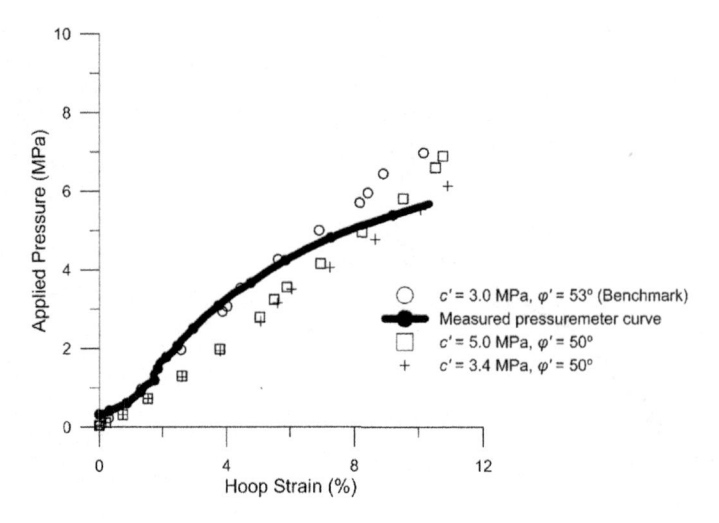

Fig. 2.17 Sensitivity of the benchmark numerical pressuremeter curve to changes in c

Figure 2.16 shows that the equivalent φ' affects the numerical pressuremeter curve at both small-strain and large-strain levels. This shows that the development of the numerical pressuremeter test curve is sensitive to changes in φ'. Consequently, it is possible to suggest that a unique value of φ' can be interpreted. However, since c' also has influence over the numerical pressuremeter curve, one may argue that the effects of φ' can be compensated by changing c'. Therefore, it is necessary to study the response of c' toward the changes in φ'. This was done by trying to reproduce the same numerical pressuremeter curve with different combinations of c' and φ'.

Investigation on the Response of c' Subject to the Changes in φ'

Figure 2.17 shows the attempt to reproduce the same numerical pressuremeter curve when the value of φ' is reduced. The benchmark numerical pressuremeter curve is based on the values of c' and φ' derived in Table 2.3, i.e., $c' = 3.0$ MPa and $\varphi' = 53°$. When φ' was reduced to 50°, the value of c' was increased to 3.4 MPa (1st attempt) and 5.0 MPa (2nd attempt).

Despite the attempts, the increase in c' could not reproduce the same benchmark numerical pressuremeter curve. This was evident, as the pressure at $\varepsilon_\theta = 5\%$ was underpredicted by approximately 30% despite the magnitude of c' was increased to 5.0 MPa. From this finding, it is important to acknowledge the high sensitivity of φ', given the significant underprediction of pressure that resulted from slight reduction in φ'. Furthermore, since changing the values of c' could not replicate the same numerical pressuremeter curve, it is sufficient to conclude that the numerical pressuremeter curve could only be reproduced with a set of unique c'-φ' combination.

Summary of Sensitivity Assessment

The earlier sensitivity assessment on the strength parameters has shown that:

- The equivalent friction angle, φ', has major influence on both elastic and plastic deformations in the numerical pressuremeter curve.
- The generated numerical pressuremeter curve is governed by a unique combination of c' and φ'.
- The combination of strength parameters $c' = 3.0$ MPa and $\varphi' = 53°$ managed to produce a numerical pressuremeter curve that is in relatively good agreement with the actual pressuremeter curve. This validates the strength parameters, which was interpreted from the actual pressuremeter curve using cavity expansion theory (Yu and Houlsby 2015).

The application of a series of real-life pressuremeter tests and their numerical verifications will be discussed in detail in Chap. 3.

2.5 Geotechnical Laboratory Testing

Laboratory tests on disturbed or undisturbed samples would generate information about physical properties (density, water content, grain size distribution, and Atterberg limits), hydraulic properties, and mechanical properties (deformability and strength parameters). The investigation of these parameters is very important as they play a major role in determining the type of mTBM, mucking system, shaft construction method, and jacking system.

While undisturbed samples retain the in situ soil structure and water content, disturbed samples are reworked and, hence, are only representative of the grain size distribution. This implies that in situ soil conditions, such as the soil water content, density, stress history, compressibility, stress–strain, and strength characteristics can be only investigated by means of undisturbed samples.

Despite their importance, undisturbed samples are very difficult to obtain, particularly in cohesionless materials where the soil structure is partially or completely destroyed when driving or pushing the sampler (Lancellotta 2009). Further problems arise when larger gravel or cobbles are present, as they will not enter the sampler, causing misleading of the strata log. However, in this soil, the quality of the samples can be increased by using suited samplers (i.e., stationary piston sampler), freezing the soil before its sampling or, in case of large elements, by taking samples of big sizes to test in the laboratory.

Problems arise also in rocks, for example in fractured strata, where collected samples can be of poor quality because of the use of wash water for cooling the bit. Even in this case, the quality of the samples can be improved using well-suited samplers (i.e., double-tube core sampler) that limit the disturbance due to the coring procedure on the collected sample. Finally, it should be noted that the results obtained are those for the samples taken, not for the actual soil in place and are related to the position of the borehole. On the other hand, as highlighted previously, geophysical investigations can contribute well to optimize the layout of borehole campaign and the lateral extrapolation of the obtained results.

2.5.1 Interface Tests

An additional relevant aspect for microtunneling projects is the determination of the friction coefficient at the soil–pipe interface. This is generally estimated to vary between an upper bound of soil friction angle, φ and a lower limit of $\varphi/3$ to $\varphi/2$ (Pellet-Beaucour and Kastner 2002). With the addition of slurry lubricant, the interface friction can be reduced significantly (Reilly 2014; Reilly and Orr 2017). The inclusion of bentonite slurry can also help to keep the stability of the borehole and minimize the ground surface subsidence due to pipe jacking. Experimental efforts have been taken to investigate the degree of reduction of interface friction induced by slurry lubricant, such as the use of standard Casagrande apparatus (Milligan and Norris 1999) and the use of a vertically pushed concrete sample in soil samples (Namli and Guler 2016). A direct shear apparatus can be modified to conduct interface tests as shown in Fig. 2.18, since the soil–pipe friction occurs in the horizontal plane and the applied overburden pressure (from 50 to 250 kPa) can be adjusted to simulate different pipe burial depth (from approximately 3 to 16.5 m).

Pipe specimens of varying materials (reinforced concrete and steel pipes) can be fabricated into blocks with dimensions of 90 mm × 90 mm × 10 mm (width x length x height). The outside surface of the pipe should be protected for use as the sliding interface. Examples of these kinds of tests are reported by Ji et al. (2019).

During pipe jacking, the jacking force will also be reduced with the occurrence of overcut as time elapses. The use of the friction coefficient determined without considering the influence of overcut will provide a conservative estimation for the jacking force.

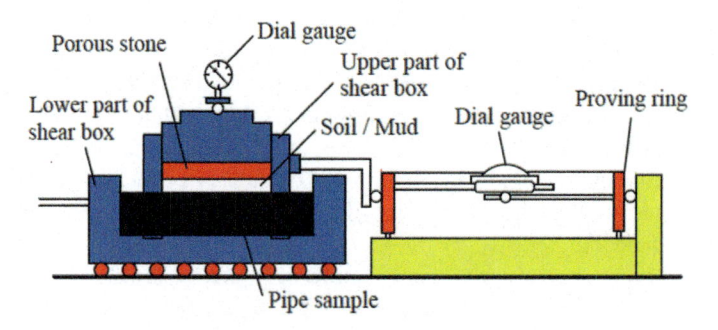

Fig. 2.18 Modified direct shear apparatus for interface tests (after Ji et al. 2019)

2.5.2 Case Study: Direct Shear Testing of Reconstituted Tunneling Spoil from Highly Fractured Rocks

2.5.2.1 Specimen Collection and Preparation

This section focuses on the development of equivalent strength parameters of highly fractured rocks. This was carried out through the use of direct shear tests on reconstituted tunneling rock spoil. Particle size distribution tests were performed in accordance with ASTM D422-63 (2002) to determine the granulometry of the collected tunneling rock spoil. Figure 2.19 shows the resulting envelopes of gradation curves before and after the reconstituted tunneling rock spoil went through the process of scalping.

It was necessary to perform scalping to fulfill the direct shear test requirements regarding the relative comparison between the tunneling rock spoil grain sizes and the shear box. To comply with the requirements from Standards Australia (1998)

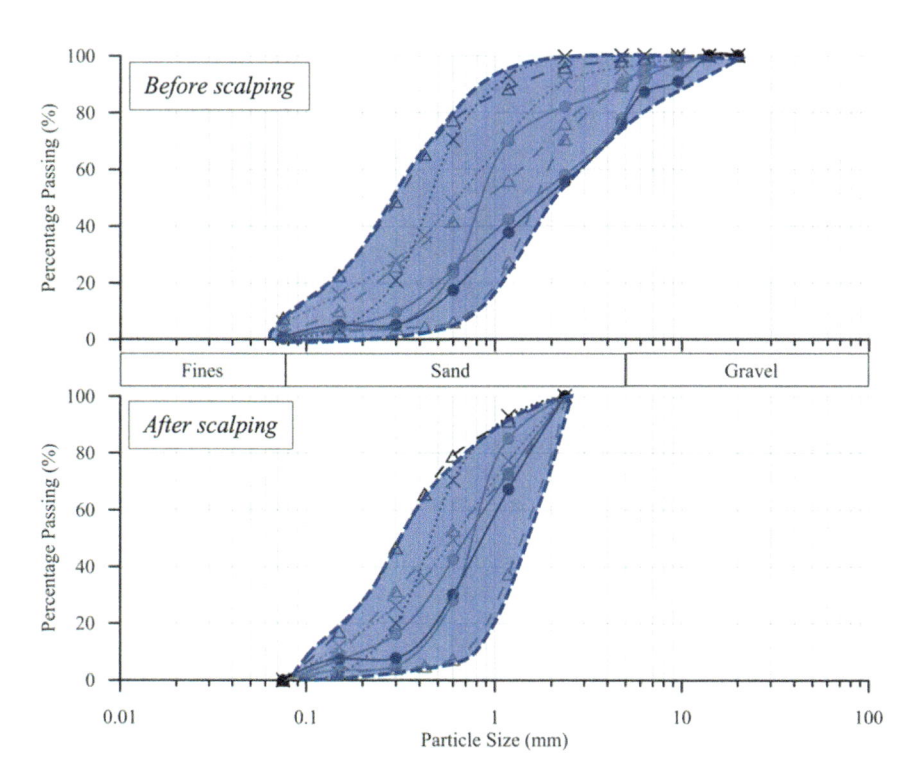

Fig. 2.19 Envelopes of gradation curves for collected specimens of reconstituted tunneling rock spoil, before and after scalping

and ASTM (2003), the particle sizes of the test specimens cannot exceed 1/10th the thickness of the test specimens and 1/6th the diameter of the test specimen. This was similarly recommended by Head (1992).

It is worth noting that the test results may be influenced by the scalping of the coarser fraction of the tunneling rock spoil. The particle shapes of the scalped samples, on the other hand, would stay identical as compared to the original unscalped spoil. This would maintain the 'blockiness' of sandstone grains due to quartz minerals, as well as the angularity of phyllite and shale spoil due to mica flakes. The tunneling rock spoil that was sieved before scalping was classified as poorly graded sand-sized spoil (SP), according to the Unified Soil Classification System (ASTM 2000).

2.5.2.2 Testing Procedure

As previously stated, the test specimens were prepared so that the highest particle size did not exceed 1/10th of the shear box height and 1/6th of the shear box diameter (ASTM 2003; Head 1992; Standards Australia 1998). Afterward, three layers of the scalped samples were poured into the shear box. After every layer was poured into the shear box, a tamper was used to properly compact the sample. A tamping plate can be used to ensure that the compaction is distributed evenly.

Direct shear tests can be performed under conditions of constant normal stiffness (CNS), constant volume (CV), or constant normal load (CNL) (Pellet and Keshavarz 2014). The direct shear tests in this study were performed under CNL conditions. This was done so that the in situ stress condition along the drive can be reflected in the test. The tests were performed based on a range of effective normal stresses. This includes the expected overburden stress at the tunnel depth. The range of various effective normal stresses at each test was used to study the strength behavior of the reconstituted tunneling rock spoil. Before the test, the specimens were saturated and subject to primary consolidation. The primary consolidation was typically over within 5 min. After the primary consolidation, the specimens were sheared at a constant rate. According to ASTM (2003), the recommended shearing rate for clean dense sands is given as

$$d_r = \frac{d_f}{t_f} \qquad (2.7)$$

where d_r is the displacement rate (mm/s), d_f is the estimated horizontal displacement rate at failure, and t_f is the total estimate elapsed time to failure (600 s for clean dense sands).

2.5 Geotechnical Laboratory Testing

The recommended shearing rate was $d_r = 0.0083$ mm/s, which was based on the recommended values of d_f and t_f. In this study, however, a lower shearing rate of $d_r = 0.0017$ mm/s was adopted to ensure possible dissipation of excess pore water pressure. All specimens were tested with a minimum of 15 mm horizontal deformation. This was sufficient for the tunneling rock spoil specimens to achieve a residual state.

2.5.2.3 Equivalent Rock Strength Parameters Using Mohr–Coulomb Failure Criterion

Mohr–Coulomb (MC) failure criterion is commonly used to characterize the strength of soils and other granular material. The MC failure criterion models a material with linear elastoplastic behavior. This criterion states that failure will occur when shear stress at any point in a soil mass becomes equal to the soil shear strength. The soil shear strength, τ_f, as defined by MC failure criterion is expressed as:

$$\tau_f = c' + \sigma_f' \tan \phi' \quad (2.8)$$

where c' is the effective cohesion of a soil, σ_f' is the effective normal stress at failure, and ϕ' = effective internal friction angle.

Soil strength tests such as triaxial tests and direct shear tests can be used to derive the values of c' and ϕ'. From the MC failure criterion equation, a linear failure envelope of the tested soil can be interpreted via a line of best fit. From this line, the value of c' is derived from the vertical axis intercept. Meanwhile, the gradient of the best fit line would give the value of ϕ'. This method is graphically illustrated in Fig. 2.20.

The interpretation of soil shear strength shown in Fig. 2.19 is also applicable to results from direct shear tests (Härtl and Ooi 2008; Nam et al. 2011). The direct shear test results can be used to generate two envelopes, each representing the peak shear

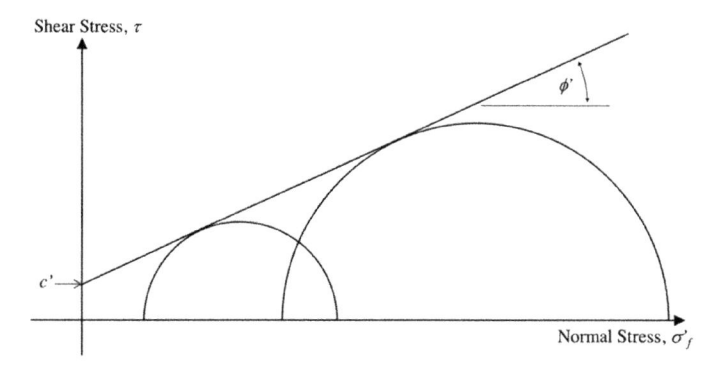

Fig. 2.20 Mohr–Coulomb failure envelope

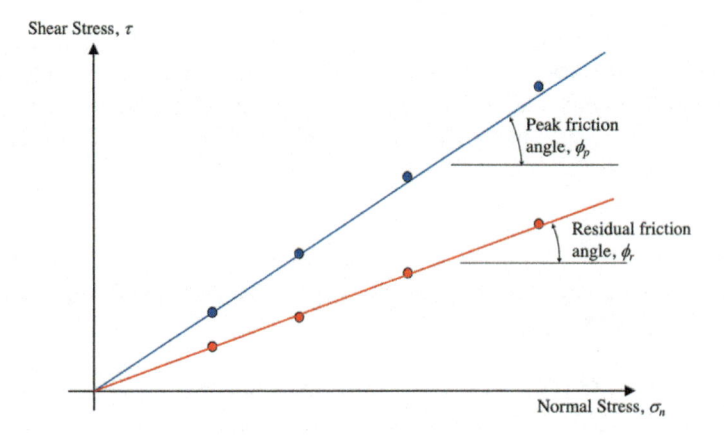

Fig. 2.21 Determination of peak and residual friction angles from direct shear tests

strength and residual shear strength, respectively. Figure 2.21 illustrates a typical interpretation of failure envelope from the direct shear test.

It is important to note that the value of cohesion, c' derived from Mohr–Coulomb failure criterion is not 'true cohesion,' but rather 'apparent cohesion. Misinterpretation of apparent cohesion (c' derived from drawing line of best fit to strength test results) against true cohesion (from interlocking) has been discussed by Schofield (1998, 2001). Furthermore, it is also worth noting that soils exhibit nonlinear strength behavior, as indicated in Fig. 2.19. As a result, the reliability of representing nonlinear soil strength using a linear best fit line would largely depend on the applied stress levels during testing.

The Mohr–Coulomb failure criterion is a convenient and simple model to interpret soil strength behavior from direct shear test results. However, it is important to realize that the assumed linear strength behavior in this model is one of the main shortcomings, since true soil behavior is nonlinear. To improve the characterization of soil strength, a power law function can be used to consider the nonlinearity in soil strength behavior. The use of this function is explored, hereinafter.

2.5.2.4 Equivalent Rock Strength Parameters Using Power Law and Generalized Tangential Method

Use of nonlinear failure envelopes in geotechnical engineering practice has been slowly growing (Anyaegbunam 2015). This is mainly due to higher quality triaxial tests at low confining pressure below 50 kN/m^2, where nonlinearity in strength is more evident at such lower normal stresses.

2.5 Geotechnical Laboratory Testing

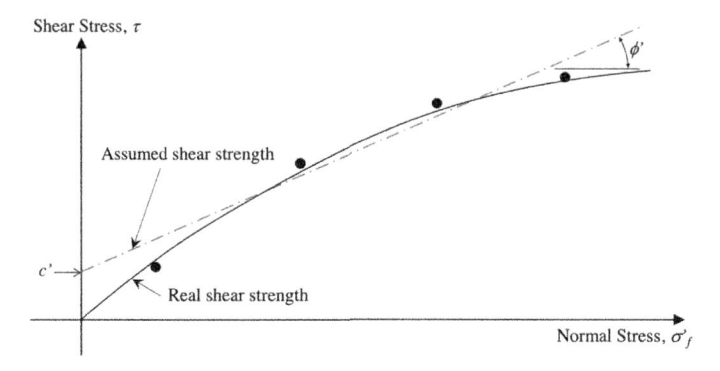

Fig. 2.22 Comparison of Mohr–Coulomb failure criterion (straight line) and power function failure criterion (curve) (after Lade 2010)

The use of a nonlinear strength envelope is motivated by the shortcomings of using linear best fit line to represent the tested material strength. Figure 2.22 shows that the linear Mohr–Coulomb failure criterion overestimates shear strength at low and high normal stresses, while it underestimates shear strength at intermediate normal stresses.

It is important to note that assessment of engineering problems such as surficial slope stability generally occurs within this stress range. Therefore, consideration of stress-dependency when evaluating soil strength is essential. The use of the power law function can be useful, as it describes the nonlinear strength in soils. This would help address the shortcomings of the Mohr–Coulomb failure criterion. The nonlinear power law function has been previously applied in embankment designs, specifically in characterizing the strength behavior of reconstituted rockfill materials (Charles and Watts 1980; De Mello 1977).

The power law function is given as

$$\tau = A \cdot (\sigma')^B \qquad (2.9)$$

where σ' is the effective normal stress, τ is the shear strength, and A & B are empirical constants.

The reported studies have shown the ability of a simple nonlinear power law in characterizing the strength of reconstituted materials. Despite this, the use of Mohr–Coulomb parameters, c and ϕ, is still commonly used in design practice. Yang and Yin (2004) used a 'generalized tangential' technique, where a tangent line was applied to the nonlinear power law failure criterion. The tangent line was used to represent the linear Mohr–Coulomb profile, as illustrated in Fig. 2.22. This technique enabled the use of Mohr–Coulomb parameters to analyze the engineering problem faced. This practice has been performed on multiple slope stability problems (Collins et al. 1988; Drescher and Christopoulos 1988) and retaining walls (Soon and Drescher 2007).

The 'generalized tangential' approach has demonstrated the ability to describe soil strength behavior from experimental data using nonlinear power law, while still using Mohr–Coulomb parameters to represent the shear strength of the soil. As a result, stress-dependency in soil strength behavior can be taken into account, while still making use of Mohr–Coulomb parameters. This can be an advantage for evaluating geotechnical engineering problems that are still reliant on Mohr–Coulomb parameters, such as the assessment of pipe jacking forces. Recent studies have shown success in using this method for the assessment of jacking forces (Choo and Ong 2015; Ong and Choo 2016, 2018; Peerun et al. 2018) and for stone column design (Ong et al 2018).

2.6 The Geological and Geotechnical Model

Based on the information collected during the investigation campaign, it is possible to describe the characteristics of the ground, the stress state at the site, the hydraulic conditions, and also to quantify the deformability and strength parameters. All this knowledge is used to define the reference geological and geotechnical model which then forms the basis of the design. The geological and the geotechnical model can be defined as a schematic representation of reality being able to describe the fundamental aspects of the behavior of a soil or a rock mass. It is obtained by combining all the information provided by the geotechnical characterization process (Fig. 2.23).

The ground characteristics, including deformability and strength properties, are considered as input data for the definition of the appropriate model. In the case of a rock mass, characterization will focus on the intact rock and the discontinuities

Fig. 2.23 Conceptual flowchart for the definition of the geotechnical model and the design approach (modified from Barla 2010)

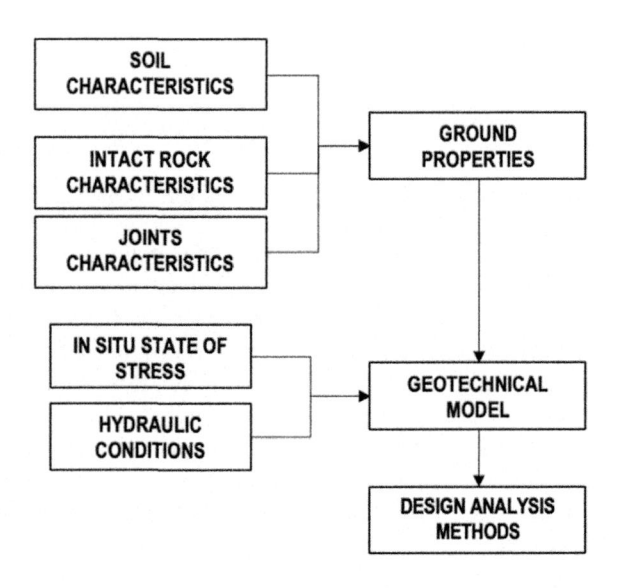

2.6 The Geological and Geotechnical Model

which will allow the characteristics of the rock mass to be determined. In addition, information on the stress state and the hydraulic conditions at the site should also be considered.

The design approaches that are to be adopted will then be based on the geotechnical model assumed. Two design approaches are of particular interest in geotechnical engineering:

- Equivalent Continuum Model, where the soil or the rock mass is treated as a continuum, where the input strength and deformability properties remain equal in all directions. This concept is then used to define the given constitutive relationship for the medium, e.g., elastic, elasto-plastic.
- Discontinuum Model, where the soil or the rock mass is represented as a discontinuum and most of the efforts are devoted to the characterization of its single components, i.e., for a rock mass, it is the rock elements and the rock joints/discontinuities. The modeling approach considers the granular or blocky nature of the system being analyzed. Each grain or block may interact with the neighboring blocks through the joints.

The equivalent continuum approach is commonly adopted for geotechnical engineering, while the discontinuum approach is more aligned to rock engineering. In the case of microtunneling, the first approach is more common.

References

Anyaegbunam AJ (2015) Nonlinear power-type failure laws for geomaterials: synthesis from triaxial data, properties, and applications. Int J Geomech 15(1). https://doi.org/10.1061/(ASCE)GM.1943-5622.0000348

ASTM D2487-00 (2000) Standard practice for classification of soils for engineering purposes (unified soil classification system). ASTM International, West Conshohocken, PA

ASTM D3080-03 (2003) Standard test method for direct shear test of soils under consolidated drained conditions, ASTM International, West Conshohocken, PA

ASTM D422-63 (2002) Standard test method for particle-size analysis of soils. ASTM International, West Conshohocken, PA

ASTM D4719-07 (2020) Standard test methods for prebored pressuremeter testing in soils. ASTM International, West Conshohocken, PA

Barla M (2010) Elementi di meccanica e ingegneria delle rocce. Celid, Torino, p 319

Barla M, Camusso M, Aiassa S (2006) Analysis of jacking forces during microtunnelling in limestone. Tunn Undergr Sp Tech 21(6):668–683. https://doi.org/10.1016/j.tust.2006.01.002

Birid KC (2015) Interpretation of pressuremeter test in rock. ISP-PRESSIO 2015, Hammamet, Tunisia, pp 289–299

Briaud JL (1992) The pressuremeter, A.A. Balkema, Rotterdam

Camusso M (2008) Barla M, Camusso M, Aiassa S (2006) Analysis of jacking forces during microtunnelling in limestone. PhD Thesis, Politecnico di Torino, 224 pp

Cao LF, Peaker SM, Sirati A (2013) Rock modulus from in-situ pressuremeter and laboratory tests. In: Proceedings of the 18th international conference on soil mechanics and geotechnical engineering, Paris, France, pp 1–4

Charles JA, Watts KS (1980) The influence of confining pressure on the shear strength of compacted rockfill. Géotechnique 30(4):353–367. https://doi.org/10.1680/geot.1980.30.4.353

Choo CS, Ong DEL (2015) Evaluation of pipe-jacking forces based on direct shear testing of reconstituted tunneling rock spoils. J Geotech Geoenviron 141(10), https://doi.org/10.1061/(ASCE)GT.1943-5606.0001348

Clarke BG, Smith A (1992) Self-boring pressuremeter tests in weak rocks. Constr Build Mater 6(2):91–96. https://doi.org/10.1016/0950-0618(92)90057-6

Collins IF, Gunn CIM, Pender MJ et al (1988) Slope stability analyses for materials with a non-linear failure envelope. Int J Numer Anal Met 12(5):533–550. https://doi.org/10.1002/nag.1610120507

Dafni J (2013) The analysis of weak rock using the pressuremeter. Dissertation, University of Washington

De Mello VFB (1977) Reflections on design decisions of practical significance to embankment dams. Géotechnique 27(3):281–355. https://doi.org/10.1680/geot.1977.27.3.281

Drescher A, Christopoulos C (1988) Limit analysis slope stability with nonlinear yield condition. Int J Numer Anal Met 12(3):341–345. https://doi.org/10.1002/nag.1610120307

Foti S, Sambuelli L, Socco LV, Strobbia C (2002) Caratterizzazione mediante onde superficiali per studi di fattibilità relativi alle perforazioni trenchless in ambito urbano. Atti del XXI Congresso Nazionale di Geotecnica, AGI, L?Aquila, 11–14 Settembre, pp 73–78

FSTT (French Society for Trenchless Technology) (2004) Microtunnelling and horizontal drilling: recommendations. Hermes Science Publishing Ltd

Gibson RE, Anderson WF (1961) In-situ measurement of soil properties with the pressuremeter. Civil Eng Public Works Rev 56(658):615–618

Haberfield CM, Johnson IW (1990) A numerical model for pressuremeter testing in soft rock. Geotechnique 40(4):569–580. https://doi.org/10.1680/geot.1990.40.4.569

Härtl J, Ooi J (2008) Experiments and simulations of direct shear tests: porosity, contact friction and bulk friction. Granul Matter 10(4):263–271. https://doi.org/10.1007/s10035-008-0085-3

Head KH (1992) Manual of soil laboratory testing, vol 2, Permeability, shear strength and compressibility tests. Halsted Press, New York

Houlsby G, Carter J (1993) The effects of pressuremeter geometry on the results of tests in clay. Géotechnique 43:567–576. https://doi.org/10.1680/geot.1993.43.4.567

Hudson JA (1989) Rock mechanics principles in engineering practice. CIRIA/Butterworths, London, p 72

Hughes JMO, Wroth CP, Windle D (1977) Pressuremeter Tests in Sands. Géotechnique 27(4):455–477. https://doi.org/10.1680/geot.1977.27.4.455

Isik NS, Ulusay R, Doyuran V (2008) Deformation modulus of heavily jointed-sheared and blocky greywackes by pressuremeter tests: numerical, experimental and empirical assessments. Eng Geol 101(3):269–282. https://doi.org/10.1016/j.enggeo.2008.06.004

Ji X, Zhao W, Ni P et al (2019) A method to estimate the jacking force for pipe jacking in sandy soils. Tunn Undergr Sp Tech 90:119–130. https://doi.org/10.1016/j.tust.2019.04.002

Lade PV (2010) The mechanics of surficial failure in soil slopes. Eng Geol 114(1–2):57–64. https://doi.org/10.1016/j.enggeo.2010.04.003

Lancellotta R (2009) Geotechnical engineering. CRC Press, p 520

Mair RJ, Wood DM (2013) Pressuremeter testing: methods and interpretation, CIRIA ground engineering report: in-situ testing. Butterworths, London

Marsland A, Randolph MF (1977) Comparisons of the results from pressuremeter tests and large in situ plate tests in London Clay. Geotechnique 27(2):217–243. https://doi.org/10.1680/geot.1977.27.2.217

Milligan GWE, Norris P (1999) Pipe-soil interaction during pipe jacking. Proc Inst Civ Eng: Geotech Eng 137(1):27–44. https://doi.org/10.1680/gt.1999.370104

Nam S, Gutierrez M, Diplas P et al (2011) Determination of the shear strength of unsaturated soils using the multistage direct shear test. Eng Geol 122(3–4):272–280. https://doi.org/10.1016/j.enggeo.2011.06.003

Namli M, Guler E (2016) Effect of bentonite slurry pressure on interface friction of pipe jacking. J Pipeline Syst Eng 8(2). https://doi.org/10.1061/(ASCE)PS.1949-1204.0000255

References

Ong DEL, Choo CS (2016) Back-analysis and finite element modelling of jacking forces in weathered rocks. Tunn Undergr Sp Tech 51(1):1–10. https://doi.org/10.1016/j.tust.2015.10.014

Ong DEL, Choo CS (2018) Assessment of non-linear rock strength parameters for the estimation of pipe-jacking forces. Part 1. Direct shear testing and backanalysis. Eng Geol 244(1):159–172. https://doi.org/10.1016/j.enggeo.2018.07.013

Ong DEL, Sim YS, Leung CF (2018) Performance of field and numerical back-analysis of floating stone columns in soft clay considering the influence of dilatancy. Int J Geomech 18(10). https://doi.org/10.1061/(ASCE)GM.1943-5622.0001261

Peerun MI, Ong DEL, Choo CS (2018) Interpretation of geomaterial behavior during shearing aided by PIV technology. J Mater Civil Eng 31(9). https://doi.org/10.1061/(ASCE)MT.1943-5533.0002834

Pellet FL, Keshavarz M (2014) Shear behavior of the interface between drilling equipments and shale rocks. J Pet Explor Prod Technol 4:245–254. https://doi.org/10.1007/s13202-014-0108-z

Pellet-Beaucour AL, Kastner R (2002) Experimental and analytical study of friction forces during microtunneling operations. Tunn Undergr Sp Tech 17(1):83–97. https://doi.org/10.1016/S0886-7798(01)00044-X

Phangkawira F, Ong DEL, Choo CS (2016) Characterisation of highly fractured and weathered rock mass using pressuremeter test. In: Chan SH, Ooi TA, Ting WH, Chan SF, Ong DEL (eds.) Deep Excavation and Ground Improvement – Proceedings of the 19th SEAGC and 2nd AGSSEAC, Subang -Jaya, Malaysia, 31 May – 3 June 2016, ISBN 978-983-40616-4-7

Phangkawira F, Ong DEL, Choo CS (2017a) Numerical verification of tensile stress development in highly fractured phyllite characterised via pressuremeter test. In: Proceedings of the 6th international young geotechnical engineers' conference (iYGEC6), Seoul, South Korea

Phangkawira F, Ong DEL, Choo CS (2017b) Numerical prediction of plastic behavior of highly fractured and weathered phyllite subjected to pressuremeter testing. In: Proceedings of the 19th international conference on soil mechanics and geotechnical engineering, Seoul, South Korea

Reilly CC (2014) The influence of lubricant slurries on skin friction resistance in pipe jacking. Dissertation, Trinity College Dublin

Reilly CC, Orr TLL (2017) Physical modelling of the effect of lubricants in pipe jacking. Tunn Undergr Sp Tech 63:44–53. https://doi.org/10.1016/j.tust.2016.11.005

Schofield AN (1998) Technical Report No. 35—The "Mohr-Coulomb" error, Cambridge University

Schofield AN (2001) Re-appraisal of Terzaghi's soil mechanics. Paper presented at the International Society of Soil Mechanics and Geotechnical Engineering Conference, Istanbul

Sharo AA, Liang RY (2012) Numerical study of rock identification number β in pressuremeter test. The Art of Foundation Engineering Practice Congress 2012, Florida, pp 717–732

Soon SC, Drescher A (2007) Nonlinear failure criterion and passive thrust on retaining walls. Int J Geomech 7(4):318–322. https://doi.org/10.1061/(ASCE)1532-3641(2007)7:4(318)

Standards Australia (1998), AS 1289.6.2.2 Soil strength and consolidation tests—determination of the shear strength of a soil—direct shear test using a shear box, Standards Australia, Australia

Yang XL, Yin JH (2004) Slope stability analysis with nonlinear failure criterion. J Eng Mech 130(3):267–273. https://doi.org/10.1061/(ASCE)0733-9399(2004)130:3(267)

Yu HS (2000) Cavity expansion in geomechanics. Springer, Dordrecht

Yu HS, Houlsby GT (2015) Finite cavity expansion in dilatant soils: loading analysis. Géotechnique 41(2):173–183. https://doi.org/10.1680/geot.1991.41.2.173

Chapter 3
Complex Soil–Pipe Interaction: Challenges in Geological Characterization and Construction

3.1 Background

With the advancement of pipe jacking technology, there has been a growing demand for research on aspects of the pipe jacking method. For instance, the need to assess frictional loads during pipe jacking was emphasized by Ripley (1989). The University of Oxford contributed to one of the pioneering works in the assessment of pipe jacking works (Norris and Milligan 1991). This involved equipping jacked pipes with instruments to measure deformation and pressures acting on the jacked pipes. The instrumentation scheme is shown in Fig. 3.1.

The instrumentations were successful in obtaining useful data during pipe jacking, such as pipeline misalignment, load transfer at pipe joints, interface stresses, frictional resistance, pipe–soil contact points, and lubrication (Norris and Milligan 1992a, b, Milligan and Marshall 1995). The developed technology managed to give essential insights on the behavior of the jacked pipes, as well as the changes in stresses during pipe jacking. This contributed in driving further studies on the assessment of jacking forces.

Until recently, the assessment of jacking forces has been focused on factors that can help to increase the span of pipe jacking drives (Bergeson 2014). This led to many developments in the understanding of jacking forces, where various methods to describe the development of jacking forces were developed. Jacking forces are often crucial in the span of a pipe jacking drive. While excessive jacking forces can lead to damaged pipes during jacking, low jacking forces can allow for longer drive spans. This necessitated the understanding of soil–pipe interaction in the investigation into jacking forces.

Geotechnical investigations appear to be of primary importance for the assessment of jacking forces required by the microtunneling boring machine (mTBM) and the pipeline to advance into the ground. These jacking forces are generated along the longitudinal axis of the pipe, due to the driving of pipes into the soil or rock mass via hydraulic jacks. As shown in Fig. 3.2, the total thrust required for jacking, P_{total}, is the sum of the thrust exerted at the head of the boring machine, R_p and the friction

© The Author(s), under exclusive license to Springer Nature Singapore Pte Ltd. 2022
D. E. L. Ong et al., *Sustainable Pipe Jacking Technology in the Urban Environment*, Cities Research Series, https://doi.org/10.1007/978-981-16-9372-4_3

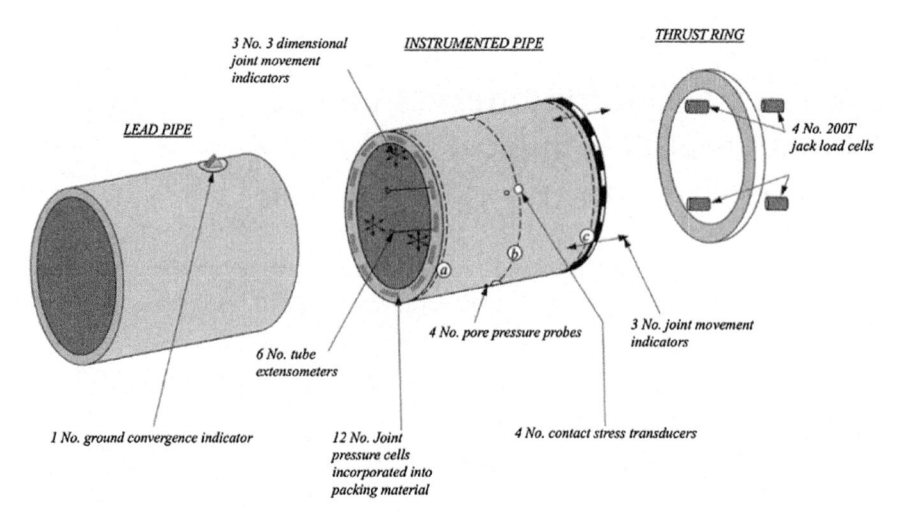

Fig. 3.1 Instrumentation scheme for the jacked pipe developed at the University of Oxford (after Norris and Milligan 1991)

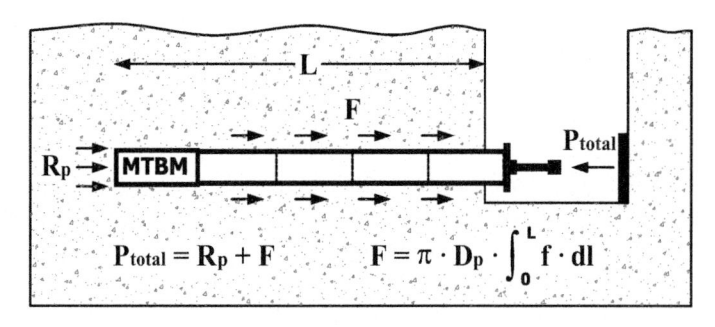

Fig. 3.2 Forces involved during microtunneling operation

forces, F developed at the soil–pipe interface (Stein 2005). As both the head of the boring machine and the pipe string are in contact with the surrounding soil, R_p and F are strongly related to soil properties.

A reliable assessment of thrust head and friction forces developed during microtunneling operations is very important, as they influence the choice of mTBM and its equipment (thrust frame, dimension of thrust wall, dimension of overcut, and lubrication), the choice of pipe to be jacked and geometry of the drive (maximum length of installation, number of shafts, and position of intermediate jacking stations).

Generally, the thrust head, R_p, does not increase with the jacking distance, since the depth of soil cover does not change significantly throughout the pipe jacking drive. Changes in thrust head could fluctuate with variations in soil type along the pipe jacking drive. Regardless, these fluctuations do not contribute to the total jacking forces as much as the frictional resistance, F.

3.1 Background

Friction forces usually constitute the main component of jacking loads (Norris and Milligan 1992b, Pellet-Beaucour and Kastner 2002). As a result of their increase in jacking length, these forces are responsible to limit the drive length. In the following part of this chapter, experimental values and empirical correlations to compute friction forces at the soil–pipe interface are taken into account and described. Some considerations about the thrust exerted at the head (face resistance) will also be presented toward the end of the chapter.

3.2 Pipe Jacking Forces

As the drive span increases, the contact area between the outer surface of the pipe and the surrounding geology increases proportionately. Thus, it can be readily surmised that the frictional resistance increases almost linearly as the drive span increases and ultimately becomes significantly larger than the penetration resistance.

The magnitudes of the friction forces developed at the soil–pipe interface depend on the pipe size and material, type of soil, its moisture content and grading, depth of cover and the detail of construction equipment and procedures employed. Factors such as the amount of overcut, misalignment of the pipes, duration of work stoppages, and whether or not a bentonite injection system is used will also affect the amount of friction (Norris and Milligan 1992b).

In general, it is possible to define the friction force F as:

$$F = \pi D_p \int_0^L f \, dL \tag{3.1}$$

with D_p the outer diameter of the jacked pipes, L the pipeline length, and f the frictional stress acting on the pipe outer surface. This stress, commonly known as unit friction, depends on the normal forces build-up on the pipe (due to the weight of the pipes and the unstable soil area around the excavation zone) and the friction at the soil–pipe interface which in turn depends on soil and pipe properties.

Chapman and Ichioka (1999) used a statistical approach to predict frictional jacking forces. This was done by studying the data obtained from 398 microtunneling projects. The data was classified based on soil type, and the frictional resistance was represented using lines of best fit. From the study, the frictional resistance is expressed as

$$F_f = a + 0.38 D \tag{3.2}$$

where F_f is the frictional resistance along the pipe, D is the external pipe diameter, and a is a soil-dependent parameter (see Table 3.1).

Table 3.1 Values of a for different types of soils (Chapman and Ichioka 1999)

Soil type	a
Clay	0.153
Sand	0.243
Sand and gravel	0.343

It is important to note that Chapman and Ichioka (1999) approach was a seminal development, offering predictive jacking force equations for different microtunneling methods traversing different soils. With the use of fairly simplistic correlation methods, along with the earlier work by the Oxford team, it became readily apparent that the traversed soil types had an (as yet unknown) effect on the accrual of pipe jacking forces. There was a need to develop predictive jacking force models using soil mechanics principles, some of which are shown in Table 3.2.

Table 3.2 Some predictive models for frictional jacking forces

Reference	Jacking force model	Definition
Staheli (2006)	$\mu_{int} \dfrac{\gamma r \cos\left(45 + \frac{\phi_r}{2}\right)}{\tan\varphi_r} \pi d l$	μ_{int} = pipe–soil interface frictional coefficient γ = soil unit weight ϕ_r = residual friction angle d = pipe diameter r = pipe radius l = pipe length
Osumi (2000)	$\beta(\pi D_e q + w)\tan\frac{\phi'}{2} + \pi B_c C'$	β = jacking force reduction factor D_e = pipe outer diameter q = normal force ϕ' = interface friction angle w = pipe weight C' = pipe–soil adhesion
Iseki (n.d.)	$\mu(q + W_s) + C$	q = vertical load to pipe axis μ = frictional coefficient W_s = pipe weight C = soil cohesion
ATV-A 161E (1990)	$\mu \gamma h \kappa [b_a + (K_2 d_a)] + K_2 \dfrac{d_a^2}{2h(b_a + d_a)}$ where: $\kappa = \dfrac{1 - e^{\frac{h}{b}\tan(0.5\phi)}}{\frac{h}{b}\tan(0.5\phi)}$; $b_a = \dfrac{2 d_a}{\sqrt{3}}$	μ = frictional coefficient γ = soil unit weight h = soil cover depth K_2 = coefficient of lateral active earth pressure d_a = pipe outer diameter b = influencing soil width above pipe ϕ = soil internal friction angle

3.2.1 Experimental Results Relating to Frictional Stress

In Table 3.3, a collection of experimental values of frictional stress noted in the literature is given.

They come from the analyses of different microtunneling work throughout the world in different types of soil and for different operation conditions. It is possible to distinguish:

- Dynamic friction f: frictional stress developed during jacking, when the pipeline is advancing through the soil;

$$f = \frac{F}{\pi D_p L} \quad (3.3)$$

- Static fiction f_{stat}: frictional stress developed after a stoppage in jacking, generally greater than dynamic friction;

$$f_{stat} = \frac{F_{stat}}{\pi D_p L} \quad (3.4)$$

- Lubricated friction f_{lub}: frictional stress developed during jacking in the presence of lubricate products injected between the ground and the pipe;

Table 3.3 Summary of the minimum, maximum and average (bold in parentheses) values of frictional stress, mentioned in the literature (FSTT 2004)

Soil type	Frictional stress [kPa]							
	French National Research Project 'Microtunnels'		Japan Society of Trenchless Technology (JSTT)		Geological Laboratory of US Army Corps of Engineers	Norwegian Geotechnical Institute	
	f	f_{lub}	Slurry type f_{conv}	Auger type f_{conv}	f_{app}	f_{app}	$f_{app,stat}$
Clay	1.4–5.8 (**3.25**)	0.65–3.3 (**2.25**)	0.7–16 (**4.9**) (**3.5**)	(**2.1**)	8.3–13 (**4.6**)	–	–
Sand	4.5–7.3 (**5.5**)	0.65–4.9 (**2.0**)	1–19 (**5.1**) (**4.0**)	(**2.8**)	2.7–12 (**6.1**)	2–11 (**5.7**)	4.3–11 (**7.8**)
Sand and gravel	1.8–17 (**7.4**)	3–10 6.9	5–6.9 6 4.8	**3.6**	–	12–30 15.9	16–30 (**20.8**)

$$f_{\text{lub}} = \frac{F_{\text{lub}}}{\pi D_p L} \tag{3.5}$$

In addition, due to the difficulties in knowing the value of the thrust at the head, certain empirical results include this force in the friction calculation or adopt conventional values for it (i.e., first load):

- Apparent average friction f_{app}: frictional stress equals the total jacking force solicited by the surface of jacked pipe, i.e.,

$$f_{\text{app}} = \frac{P_{\text{total}}}{\pi D_p L} \tag{3.6}$$

- Average conventional friction f_{conv}: frictional stress computed considering a conventional value $R_{p,\text{conv}}$ for the thrust head, i.e.,

$$f_{\text{conv}} = \frac{P_{\text{total}} - R_{p,\text{conv}}}{\pi D_p L} \tag{3.7}$$

The values listed in Table 3.3 refer to 4 statistical studies on friction stress involved during microtunneling in France (Pellet-Beaucour 1997, Phelipot 2000), Japan (Working Group No. 3 1994), USA (Collers et al. 1996), and Norway (Lauritzen et al. 1994). The comparison among these values is quite tricky due to the different approaches utilized for their calculation (FSTT 2004):

- Firstly, due to the approximations made for the frictional stress: while the French work considers the dynamic friction f, the Norwegian and American works calculate the average apparent friction f_{app} and the Japanese studies take into account the average conventional friction f_{conv}, assuming the thrust at the head equals to the first value of the jacking thrust.
- Secondly, the Japanese and American works do not distinguish dynamic friction from static friction.
- Finally, the lubrication, which plays a predominant role on the magnitude of friction forces, are only described in the French study.

However, by considering that taking into account the average apparent friction f_{app} leads to an overestimation of 1 kPa in frictional stress, as estimated by the JSTT (Working Group No. 3 1994), values mentioned by different studies appear to be very close. They are in the range 3.25–3.6 kPa in clay, and in the range 4–5.5 kPa in sand. The results are much more scattered in sands and gravels. High frictional stress values mentioned by the Norwegian Institute can be partly explained by an average pipe diameter of 3.2 m, much larger than the average taken into account by the French and Japanese studies (Pellet-Beaucour and Kastner 2002).

3.2 Pipe Jacking Forces

3.2.2 Calculation Methodology for Frictional Stress

The calculation of frictional stress developed at soil–pipe interface needs to rely on the stability condition of excavation made by the boring machine (Fig. 3.3):

- In case of unstable excavation or stable excavation with ground convergence greater than the overcut size, the ground comes in contact with the entire pipeline. The friction forces are then calculated by multiplying the normal forces build-up on the pipe by the soil–pipe friction coefficient.
- In case of stable excavation and ground convergence smaller than the overcut size, the pipeline slides on its base in obvious contact with the geology (assuming a non-suspension condition), inside the open annular space. The friction forces are related to the weight of pipes and to the soil–pipe friction coefficient.

3.2.2.1 Excavation Stability

In this paragraph, a possible approach to the study of excavation stability is described (Milligan and Norris 1994). The excavation stability problem can be evaluated with reference to the confining pressure σ_T that is required in order to ensure the stability of the realized bore. If this pressure results to be positive, the excavation will be unstable unless fluid pressure is provided (excavation with compressed air or pressurized slurry).

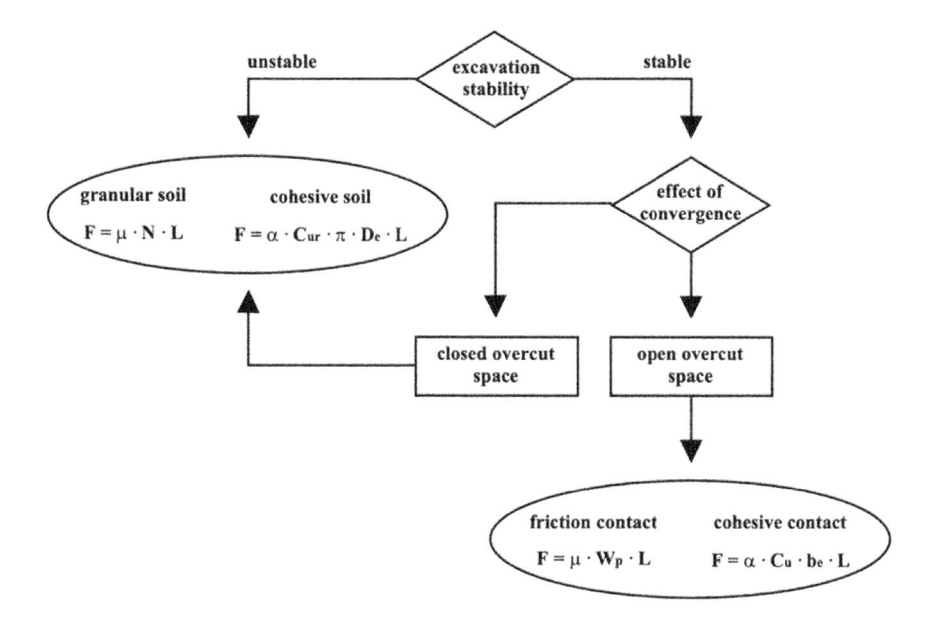

Fig. 3.3 Calculation methodology for friction forces (FSTT 2004)

In case of cohesive soil, the short-term stability is linked to the undrained cohesion c_u and to the critical stability ratio N_c (stability ratio at collapse) introduced by Broms and Bennermark (1967). The confining pressure required in order to assure the stability is given by the following relation:

$$\sigma_T = \gamma \cdot \left(h + \frac{D_e}{2}\right) - N_c \cdot c_u \tag{3.8}$$

where γ is the unit weight of soil, D_e the excavation diameter, and h the cover.

The graph in Fig. 3.4 shows the values of N_c obtained from centrifuge model tests, in terms of dimensionless ratios h/D_e and P/D_e (Mair 1979; Kimura and Mair 1981), where P represents the distance from the face to the point where stiff support is provided. The value of P can be considered equal to the drilled length for microtunneling installation, and this implies that the ratio P/D_e goes toward infinity.

In the case of granular soil (cohesionless soil) drained stability should be considered. The stability depends on the internal angle of friction of the soil φ. The required pressure σ_T for the excavation stability can be calculated with the following equation (Atkinson and Mair 1981):

$$\sigma_T = \gamma \cdot D_e \cdot T_\gamma + q_s \cdot T_s \tag{3.9}$$

where q_s is the surface surcharge and T_γ and T_s are the stability numbers, respectively, for soil load and surface surcharge. They can be determined from the graphs given

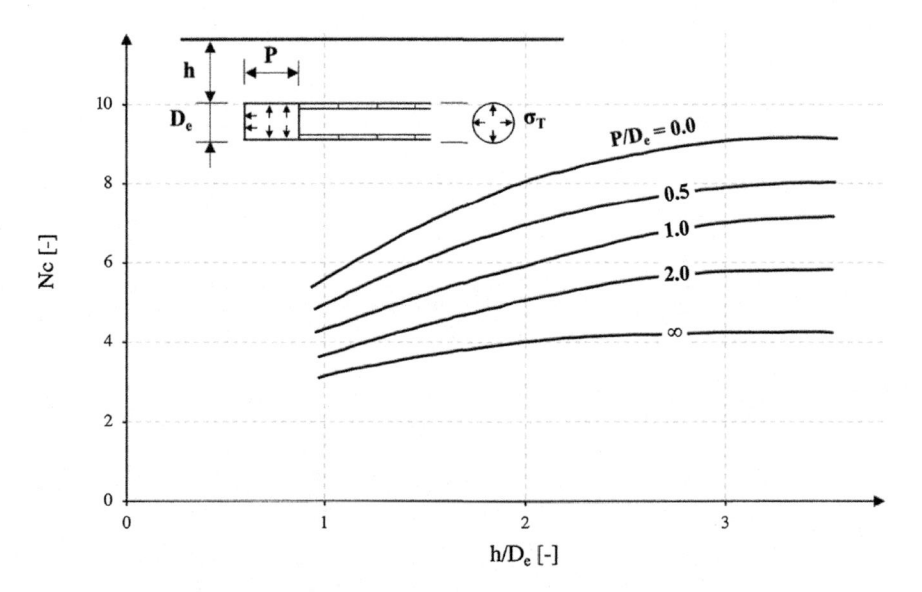

Fig. 3.4 Critical stability ratio N_c (after Mair 1979, Kimura and Mair 1981)

3.2 Pipe Jacking Forces

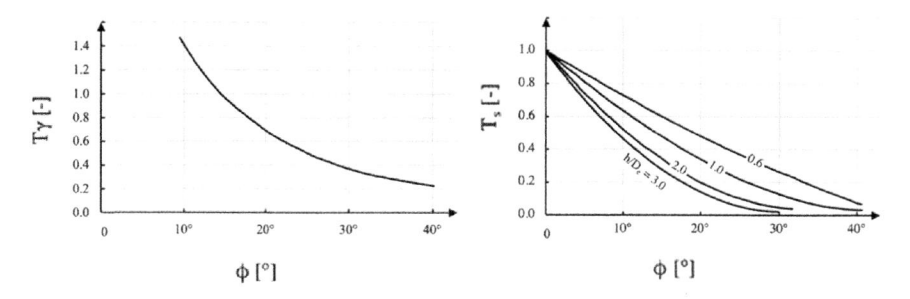

Fig. 3.5 Tunnel stability numbers T_γ and T_s (after Thomson 1993)

in Fig. 3.5. In particular, the graph of T_s shows that for sands and gravels, which typically have values of φ greater than 30°, the influence of T_s can be ignored, since the cover to excavation diameter ratio h/D_e is generally greater than 3.0 for microtunneling installations (Thomson 1993).

Based on Eq. 3.9 when the soil is purely frictional, the required pressure σ_T is always positive, and hence, the excavation is always 'unstable.'

Besides the given approach, other approaches are available today for the excavation stability analysis, both analytical (Panet 1995; Ribacchi and Riccioni 1977) and numerical. In particular, the latter method allows the soil to be modeled as a continuum or a discontinuum and takes into account the real conditions of the problem (i.e., the soil properties, the geometry of the excavation, the boundary conditions, the state of stress) which can be only considered in a simplified manner in the analytical approaches.

3.2.2.2 Ground Convergence Effect

Even in the case of stable excavation in cohesive soils, the inward movements of the tunnel walls due to elastic stress relief may be greater than the overcut size and give rise to normal forces build-up on the pipe. In order to evaluate the reduction in horizontal and vertical diameter, the following equations, based on the elastic solution for unlined tunnel and the initial state of stress (i.e., the horizontal and vertical stresses, σ_h and σ_v), can be used (Kirsch 1898):

$$\delta_h = \frac{1+\nu_s}{4 \cdot E_s} \cdot D_e \cdot [(\sigma_v + \sigma_h) - (\sigma_v - \sigma_h) \cdot (3 - 4 \cdot \nu_s)] \quad (3.10)$$

$$\delta_v = \frac{1+\nu_s}{4 \cdot E_s} \cdot D_e \cdot [(\sigma_v + \sigma_h) + (\sigma_v - \sigma_h) \cdot (3 - 4 \cdot \nu_s)] \quad (3.11)$$

where E_s and ν_s are, respectively, Young's modulus and Poisson's ratio of the soil.

If a pressure σ_p is applied inside the overcut (i.e., injection of lubricant in order to reduce friction forces), this leads to a uniform increase δ_p in the excavation radius that can be calculated on the basis of Lamé elasticity theory (Lamé 1852):

$$\delta_p = \frac{1 + \upsilon_s}{2 \cdot E_s} \cdot D_e \cdot \sigma_p' \qquad (3.12)$$

where σ_p' is the applied pressure σ_p reduced by the pore pressure of the soil.

According to the size of the overcut O_{ep}, two different conditions may be considered:

- If δ_v (and δ_h) $- 2 \cdot \delta_p < O_{ep}$, the overcut remains open and friction forces depend on the weight of the pipes.
- If δ_v (and δ_h) $- 2 \cdot \delta_p \geq O_{ep}$, the overcut is closed and the normal forces build-up on the pipe give rise to additional friction forces, which can be computed in the same manner of friction forces for unstable excavation.

3.2.2.3 Calculation of Friction Forces for Unstable Excavation in Granular Soil

The friction forces can be computed based on the normal forces build-up on the pipe N (normal force for a unit length of pipe) and the soil–pipe friction coefficient μ:

$$F = \mu N L = \mu \left(2 \int_{-\pi/2}^{\pi/2} \sigma_n \frac{D_p}{2} d\theta \right) L \qquad (3.13)$$

This approach assumes that the ground is in contact with the whole outside area of the pipe and only takes into account stresses due to ground pressure. The normal force N is gained by integrating the normal stress σ_n acting on an element of the pipe surface, along the pipe circumference. Considering the horizontal and vertical stresses as principal stresses, this stress can be determined as:

$$\sigma_n = \frac{\sigma_v + \sigma_h}{2} + \frac{\sigma_h - \sigma_v}{2} \cdot \cos(2 \cdot \vartheta) \qquad (3.14)$$

where ϑ is the direction angle of the pipe surface element (measured counterclockwise from the positive x-axis).

Since the excavation of a microtunnel causes stress relief in the surrounding ground, the horizontal and vertical stresses of Eq. 3.14 have to be determined by a model. At present, by far the most used model is Terzaghi's silo model that gives the vertical stress σ_{EV} acting on the crown of the excavation (Terzaghi 1943). Therefore, the horizontal and vertical stresses at a given point of the pipe can be written as:

$$\sigma_v = \sigma_{EV} + \gamma \cdot \left(\frac{D_e}{2} - y_p \right) \qquad (3.15)$$

3.2 Pipe Jacking Forces

Fig. 3.6 State of stresses around the pipe

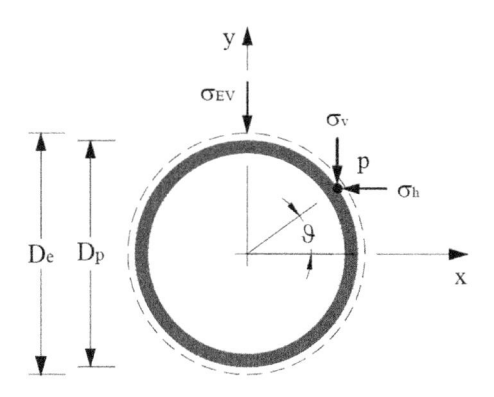

$$\sigma_h = K_2 \cdot \left[\sigma_{EV} + \gamma \cdot \left(\frac{D_e}{2} - y_p\right)\right] \quad (3.16)$$

with y_p the ordinate of the considered point with respect to the center of the pipe and K_2 the coefficient of earth pressure of the soil around the pipe. This coefficient can be assumed to be equal to 0.3 (Stein et al. 1989) or to the active earth pressure coefficient, computed according to the Rankine formula (FSTT 2004) (Fig. 3.6).

By substituting Eqs. 3.15 and 3.16 into Eqs. 3.14 and 3.13, the total normal force build-up on a unit length of pipe during microtunneling installation in unstable granular soil can be expressed as:

$$N = D_p \cdot \frac{\pi}{2} \cdot \left(\sigma_{EV} + \gamma \cdot \frac{D_e}{2}\right) \cdot (1 + K_2) \quad (3.17)$$

3.2.2.4 Calculation of the Vertical Stress, σ_{EV}

One such equation was developed by Pellet-Beaucour and Kastner (2002), which considered the effects of soil arching. Soil arching occurs when there is a redistribution of soil stresses due to a stress relaxation around an excavated cavity, e.g., tunnels. Think of this as an arch walkway, passing through a wall, as in Fig. 3.7.

As shown in Fig. 3.7, the weight of the wall (represented by solid vertical arrows) would be oriented in the vertical direction. However, with the introduction of a gap in the wall, the weight of the wall would become unsupported, resulting in the collapse of the wall. The arrangement of stones into an arch walkway helps to channel (or redistribute) the weight of the wall around the gap while ensuring the stability of the wall. The soil arching phenomenon works in a similar way. For pipe jacking work as shown in Fig. 3.7, think of the arch walkway as the tunnel through which the pipe is being jacked into. When applied to pipe jacking context, there are less stresses

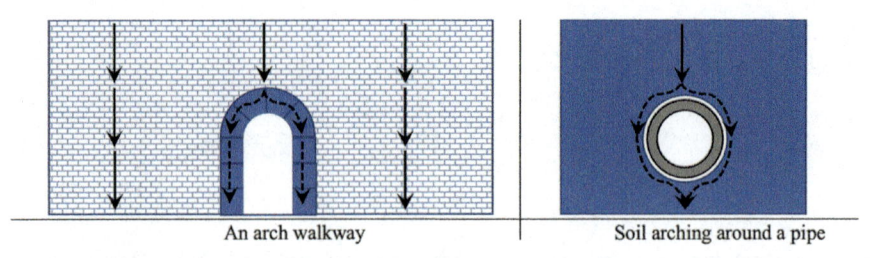

Fig. 3.7 Redistribution of stresses as represented by an arch walkway, and as depicted in soil arching. *Note* Continuous arrows represent the flow of gravitational normal stresses, with dashed arrows representing the redistribution of normal stresses due to arching

pressing (or squeezing) onto the pipe. With effective soil arching, this can reduce the amount of jacking forces.

The Pellet-Beaucour and Kastner (2002) jacking force equation is expressed as

$$F_f = \pi L D \mu \times 0.5 \left[\sigma_{ev} + 0.5\gamma D + K_2(\sigma_{ev} + 0.5\gamma D) \right] \quad (3.18)$$

where F_f is the frictional resistance along the pipe, D is the pipe outer diameter, L is the jacking distance, γ is the soil unit weight, and K_2 is the coefficient of soil thrust on the pipe. Soil arching is considered in the computation of σ_{ev} which is the vertical stress at pipe crown and given in Eq. 3.19.

$$\sigma_{ev} = \frac{b\left(\gamma - \frac{2c}{b}\right)}{2K \tan\delta} \times \left(1 - e^{\frac{2Kh}{b} \tan\delta}\right) \quad (3.19)$$

where b is the ideal silo width, c is the apparent soil cohesion, and h is the height of soil cover at the pipe crown.

In Terzaghi (1943), the silo theory assumed that the ground located above the excavation (and the pipeline) is settling along two vertical planes, whose distance b depends on the excavation diameter and mechanical properties of the soil. The displacements are significant enough to produce sliding planes and the creation of an arching mechanism (Fig. 3.8).

If ground water or surface load is present, Eq. 3.19 should be modified, in order to take into account their contributions. In cohesionless soils, Eq. 3.19 can also be written as:

$$\sigma_{EV} = \gamma \cdot h \cdot k \quad (3.20)$$

where k is a coefficient (smaller than unity) that lowers the soil weight $\gamma \cdot h$ and represents the arching effect taking place in the soil above the pipeline:

3.2 Pipe Jacking Forces

Fig. 3.8 Ground loading from Terzaghi model

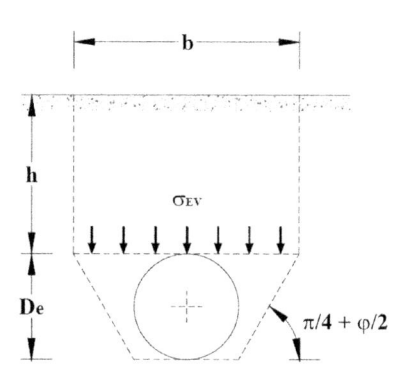

$$k = \frac{1 - e^{-2 \cdot K \cdot (\tan \delta) \cdot h/b}}{2 \cdot K \cdot \tan \delta \cdot h/b} \quad (3.21)$$

Equation 3.19 can be used for pipes having a depth varying from b to $2.5 \cdot b$. In fact, in case of smaller covers, the decompression displacements after boring operation affect the whole soil above the pipe and the vertical stress σ_{EV} can be assumed equal to the soil weight $\gamma \cdot h$ (Szechy 1970; AFTES 1982). Contrarily, as the arching effect only extends to a height of $2.5 \cdot b$, in case of cover greater than this value, the vertical stress σ_{EV} can be calculated as (FSTT 2004):

$$\sigma_{EV} = \frac{b \cdot (\gamma - 2c/b)}{2 \cdot K \cdot \tan \delta} \quad (3.22)$$

The calculation of the σ_{EV} requires some physical parameters which can be determined with reasonable accuracy (i.e., the height of cover h and the unit weight of soil γ) as well as some empirical parameters (i.e., the coefficient of earth pressure K, the width b, and the friction angle δ) whose definition varies from one author to another. In Table 3.4, three different approaches, taken from the literature, are given.

Considering a cohesionless soil, a comparison of the coefficient k obtained by these three different approaches shows that the ATV-A 161 model leads to the most cautious assumption (Fig. 3.9). In fact, contrary to the other two models, this method

Table 3.4 Parameters of Terzaghi's silo theory from different Authors (Pellet-Beaucour and Kastner 2002)

Parameter	Terzaghi (1943)	ATV-A 161 (1990)	Pipe Jacking Association (Milligan and Norris 1994)
b	$D_e[1 + 2\tan(\pi/4 - \phi/2)]$	$D_e \cdot \sqrt{3}$	$D_e \tan(3\pi/8 - \phi/4)$
δ	ϕ	$\phi/2$	ϕ
K	1	0.5	$\tan^2(\pi/4 - \phi/2)$

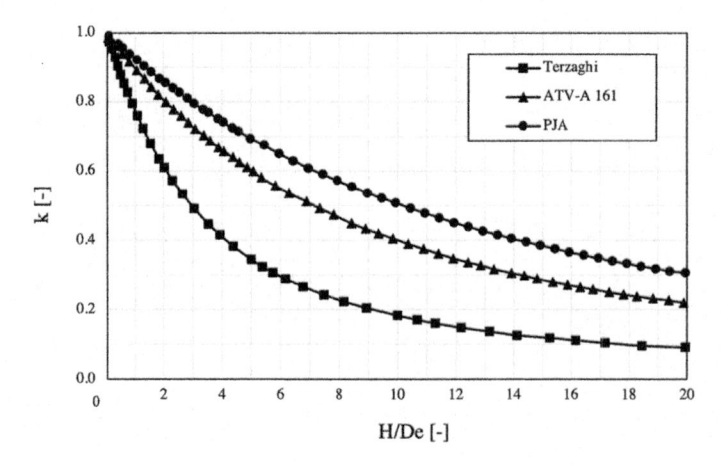

Fig. 3.9 k coefficient versus h/D_e according to the three different approaches of Terzaghi's silo theory for a φ equal to 30° (after Pellet-Beaucour and Kastner, 2002)

does not class shear planes as perfectly rough, which leads to decreasing importance of the arching effect. Consequently, the ATV-A 161 approach will influence vertical stress, and therefore, friction forces are higher than those calculated by the PJA, which in turn produces higher values than the Terzaghi model (Pellet-Beaucour and Kastner, 2002).

It is worth noting that comparison of the experimental results, mentioned in the literature, with various analytical models has shown that it is Terzaghi's approach which is the closest to reality while the two other approaches (ATV-A and PJA) lead to far higher values of friction stresses (Kastner et al. 1997, Pellet-Beaucour 1997).

3.2.2.5 Determination of the Friction Coefficient

The friction coefficient μ is related to the angle of friction δ_{sp} at the soil–pipe interface:

$$\mu = \tan\delta_{sp} \tag{3.23}$$

In soil–structure interaction calculations, the angle of friction δ_{sp} is assumed to be between an upper limit, equal to the internal friction angle of soil φ, and a lower one, usually comprised in the range $\varphi/3$ to $\varphi/2$, depending on the roughness of the soil–structure interface. In the case of microtunneling installations, the friction coefficient is also dependent on the friction nature (i.e., dynamic or static friction) and the use of lubricants (Pellet-Beaucour and Kastner 2002).

3.2 Pipe Jacking Forces

Table 3.5 Standard values for friction coefficient (Stein et al. 1989)

Friction state	Pipe and soil materials	Friction coefficient
Static friction	Concrete on gravel or sand	0.5–0.6
	Concrete on clay	0.3–0.4
	Centrifuged concrete on gravel or sand	0.3–0.4
	Centrifuged concrete on clay	0.2–0.3
Dynamic friction	Concrete on gravel or sand	0.3–0.4
	Concrete on clay	0.2–0.3
	Centrifuged concrete on gravel or sand	0.2–0.3
	Centrifuged concrete on clay	0.1–0.2
Lubricated friction		0.1–0.3

Values of dynamic friction coefficient can be assessed equal to 0.3–0.4 for sand and gravel, and 0.2–0.3 for clay (Stein et al. 1989). These ranges correspond to a soil–pipe friction angle equal approximately to half the soil internal friction angle (Table 3.5).

Furthermore, according to Stein, the coefficient of static friction is equal to 1.5 times the coefficient of dynamic friction while, if lubricants are used during installation, a coefficient ranging from 0.1 to 0.3 should be adopted, depending on the type and volume of injections (in this case the coefficient of friction no longer depends on the nature of the soil but mainly on the liquid limit of the lubricant).

According to Stein (2005), the frictional coefficient, μ, is a function of wall friction angle, which can be determined through tests on interface between the pipe material and surrounding geology (Iscimen 2004). However, the frictional coefficient derived from this assumes full contact between the pipe material and the surrounding geology. This may not always be the case during actual pipe jacking works, given the effects from other factors such as lubrication, pipe alignment deviations, and the difference between the diameters of the pipe and the bored tunnel. Without specialized instruments, the contact area between the pipe and the surrounding soil or rock mass can be difficult to assess.

Alternatively, Stein (2005) recommended some typical ranges of frictional coefficients for different interface materials and conditions, which are expressed in Table 3.5.

3.2.2.6 Calculation of Friction Forces for Unstable Excavation in Cohesive Soil

The friction forces developed at the soil–pipe interface depend on the undrained cohesion c_u of the soil and the total lateral surface of the pipes. In the case of relatively large soil–pipe displacements (such as the displacements caused by jacking operations), the clay in contact with the pipeline is greatly reworked and a better assessment of the friction forces can be obtained referring to the undrained cohesion of reworked clay c_{ur}:

$$F = \alpha \cdot c_{ur} \cdot \pi \cdot D_e \cdot L \tag{3.24}$$

where α is a coefficient introduced in order to take into account the adhesion effect at the interface (α can be considered equal to 0.6 for concrete pipes and 0.5 for steel pipes, according to the results obtained for drilled piles of large diameter) (FSTT 2004).

The undrained cohesion of reworked clay can be estimated from the abacus of Fig. 3.10 (Leroueil et al. 1983). However, this approach assumes that the natural water content of clay in contact with the pipeline is not influenced by percolations of mucking or injection fluids, usually adopted in microtunneling work.

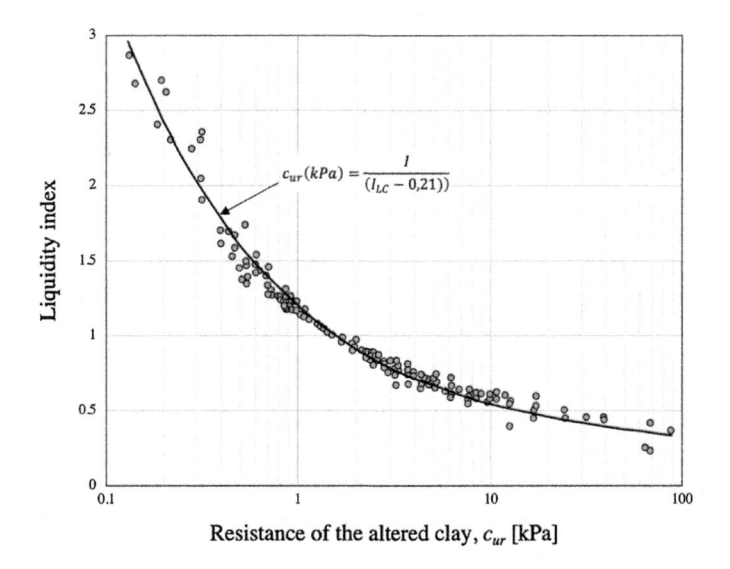

Fig. 3.10 Estimation of the undrained cohesion of reworked clay (after Leroueil et al. 1983)

3.2.2.7 Calculation of Friction Forces for Stable Excavation

If the excavation remains stable and the convergence is less than the overcut size, the pipeline slides on its base inside the bore. This is the case of most microtunnels in clayey soil which are stable during the construction phase, due to the relatively small diameters of excavation (except those in very soft clay) and microtunnels in damp fine or silty sand where the capillary suctions may be sufficient to maintain temporary stability (Milligan and Norris 1999).

The calculation of friction forces depends on the type of contact between the soil and the pipe. If this can be assumed as purely frictional (such as in dense sand or rock), the resistance to sliding can be assessed with the following formula:

$$F = \mu \cdot W_p \cdot L \quad (3.25)$$

where μ is the friction coefficient at the soil–pipe interface and W_p is the weight of the pipe per unit length (when the bore is filled with slurry, the pipes become buoyant).

However, in the case of microtunneling in rock, O'Reilly and Rogers (1987) have suggested computing the friction forces taking into account an excavation that is not perfectly circular (Fig. 3.11), as follows:

$$F = \frac{\mu \cdot W_p \cdot L}{\cos\zeta} \quad (3.26)$$

where ζ is the offset angle of reaction from the vertical, comprised in the range from 10 to 60°, leading to higher forces, up to twice (Phelipot 2000).

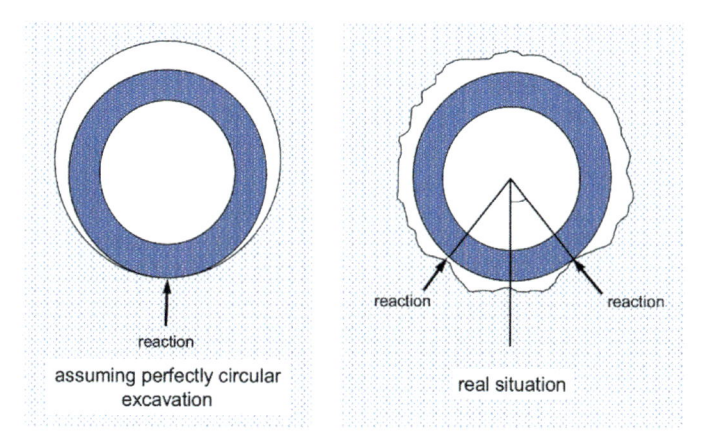

Fig. 3.11 Pipe seated in an oversized cylindrical void showing (left) perfectly circular excavation, and (right) real situation (after O'Reilly and Rogers 1987)

Finally, in the case of soft clay, Haslem (1986) has proposed a different formula for the calculation of friction forces, based on the undrained cohesion c_u and the width of contact area b_e between the soil and the pipe:

$$F = \alpha \cdot c_u \cdot b_e \cdot L \tag{3.27}$$

The width b_e is obtained considering an elastic contact between two curved surfaces and can be expressed with the following equation:

$$b_e = 1.6 \cdot \sqrt{P_u \cdot k_d \cdot C_e} \tag{3.28}$$

where P_u is the contact force per unit length (or the weight of the pipe per unit length) and k_d and C_e are two parameters defined as follows:

$$k_d = \frac{D_e \cdot D_p}{D_e - D_p} \tag{3.29}$$

$$C_e = \frac{1 - v_s^2}{E_s} - \frac{1 - v_p^2}{E_p} \tag{3.30}$$

with E_p and v_p being Young's modulus and Poisson's ratio of the pipe material, respectively. Experimental results obtained from microtunneling installations in soft clay soils (Milligan and Norris 1994) have shown that this approach gives values of friction forces very close to reality.

3.3 The Influence of Construction Operational Parameters

The above-mentioned studies have shown that to this date, the use of soil mechanics in understanding jacking forces has been well-developed. These equations are particularly useful during the planning stages of a pipe jacking project. However, during the construction phase of the project, these equations can be limited in their usage, since decisions are primarily made based on pipe jacking operational parameters, e.g., jacking forces, jacking speed, lubrication, stoppages, cutter torque, etc. Many of these operational parameters can be plotted against a common horizontal axis for the pipe jacking drive length, in order to better understand the relationships between these operational parameters. An example of this plot is shown in Fig. 3.12.

Table 3.6 summarizes some of the operational parameters and their effects on jacking forces.

Many of these relationships were based on field observations; pipe jacking contractors and mTBM operators also tend to base these relationships on their experiences, which influence the decisions made during the pipe jacking process. In recent years, there have been efforts to develop quantifiable models for the prediction of

3.3 The Influence of Construction Operational Parameters

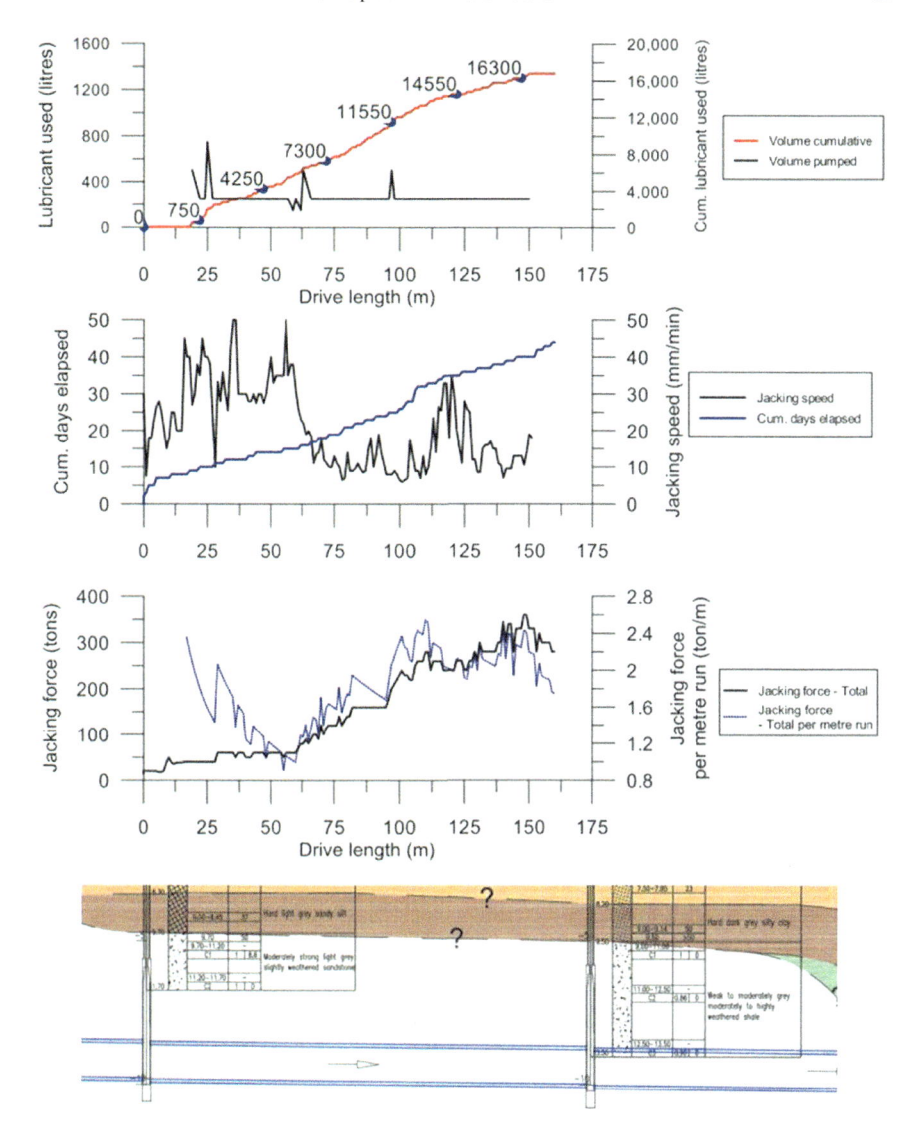

Fig. 3.12 Plots of pipe jacking operational parameters (lubricant, days elapsed, jacking speed, and jacking forces) against drive length

jacking forces based on operational parameters (Wei et al. 2021; Cheng et al. 2020; Chen et al. 2019). These methods typically employ the use of machine learning and deep learning and are currently a growing area of interest. Some specific application of machine learning in pipe jacking works can be found in Chap. 7 of this book.

Still, documented studies on pipe jacking works in rocks remain limited. It is important to note that the methods to assess jacking forces in soils (whether based on soil mechanics or operational parameters) cannot be directly applied to pipe jacking

Table 3.6 Some operational parameters and their effects on jacking forces

Operational parameters	Influence on jacking forces	References
Lubrication	Reduces frictional force	Choo and Ong (2012, 2015), Ong and Choo (2018), Shao et al. (2009), Shou et al. (2010)
Stoppage	Increases frictional static resistance	Cheng et al. (2017), Choo and Ong (2012, 2015), Shao et al. (2009)
Progress drive length	Increase in the frictional force acting on the surface of the pipeline	Choo and Ong (2017), Ji et al. (2019), Pellet-Beaucour and Kastner (2002)
Jacking speed	Increase in face pressure force	Cheng et al. (2017), Hadri and Mohammad (2020), Pellet-Beaucour and Kastner (2002), Staheli (2006)

in rocks. This is because the behavior and properties of rocks are different from soils, especially the relatively big differences in their strengths and stiffnesses. As a result, rock properties and the way they are characterized are usually different from those for soils. Furthermore, the presence of joints and discontinuities in rocks can affect pipe jacking operational parameters in ways that are not usually experienced for drives in soils. Various refined methods to characterize soils and rocks have been presented in Chap. 2 earlier. Recent developments in the assessment of jacking forces in different rock conditions (intact or highly weathered and fractured) will be discussed in the 2 case studies found in this chapter.

3.3.1 Influence of Stoppages

After a stoppage in the jacking exercise, larger jacking forces are required to advance the pipeline than are required to keep it moving. Assuming a constant thrust at the head, this phenomenon can be explained by soil creep, which leads to tightening around the pipe section. Furthermore, if lubrication is used, the dissipation of induced interstitial overpressures in the bentonite film (due to the stopping of bentonite pumping) leads to an increase of the effective stress acting on the pipe and friction forces at pipeline restart.

It is therefore recommended that a capacity of twice the base force be available to advance a pipe string following a substantial stoppage (Thomson 1993). Although this effect is negligible in soft clay and cohesionless soil (Pellet-Beaucour 1997, Milligan and Norris 1999), it is particularly noticeable in highly plastic clay. In this case, as shown in Fig. 3.13, increases in jacking forces are detectable after only a few minutes of a stoppage episode and can lead to increase of 50% or more after a few hours (Milligan and Norris 1999).

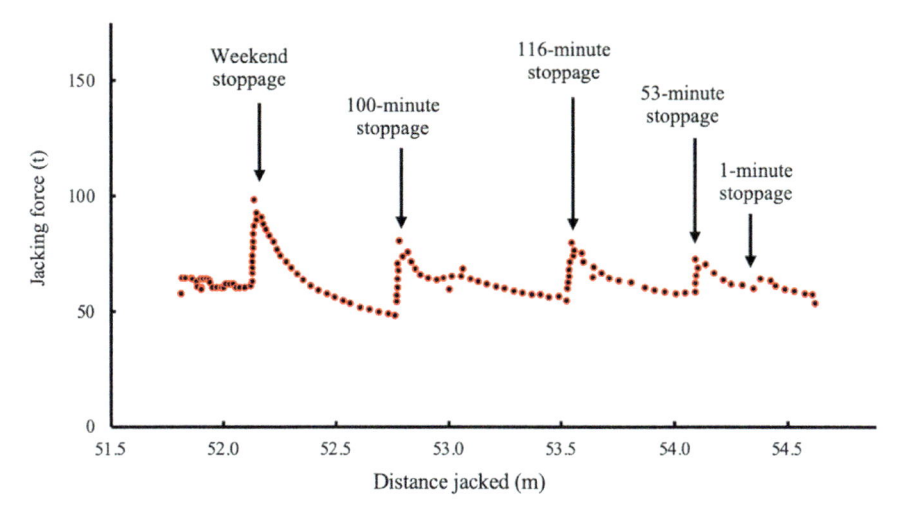

3.3 The Influence of Construction Operational Parameters

Fig. 3.13 Effect of stoppages on jacking forces (after Milligan and Norris 1999)

Experiment data collected from a microtunneling installation in silty marl, carried out as part of the French National Project 'Microtunnels' (Pellet-Beaucour and Kastner 2002), has shown that the increases in jacking forces after stoppages are mainly related to the drive length and the duration of work interruptions and can be evaluated to be equal to 0.8 kPa for short-length interruptions (shorter than three hours), 2.0 kPa for daily interruptions, and 2.4 kPa for weekend stoppages.

Further research studies carried out by the authors and Milligan and Norris (1999) have highlighted a linear relationship between the percentage of increase in jacking forces and the stoppage duration on a logarithmic scale, multiplied by a coefficient varying between 6 and 8.

3.3.2 Influence of Lubrication

Lubrication and lubrication system can take many forms according to the job and the nature of soil conditions. Theoretically, if an appropriate amount of lubricant is introduced and can be maintained in the void space around the pipe (which depends on the overcut size), the pipeline being jacked can be suspended in this fluid. The force required to jack the pipe will then only be that required to overcome the shear value of the fluid.

In practice, difficulties arise in maintaining a void space around the pipe and a suitably pressurized fluid in it, due to the inward movements of material onto the pipe (e.g., in case of excavation in expansive clays, in stress-relieved strata and generally in weaker clays and silts) or to the loss of slurry into permeable soils (granular soils).

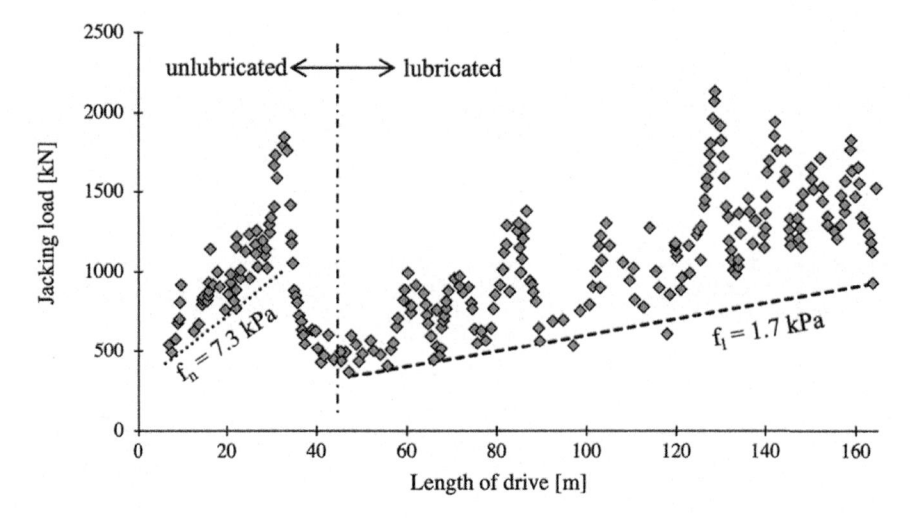

Fig. 3.14 Lubrication influence on jacking forces (after Pellet-Beaucour and Kastner 2002)

However, results of different research studies (Pellet-Beaucour and Kastner 2002, Norris and Milligan 1992b) have highlighted the effectiveness of lubricants in reducing dynamic frictional stress. Figure 3.14, obtained from the analysis of a microtunneling installation in fine sand, shows a reduction of about 77% of frictional stress when lubrication is adopted.

Besides ground conditions, lubrication efficiency appears to be linked to equipment and procedures of injection (i.e., nature and volume of injected slurry, continuum or discontinuum injections, distribution of injection points along the pipeline).

A further effect of lubrication is linked to the reduction of additional frictional stress, due to soil creeping, at pipeline restart after stoppages. In fact, the consolidation time of the slurry is proportional to its thickness, it is supposed that beyond a critical amount, part of the bentonite slurry remains unconsolidated at the soil–pipe interface during the stoppage, thus leading to friction reduction at the restart (an increase of frictional stress at the restart from 2.0 to 0.6 kPa has been noted for an increasing injection amounts from 26 to 200 L/m) (Pellet-Beaucour and Kastner 2002).

The introduction of lubricant into the overcut annulus is performed by connecting the lubricant injection unit to inlet valves located on the inner wall of the jacking pipes. With reference to the Kuching Wastewater Management System (Phase 1) (KWMSP1), these valves (subsequently known as lubricant injection ports) are positioned at three locations in the jacking pipes, i.e., at the pipe crown and pipe knees (Fig. 3.15).

3.3 The Influence of Construction Operational Parameters 65

Fig. 3.15 Positions of lubricant injection ports

Lubrication is not carried out at each pipe. The general practice was to designate two numbers of pipes as lubricant injection points, with Pipe 1 as the primary lubricant injection point and other pipes identified through on-site inspection of lubricant sufficiency.

The assessment of lubricant sufficiency was made by the qualified site personnel traversing the pipeline during jacking and checking every lubricant injection port for the presence of a lubricant. As lubrication was pressurized during injection at approximately 2 bars, the adequacy of lubrication was immediately apparent upon opening the lubricant ports. Should lubricant spray into the pipe from the overcut annulus, the indication was that lubrication efforts had been sufficient. When no seepage of lubricant was encountered through the injection port, it was probable that there was insufficient lubrication of the overcut annulus at the said port.

Figure 3.16 charts the active lubricant injection ports (by pipe number) with respect to drive progress for a particular stretch of one pipe jacking drive traversing from phyllite into graywacke. The numbers on the table show the pipe number, with active lubricant ports highlighted in yellow. This chart was derived from Staheli (2006) as a means of visualizing the position of active lubricant injection ports as pipe jacking progressed.

Pipe 1 is the primary injection port for replenishing lubricant in the newly excavated overcut as the drive progresses. The red arrows follow the positions of active ports as they shift for each lubricant injection unit. For example, one of the lubricant injection units was initially connected to Pipe 16, before shifting back to Pipe 22 and then to Pipe 25. For the other lubricant injection unit, it was initially connected to Pipe 24, before shifting back to Pipe 29 and then to Pipe 33.

Active injection ports would be at the same pipe throughout the drive should there be no losses of lubricant material. Also, prolonged lubrication at a particular chainage could imply a loss of lubricant. The 'shifting back' of the active lubricant

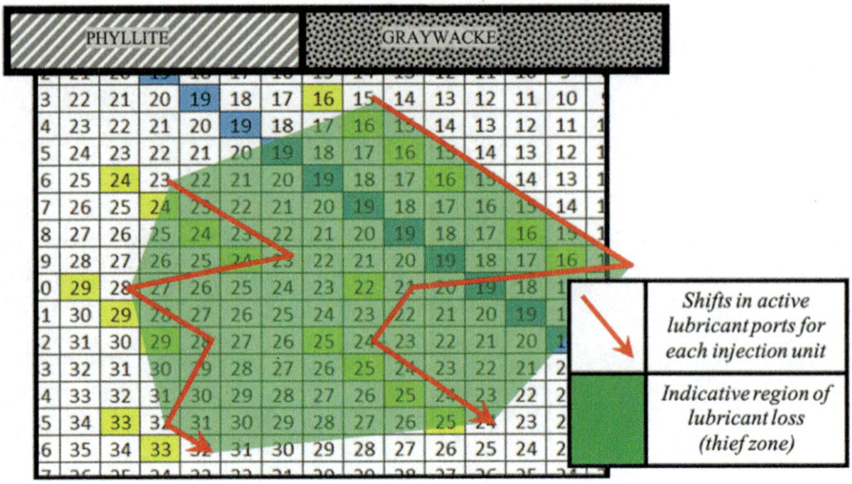

Fig. 3.16 Shifts in active lubrication ports

ports seemed to suggest that there was significant lubricant loss, possibly in the zone between these shifts in the active lubricant port (shaded in green).

3.3.3 Influence of Overcut Size

The excavation diameter is often larger than the pipe external diameter, in order to reduce the friction forces developed at the soil–pipe interface during microtunneling installation. The difference between these two dimensions, called overcut, is typically in the range 10–30 mm (FSTT 2004; Milligan and Norris 1999; Thomson 1993) although larger values, up to 300 mm, have been used (Thomson 1993). In this case, the void around the pipe has to be backfilled by grouting once the driving is complete in order to limit settlements.

A research work carried out at Loughborough University of Technology indicated that the optimum overcut ratio (i.e., the ratio between the difference in diameter of excavation and pipe and the diameter of pipe) is 0.04. The friction forces remained low as the overcut ratio was increased above this value but rose sharply as the overcut ratio reduced to zero (Rogers and Yonan 1992). This result refers to sand above the water table and relates to the generation of arching, while there are no data to suggest that an overcut of this magnitude is also an optimum value in cohesive soils (Thomson 1993).

As shown in Fig. 3.17, experimental results of the French National Project 'Microtunnels' for a microtunnel installation in sand and gravel (Pellet-Beaucour and Kastner 2002) confirm the previous research work, showing an increase in frictional

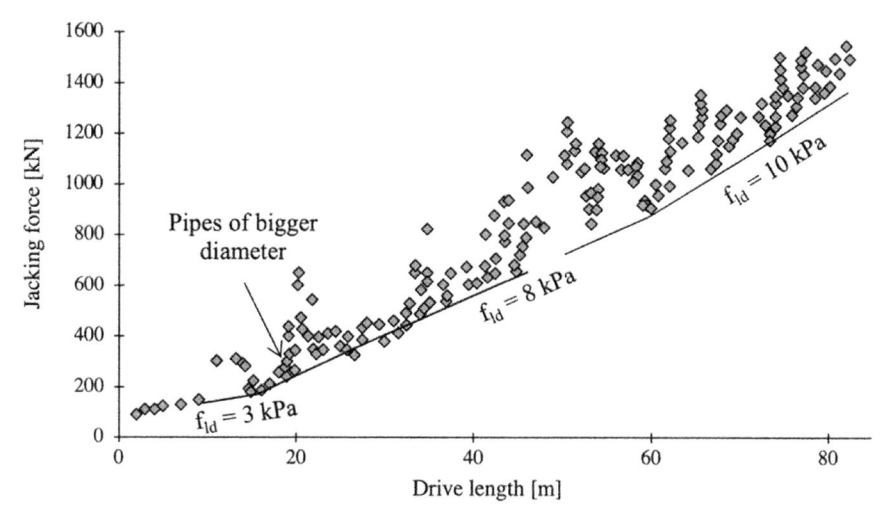

Fig. 3.17 Overcut influence on frictional stress (after Pellet-Beaucour and Kastner 2002)

stress of about three times for an increase of 20 mm in pipe external diameter and an excavation diameter of 660 mm (the overcut ratio decreases from 0.05 to 0.02).

3.3.4 Influence of Misalignment

An increase in jacking forces results from increased radial stress on misaligned pipes. Experiment works on jacked pipes that were maintained in deflection positions explained this phenomenon as the attempt of pipes to realign when axial force is applied, resulting in large radial forces on the inside surface of a curve.

In instrumented field trials on misalignment (Norris and Milligan 1992a), good agreement was found between the maximum angular deviation of the pipe train and the position of peak radial stress (several times greater than the average radial stress), confirming that the pipeline tried to straighten at position of maximum misalignment.

Predicting the forces caused by misalignment is tricky, due to the difficulty in establishing its exact contribution. However, from the limited data available, it seems to be correct to consider a threefold increase in jacking force for every $0.1°$ of relative angular deviation at a pipe joint in dense or heavily over-consolidated soils (Thomson 1993). A significant lower value would apply in less competent soils.

Furthermore, Milligan and Norris (1999) have analyzed the effect of vertical and horizontal misalignments, showing a greater influence of the latter on the jacking force increase. In particular, while in the former case the pipeline was observed to span between the high points of the bored tunnel (without significant increment of contacts at soil–pipe interface), horizontal deviations were noticed to create additional friction forces due to the contacts made by the tunnel sides and the inside of each pipeline bend, at the point of maximum curvature.

3.3.5 Influence of Steering

It can be demonstrated that steering causes temporary increases in jacking forces. The jacking forces from a particular drive in the KWMSP1project are shown in Fig. 3.18.

Normally, steering of the microtunneling boring machine (mTBM) (Iseki TCS1500) is carried out only by halting the main jacks. The normal practice is to segregate operations of the steering jacks and main jacks to mitigate sustained excessive jacking forces. Toward the end of the observation period, 'mixed steering' was utilized. The term 'mixed steering' refers to simultaneous use of both steering jacks and main jacks. Mixed steering was utilized as the operator had difficulties realigning the TBM under normal steering practice.

Additionally, between 1 and 5 min were spent steering during normal steering procedure. The cutter revolution speed for the TCS1500 model was rated at 8.3 rev/min, implying that the overcut roller cutters made up to 120 passes against the rock face due to steering. Figure 3.19 shows some of the roller cutters used.

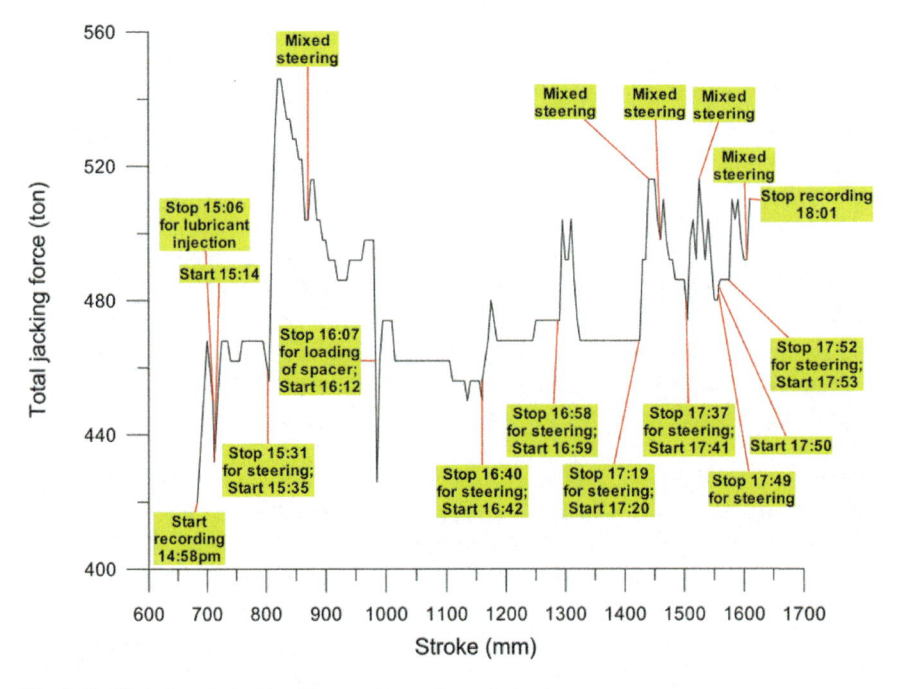

Fig. 3.18 Variations in jacking forces with stroke and steering

3.3 The Influence of Construction Operational Parameters 69

Fig. 3.19 (clockwise from top left) Roller cutters from Iseki Super Unclemole TCS1500, roller cutters from Iseki Unclemole TCS450, disk cutters from Herrenknecht AVN1200TC, worn out roller cutters

Any forces exerted during this time were solely from the steering jacks as the main jacks were halted under normal procedures. The steering jacks for the TCS1500 model were rated at four numbers of 820 kN each. Steering only utilizes two of the four hydraulic steering jacks, depending on the direction of steering, resulting in a maximum of 1640 kN. Steering dials were usually turned intermittently, alternating between 'steer' and 'stop' for a few seconds at a time.

Generally jacking forces increased temporarily by 20–50 tons after each steering effort. The reasons for these 'spikes' in jacking forces can be explained in the way mTBMs are steered, as opposed to steering of a bicycle, as depicted in Fig. 3.20.

Steering a TBM is unlike steering a bicycle or a car, which re-centers itself upon exiting a turn due to the caster of the suspension system. The TBM has no re-centering feature owing to the following reasons:

- The steering jacks are extended or retracted along an imaginary line which is almost parallel with the direction of motion of the TBM, creating no caster angle.
- The differences in earth pressure against the direction of TBM motion are mostly minimal across the cutter face, owing to the small TBM diameter.
- The hydraulic steering jacks do not displace significantly after extending or retracting.

Furthermore, additional friction is initiated during steering by exerting the TBM shield against the wall of the bored tunnel. This restrictive bore is created by the slight protrusion of the overcut roller cutters beyond the shield body. Concentrated overcutting has to take place at the rock face before drive progress can be made along the new alignment.

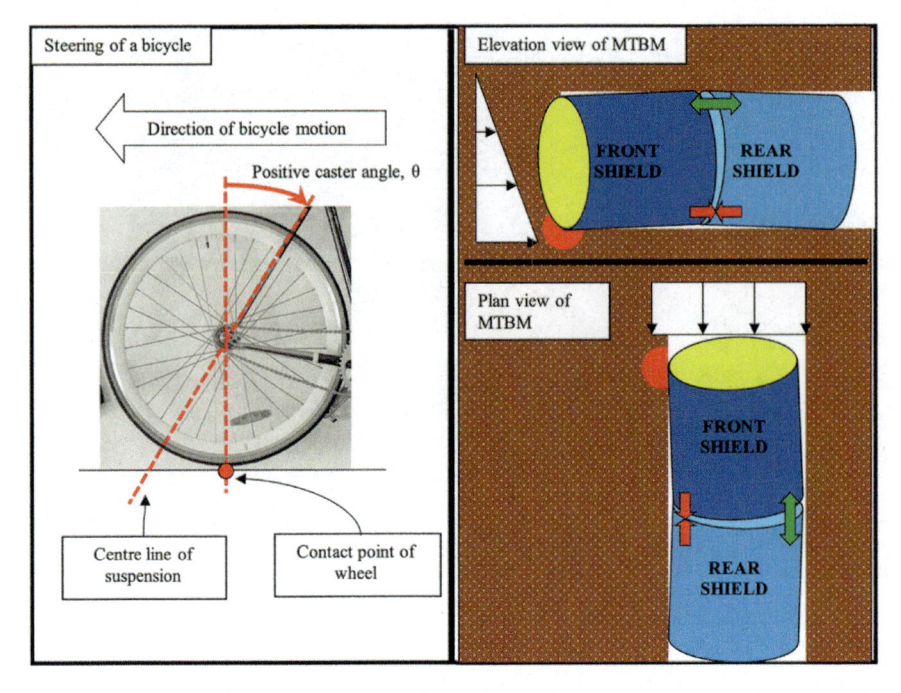

Fig. 3.20 Differences between steering of a bicycle with a caster angle, as opposed to steering of MTBM. Concentrated overcutting of MTBM shown as red circles

3.4 Case Study 1: Pipe Jacking in the Altamura Limestone

Barla et al. (2006) presented a case study on microtunneling works in a limestone rock mass. The project was in the city of Martina Franca, Italy, involving the installation of a 200 m sewer pipeline under the central area of the city. The sewer pipeline was installed via the pipe jacking method. This case study was motivated by the high measured jacking forces, which stopped the pipe jacking works from proceeding. After the stoppage, the rest of the pipeline installation had to be done via cut and cover method. Figure 3.21 shows the location of the studied pipe jacking drive.

3.4.1 Characterization of the Altamura Limestone

Geological mapping was done to characterize the limestone rock mass that was traversed by the microtunneling works. The structure of the intact rock mass was mapped from rock outcrops along the road near the site, as well as within the jacking shaft. The mapping was done to characterize the joint characteristics (e.g., joint length and spacing, dip direction, bedding plane directions) in the intact limestone rock

3.4 Case Study 1: Pipe Jacking ... 71

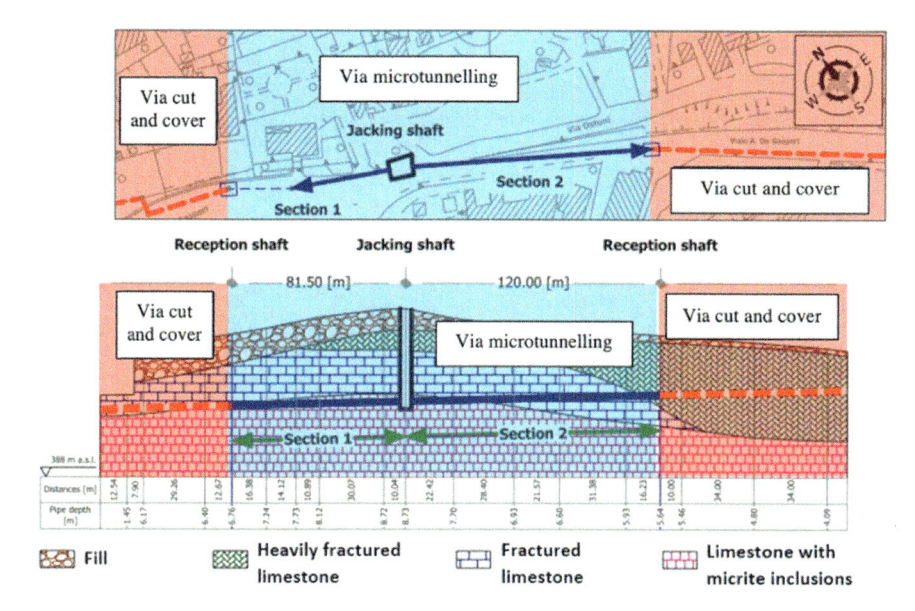

Fig. 3.21 Location of the pipe jacking drive in the case study (after Barla et al. 2006)

mass. This information is essential, as the behavior of the intact rock mass is predominantly governed by its joint characteristics. Stereonet analyses were performed using the data obtained from field observation, as shown in Fig. 3.22.

This was done to identify the number of joint sets in the rock mass, from which three sets were found. Apart from geological mapping, laboratory tests were also performed on extracted samples to characterize the intact rock properties, such as its unit weight, mineral contents, hardness, and strength.

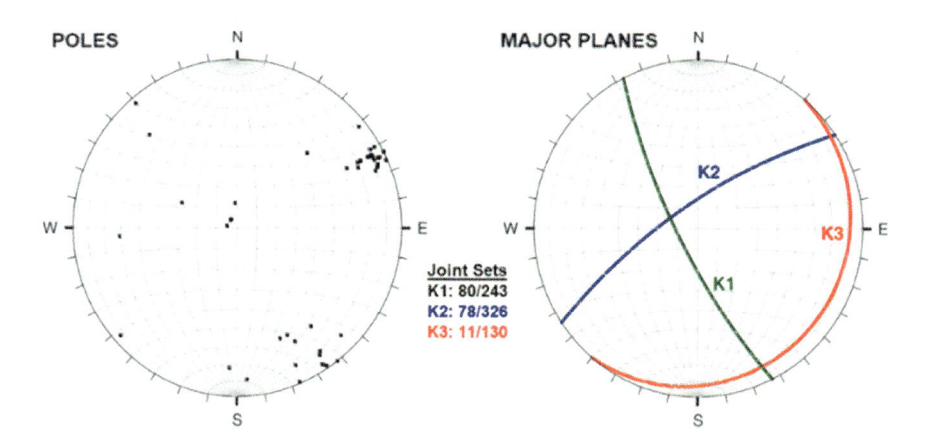

Fig. 3.22 Stereonet analysis on the observed limestone outcrop (Barla et al. 2006)

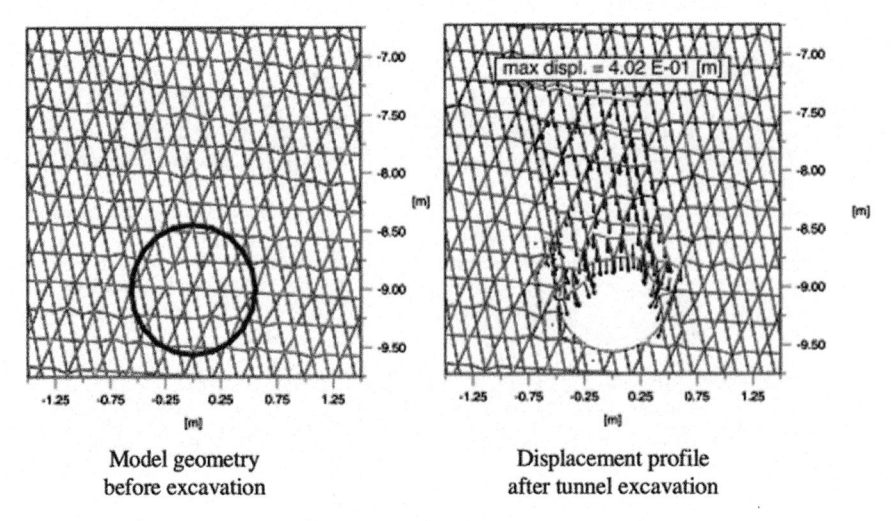

| Model geometry before excavation | Displacement profile after tunnel excavation |

Fig. 3.23 Discontinuum numerical model of tunnel excavation within limestone rock mass (after Barla et al. 2006)

The nature of discontinuities in the rocks can affect the behavior of the rock mass significantly. Barla et al. (2006) demonstrated this via discontinuum numerical method. This method allows for the simulation of rock mass behavior while considering the effects of the rock joint characteristics. This was possible, since the method models the rock mass as discontinuous 'blocks,' as opposed to the conventional finite element method that models materials as one continuous mass. The discontinuum numerical method was used to back-analyze measured jacking forces during the pipe jacking works across the limestone rock mass. This was done by simulating a tunnel excavation within the modeled limestone (together with its joint properties). Changes in stresses and deformations around the modeled tunnel were observed. Figure 3.23 shows an example of the discontinuum numerical model that was generated in the case study, as well as the resulting deformations from the tunnel excavation.

The numerical model showed higher displacements along the modeled joints. Regions of unstable 'blocks' were also observed at the modeled tunnel crown and sidewall areas. The back-analysis has managed to show that:

- Jacking forces will increase when traversing more fractured rock masses. This is because the fractured region of the rock mass is likely to collapse onto the pipes during pipe jacking works, hence increasing the pipe-rock friction.
- Characterizing the discontinuum behavior of rock mass (i.e., joint characteristics) can improve the prediction of jacking forces, which is essential at the design stage.

The findings presented in the case study highlight the importance of performing a comprehensive geotechnical investigation of rock mass characteristics in pipe jacking design. Information such as the joint characteristics in a rock mass can help geotechnical engineers to predict the anticipated loads during pipe jacking works more accurately. That way, these loads can be designed for and mitigation plans can be

3.5 Case Study 2: Pipe Jacking in the Highly Weathered 'Soft Rocks' of the Tuang Formation

made early before the pipe jacking works start. With proper planning and designs, unwanted stoppages such as that presented in the case study can be reduced, if not avoided completely.

Pipe jacking through the highly weathered 'soft rocks' of the Tuang Formation was challenging. The key challenge was in the strength characterization of these highly weathered rocks, which showed predominantly low RQD values. Ong and Choo (2011) reported the RQD values from a site overlying the Tuang Formation. Some of the extracted rock cores from this site are shown in Fig. 3.24, with the RQD values from the site as shown in Fig. 3.25 and Table 3.7.

This presented difficulties in extracting sufficient lengths of intact rock cores for conventional rock strength testing. Despite the highly weathered and highly fractured nature of rocks from the Tuang Formation that yields 73% of the total core lengths to have RQD 0%, the bored tunnels were observed to be able to self-stand, which is counterintuitive! One of these bored tunnels is shown in Fig. 3.26, which was formed after the mTBM was pulled out.

This then became an imperative research question that needs to be solved in order to develop the 'equivalent' design rock strength parameters for estimating pipe jacking forces, especially for frictional resistance. To address this challenge in characterizing the highly weathered and highly fractured traversed rock, two novel methods are presented to overcome this issue. The first method utilized reconstituted tunneling rock spoil subject to direct shear testing (DST), while the second method made use of in situ pressuremeter testing (PMT) of the less-than-perfect rock mass. The former method may be contradicting to general understanding because PMT is often used to measure the in situ stiffness of stiff clay (its operation is limited by

Fig. 3.24 Core boxes of highly fractured highly weathered rocks from a site founded on the Tuang Formation, with joints and discontinuities indicated (after Ong and Choo 2011)

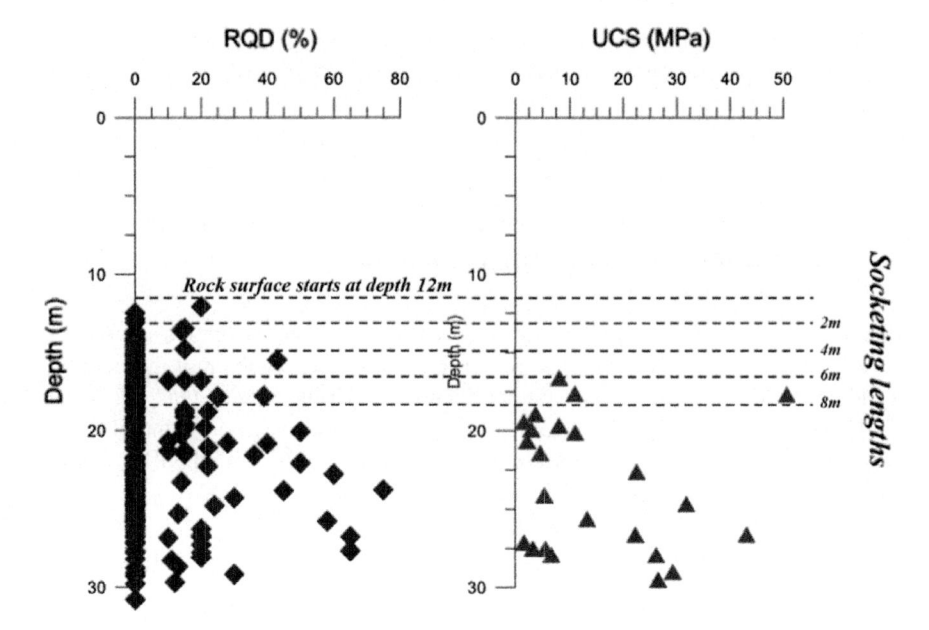

Fig. 3.25 Variation of RQD and UCS values from a site founded on the Tuang Formation (after Ong and Choo 2011)

Table 3.7 RQD values from 194 phyllite cores extracted from a site founded on the Tuang Formation (Ong and Choo 2011)

Borehole number	No. of cores extracted	No. of cores with RQD = 0%
1	10	8
2	0	–
3	1	0
4	10	6
5	5	4
6	16	11
7	11	7
8	9	9
9	13	6
10	15	12
11	12	2
12	16	16
13	14	13
14	15	13
15	15	9
16	15	13
17	17	12
Total	194	141 (73%)

3.5 Case Study 2: Pipe Jacking … 75

Fig. 3.26 Produced tunnel after extraction of the MTBM

the maximum pressure that can be applied) and not weathered rocks. As such, this method must be complemented by FEM simulation to obtain the yield point of the tested rock whose value is out of the testing range of the PMT. The testing principles and details of these two methods have previously been discussed in Chap. 2. The practical applications of these novel testing methods will now be discussed in detail, hereinafter.

3.5.1 Developing Geotechnical Strength Parameters Using Direct Shear Testing of Reconstituted Tunneling Rock Spoil

3.5.1.1 Collection of Rock Spoil for Testing

In the pipe jacking process, crushed excavated rock spoil is transported from the tunnel face to desanders located at the ground surface. The tunneling rock spoil was transported via pressurized slurry and segregated by the desanders (Fig. 3.27).

The tunneling rock spoil was collected from several drives across the KWMSP1 project. In order to assess the pipe jacking forces, these spoil was subject to direct shear testing (DST) to develop somewhat of lower bound strength characteristics of the geology surrounding the driven pipe string. The direct shear tests were carried out on the scalped tunneling rock spoil (see Sect. 2.5.2.1) acquired from the traversed geology in order to characterize the strength of the rock surrounding the driven pipe string. The Geocomp Shear-Trac II system, which is completely automated, was used to conduct the direct shear testing (Geocomp Corp. 2010). The fully automated direct shear system was used to ensure exact control and automation of the direct

Fig. 3.27 Desanders or decantation chamber where tunneling rock spoil are segregated from transport slurry

shear tests. From these tests, strength parameters were developed for the assessment of frictional jacking forces.

3.5.1.2 Development of Equivalent Rock Strength Parameters

Many well-established jacking force models make use of Mohr–Coulomb (MC) strength factors, such as cohesion (c') and friction angle (ϕ'), to calculate the frictional jacking forces. As a result, it is practical to define the strength of the tunneling rock spoil in terms of the linear MC failure criterion because it is straightforward. The direct shear test results provide the data points from which interpretation of strength parameters could be performed in a few ways, including the linear MC failure criteria, a nonlinear power law function, and a generalized tangential approach to the power law (see Sect. 2.5.2.4). These failure envelopes are shown in Fig. 3.28 for direct shear tests performed on the reconstituted tunneling spoil of sandstone from the Tuang Formation.

The tunneling rock spoil generally demonstrated nonlinear shear strength profiles. It was necessary to acknowledge this nonlinearity, while developing MC parameters for use in jacking force equations. It is important to interpret MC strength characteristics from the produced nonlinear power law envelopes in order for the power law strength characterization to be effective in the assessment of jacking forces. Through the use of a generalized tangential approach, MC strength parameters could be developed while acknowledging the nonlinearity of the tested tunneling rock spoil.

Thereafter, a generalized tangential approach was utilized to estimate the required MC parameters from the nonlinear power law failure criteria as the input (Yang and Yin 2004). The generalized tangential method together with the nonlinear power law envelope takes into consideration the variation in strength behavior as a function

3.5 Case Study 2: Pipe Jacking ...

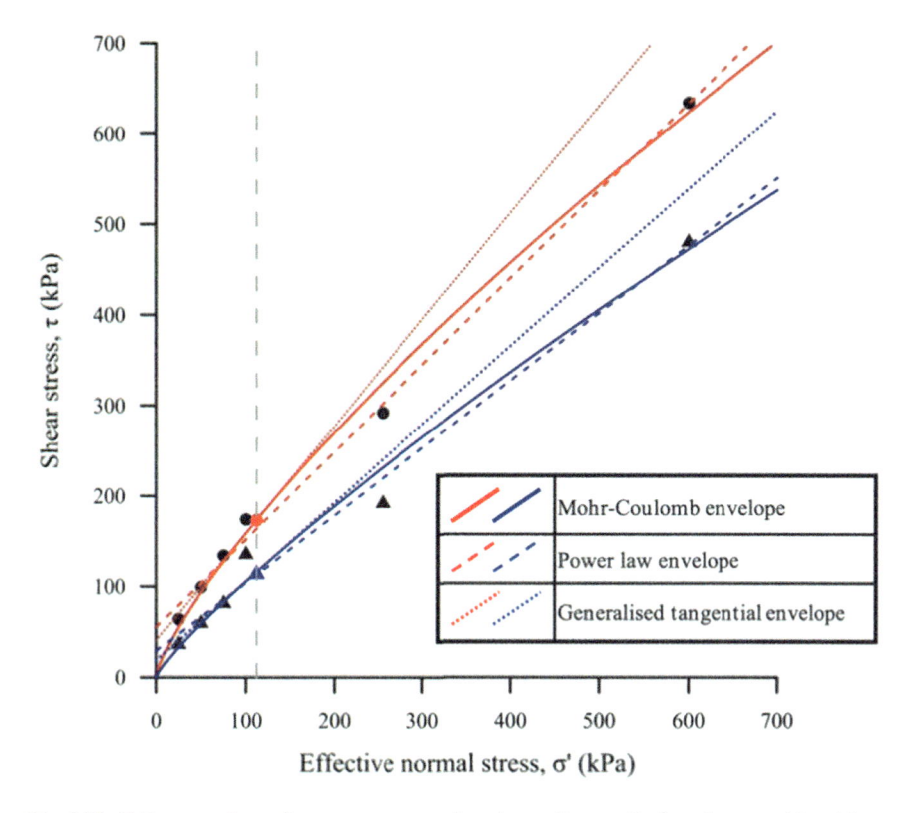

Fig. 3.28 Failure envelopes for a test on reconstituted tunneling spoil of sandstone, with red lines showing peak envelopes and blue lines showing residual envelopes

of applied stress. Similar approaches have been used in past research to understand nonlinearities in the strength behavior of different particulate materials (Collins et al. 1988; Drescher and Christopoulos 1988; Soon and Drescher 2007). The generalized tangential technique varies from the MC and power law failure criteria, which consisted of simply fitting a regression line (linear or nonlinear) across the points in the data.

In order to simplify the nonlinearity in strength behavior for practical use, a linear MC strength envelope that is tangential to the nonlinear power law envelope is created. Considering that power law functions are stress-dependent, it was necessary to identify and specify adequate effective stress levels before applying the tangent to power law curves could be performed. The effective overburden pressures at the tunnel crowns for the individual pipe jacking drives were selected for applying the tangent to the power law curves.

The interpreted strength parameters (for both peak and residual states) are shown in Table 3.8 for three highly weathered rocks commonly found in the Tuang Formation, namely sandstone, shale, and phyllite.

Table 3.8 Summary of some equivalent design strength parameters for various 'soft rocks'

	Geology		Sandstone	Shale	Phyllite
MC parameters	Cohesion [kPa]	Peak, c'_p	55.3	19.8	49.9
		Residual, c'_r	29.3	0.6	44.2
	Friction angle [°]	Peak, ϕ'_p	44	37.4	44.3
		Residual, ϕ'_r	36.7	35.2	39.1
Power function parameters	A [-]	Peak, A_p	4.68	1.87	3.86
		Residual, A_r	2.24	0.97	3.37
	B [-]	Peak, B_p	0.76	0.86	0.79
		Residual, B_r	0.84	0.95	0.78
Generalized tangential MC parameters	Cohesion [kPa]	Peak, c'_p	40.7	21.7	45.1
		Residual, c'_r	18.9	6.6	39.3
	Friction angle [°]	Peak, ϕ'_p	49.8	38.7	46.2
		Residual, ϕ'_r	40.9	35.2	41.1

In the case of soils or other particulate materials, the intercept with the vertical axis is referred to as 'apparent cohesion' when linear regression lines are used to depict the strength behavior of the materials. This apparent cohesiveness in remolded soils is related to the interlocking of solid particles inside the soil matrix (Schofield 2005). It was important to obtain more clarity in order to establish if the presence of a 'apparent cohesiveness' was warranted. Therefore, the specimen spoil was subjected to low effective normal stress levels (25, 50, 75, and 100 kPa) in order to establish this nonlinearity. At low effective loads, the results revealed that the shear strengths of the reconstructed tunneling rock spoil gravitated toward the origin (Fig. 3.28), suggesting that the tested spoil behaved in a nonlinear fashion.

3.5.1.3 Interpretation of Strength Parameters Using Mohr–Coulomb Failure Criterion

Based on a linear 'line of best fit' to the test results, elastoplastic MC failure criterion was developed for each material. Generally, the apparent cohesion reduced from peak to residual, especially for the sandstone and shale specimens. A similar observation was made for the phyllite specimens, although this reduction was less pronounced. There were similar findings for the friction angles, where reductions were observed from the peak (ϕ'_p) to the residual (ϕ'_r) states.

When comparing the strength parameters across the different rock spoil, the shale spoil showed the lowest values of cohesion and friction angle. Shale had significantly lower values of cohesion (c'_p and c'_r) than other types of tunneling rock spoil. The apparent cohesion, c'_r, of shale was almost negated in the residual state. The apparent peak cohesion, c'_p, interpreted from tests on phyllite and sandstone were similar. However, the c'_r for phyllite spoil was only reduced by 5.7 kPa from c'_p, while the

3.5 Case Study 2: Pipe Jacking ...

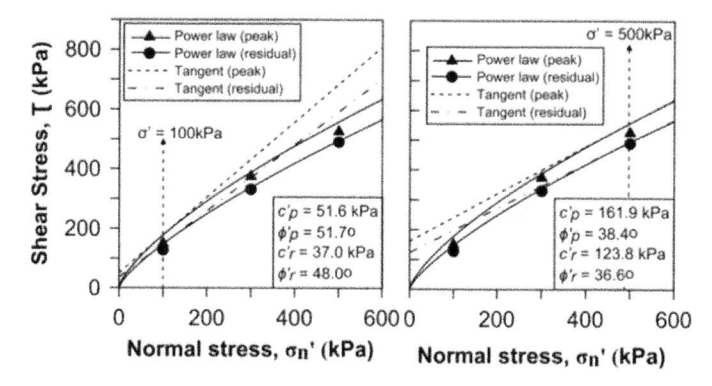

Fig. 3.29 Dependency of strength envelopes on applied normal stresses in direct shear tests (Peerun et al. 2015)

spoil of sandstone showed a significant loss in apparent cohesion of 26 kPa in the residual state. In general, both MC strength parameters, c′ and f′, declined as the system progressed from the peak to the residual states.

Although the use of the basic linear MC model for strength characterization of tunneling rock spoil had certain advantages, there were some disadvantages as well. With the use of regressive lines, such a generalization limit their applicability to the range of stresses for which the experiments was done. This can make it more difficult to characterize the strength of the tested material, particularly when using the linear MC model over a wide range of stress levels and strains. This implies that the linear MC model omits the stress-dependency of the tested material's strength, even if the material demonstrates clear nonlinearity. This results in a significant inaccuracy in the estimation of shear strengths at extremely low- or high-stress levels, in relation to the levels of normal stresses in the tests. This is illustrated in Fig. 3.29.

The incompatibility of MC parameters (based on a regressive line of best fit) to the nonlinear materials can be further emphasized by understanding the variations in normalized secant friction angle, $\tan(\phi'_{sec}) / \tan(\phi'_p)$. Figure 3.30 shows that friction angles can change significantly (up to two times!) within the range of overburden pressures for pipe jacking drives in the KWMSP1 project.

3.5.1.4 Interpretation of Strength Parameters Using Power Law Function with the Generalized Tangential Approach

When addressing the nonlinearity in the shear strength profile of the reconstituted tunneling rock spoil, a simple power law function was examined as a possible solution. Several researchers, including Charles and Watts (1980), Charles and Soares (1984), and De Mello (1977), have employed this power law function to characterize the nonlinear shear strength behavior of rock fills. This power law function is given us

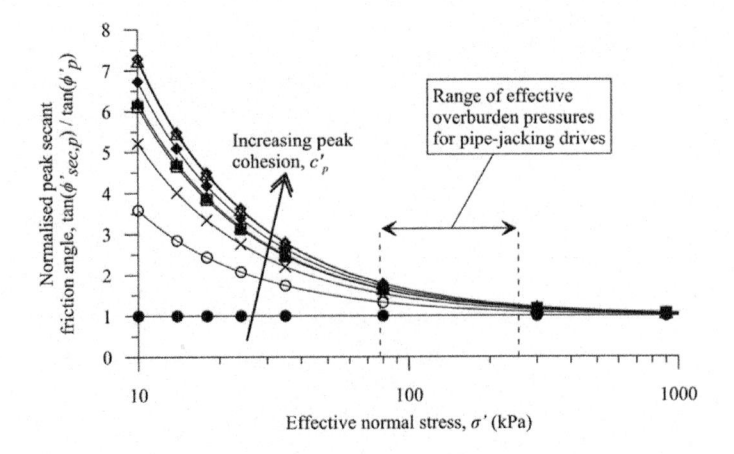

Fig. 3.30 Variation of friction angles with normal stresses

$$\tau = A \times \sigma'^{B} \tag{3.31}$$

where A and B are two dimensionless constants, respectively.

The power function is controlled by two parameters: A, which regulates the magnitude of shear strength, and B, which determines the curvature of the envelope. Values of B are limited to between 0 and 1. Lower values of B resulted in higher curvature in the power law envelope, while larger values would cause the power function to become more linear. When the power law function is reduced to the simplest form (where $B = 1$), the MC failure criterion becomes apparent.

From the earlier reported test results (Table 3.8), B was between 0.72 and 0.96, which is consistent with the power law exponent values provided by De Mello (1977) and Anyaegbunam (2015). These values are reported in Table 3.9 and illustrated in Fig. 3.31.

From the tested specimens, the A parameter's peak values (A_p) were significantly bigger than the corresponding residual values (A_r). This agrees with the earlier observations using a regressive MC envelope, where magnitudes of c' and ϕ' generally reduced with continued shearing. For the exponential B parameter, as the tests progressed from peak to residual states, B increased from its peak value, B_p to its residual value, B_r. This implied that the curvature of strength profiles decreased from peak to residual states, which is equivalent to a loss of apparent cohesion from the peak to the residual states in the linear MC interpretation.

Because of the nonlinearity in the shear strength behavior of the reconstructed tunneling rock spoil captured by the power law function, the linear MC model was found to have a severe weakness that was not captured by the power law function. However, although the power law function is capable of accurately representing the nonlinear behavior of the tested spoil, jacking force models are still significantly reliant on the MC strength parameters c' and ϕ'. An extra step in the development of

3.5 Case Study 2: Pipe Jacking …

Table 3.9 Power law parameters, A and B from tests on reconstituted tunneling rock spoil, and other previous studies

Material type	Geology	A	B	Source
Tunneling rock Spoil	Sandstone (peak)	4.68	0.76	This book
	Sandstone (residual)	2.24	0.84	
	Shale (peak)	1.87	0.86	
	Shale (residual)	0.97	0.95	
	Phyllite (peak)	3.86	0.79	
	Phyllite (residual)	3.37	0.78	
Rockfill	Sandstone	6.8	0.67	Charles and Watts (1980)
	Slate	5.3	0.75	
	Slate	3.0	0.77	
	Basalt	4.4	0.81	
	Basalt	1.54	0.821	De Mello (1977)
	Diorite	1.10	0.870	
	Conglomerate	1.27	0.846	
	Conglomerate	1.19	0.881	
	Conglomerate	1.59	0.808	

Fig. 3.31 Plot of power law parameters, A and B for rockfill materials and reconstituted tunneling rock spoil

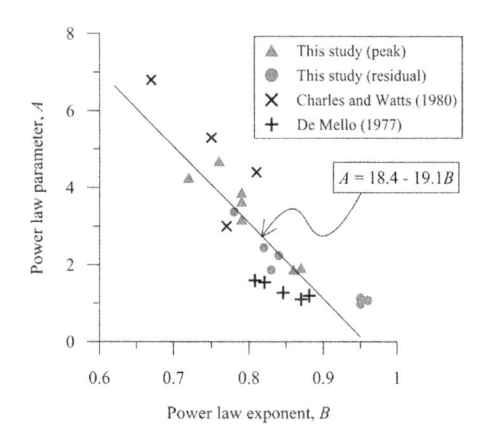

nonlinear power law strength profiles is necessary to allow for the interpretation of c' and ϕ' from A and B. This would be possible through the use of the generalized tangential approach (GTA).

The GTA can be applied by plotting a tangent to the power law envelope at designated normal stress, as shown in Fig. 3.32.

In the prediction of pipe jacking forces, the designated normal stress is the effective overburden pressure at the pipe crown. The gradient and y-intercept of the tangential function will give rise to the tangential MC parameters, ϕ'_t and c'_t, respectively.

Fig. 3.32 Tangential line for a nonlinear failure criterion (after Yang and Yin 2004)

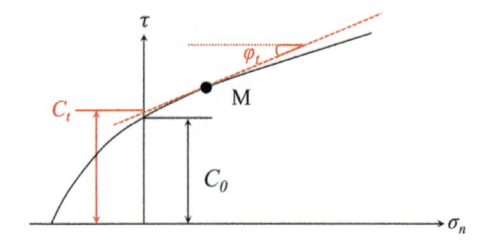

Alternatively, these tangential parameters could be obtained by differentiating the power function (Choo and Ong 2016), giving rise to the following solutions:

$$\phi'_t = \tan^{-1}\left(AB\sigma_v'^{B-1}\right) \tag{3.32}$$

$$c'_t = A\sigma_v'^{B}(1-B) \tag{3.33}$$

where σ'_v is the effective overburden pressure at the pipe crown.

These tangential MC parameters, ϕ'_t and c'_t will be used in the back-analysis of frictional jacking forces from some case studies, the details of which will be discussed, hereinafter.

3.5.1.5 Back-Analysis of Pipe Jacking Field Measurements

The majority of jacking force models are created to assess the jacking loads accumulated during pipe jacking drives that traverse soils. For drives into rock masses, only the most basic factors are taken into account in these jacking force models. The Pellet-Beaucour and Kastner (2002) (PBK) model is a well-established model that was designed for the measurement of frictional jacking forces for drives crossing different soil conditions. Because frictional forces rise with the contact area between the outer periphery of the pipes and the surrounding soils, frictional forces are the critical component of jacking loads. The length of the driven pipeline is the limiting element for frictional forces and jacking loads on the assumption that the pipe–soil contact area is uniform. In the PBK model, the soil stresses acting on the outside surface of the drive pipeline are a function of MC parameters, ϕ' and c'. From the earlier GTA method, these parameters will be substituted with ϕ'_t and c'_t, respectively.

The Pellet-Beaucour and Kastner (2002) jacking force equation is reproduced below for the reader's convenience.

$$F_f = \pi L D \mu \times 0.5\big[\sigma_{ev} + 0.5\gamma D + K_2(\sigma_{ev} + 0.5\gamma D)\big] \tag{3.34}$$

3.5 Case Study 2: Pipe Jacking …

where F_f is the frictional resistance along the pipe, D is the pipe outer diameter, L is the jacking distance, γ is the soil unit weight, and K_2 is the coefficient of soil thrust on the pipe.

The effects of soil arching on friction forces are considered in the PBK model through the computation of σ_{ev} which is the vertical stress at pipe crown due to soil arching and given in Eq. 3.35.

$$\sigma_{ev} = \frac{b\left(\gamma - \frac{2c}{b}\right)}{2K\tan\delta} \times \left(1 - e^{\frac{2Kh}{b}\tan\delta}\right) \quad (3.35)$$

δ is the internal friction angle of the soil, while c is the apparent cohesion of the soil.

σ_{EV} was developed from Terzaghi's classical trapdoor experiment, where he made observations of soil strains developing over a vertically displaced trapdoor (Terzaghi 1936, 1943). A vertical sliding block of overburden soil into trapdoor causes soil stresses to redistribute, enabling the soil to arch over the void, such as a bored tunnel during pipe jacking. Small displacements of the overlying soil frequently result in arching. Therefore, it was appropriate to use peak tangential MC parameters to assess the arching effects in σ_{EV}. Naturally, the earlier developed peak tangential MC parameters, $\phi'_{t,p}$ and $c'_{t,p}$, would be directly substituted into δ and c, respectively. The redistribution of stresses in soil arching was shown previously in Fig. 3.7. If soil arching is effectively mobilized, the normal stresses acting onto the pipeline would reduce, thus lowering frictional jacking resistance generated during pipe jacking. Therefore, arching has a substantial impact on the accumulation of jacking forces hence the importance of this knowledge in pipe jacking in rocks.

The friable geology encountered in the Tuang Formation was similar to soft rock that could act as soil. This enabled for the application of not just soil-based jacking force calculations, but also early direct shear testing on reconstructed tunneling rock spoil. Shearing of tunneling spoil could be analogous of the soil mechanics at the overburden's vertical slip planes. Arching happens at modest displacements, as previously noted. As a result, the peak results from direct shear testing on tunneling rock spoil should be utilized in conjunction with the PBK models to assess frictional pipe jacking forces.

The Kuching Wastewater Management System Phase 1 (KWMSP1) project included the installation of 7.7 km of trunk or main sewer pipelines. The gravity flow system would transfer wastewater to a 100,000 population equivalent (PE) wastewater treatment plant. The trunk sewer pipelines were made up of 3 m lengths of 1.2 m and 1.5 m diameter spun concrete pipes. Trenchless technology, notably microtunneling by pipe jacking, was used to install the trunk sewer pipelines. Pipelines were erected at depths of up to 35 m to transport graywater, under a full gravity flow system. This meant that the majority of pipe jacking operations took place within the Tuang Formation rock masses. The jacking and reception shafts were built with excavators and supported by segmental caisson rings due to the difficult geotechnical conditions and high ground water table. A case study on this can be found in Chap. 4 that discusses the challenges encountered during the construction of deep shafts.

Using the previously developed peak tangential MC parameters, $\phi'_{t,p}$ and $c'_{t,p}$ with the PBK model, along with other relevant pipe jacking geometries, the results from back-analysis are presented in Table 3.10. Pipe jacking operational parameters have also been reported in order to verify the interpreted σ_{EV} and pipe-rock coefficient of friction, μ.

One of the key outcomes from the back-analysis of jacking forces was σ_{EV}, which would indicate the degree of arching that could develop around the pipe. In the cases of the drives in sandstone and phyllite, the computed σ_{EV} values were negative, implying that tensile soil stresses were present in the rock surrounding the pipe. However, due to the presence of an overcut zone (which is effectively a gap) between the pipe's outer periphery and the surrounding geology, tensile soil forces could not be transmitted onto the pipe. For the drive in shale, there was 12.5 kPa of rock stresses acting onto the pipe, suggesting that the arching effect in shale was less pronounced compared to sandstone and shale. The other key outcome is the pipe-rock coefficient of friction, μ. These values were compared against those suggested by Stein (2005) for lubricated drives, i.e., μ of less than 0.3. For the drive in phyllite, the back-analyzed μ of 0.080 suggested effective lubrication, while the drives in sandstone and shale had μ values exceeding 0.3, suggesting that lubrication may not have been effective. Comparing the frictional resistances in these drives, the drive in shale exhibited the largest at 37.2 kN/m^2, while the drive in phyllite accrued only 4.8 kN/m^2. In order to frictional resistances in terms of the key back-analysis outcomes, it is essential to validate them against the pipe jacking operational parameters for each of these drives. Therefore, each of these drives will be studied in detail, hereinafter.

Drive in Highly Weathered Sandstone

The Herrenknecht AVN 1200TC mTBM was used to construct this pipeline. The 140-m-long concrete pipeline had to negotiate through highly weathered sandstone that was 12.5 m deep. The theoretical overcut annulus had a volumetric volume of 87 L/m, into which lubricant was injected at an average rate of 47 L/m, making the overcut only 55% filled with lubricant. This indicated that the overcut region was not completely filled, implying that the frictional force-reducing effects of lubrication were not fully mobilized. At 104 kN/m^2, the measured face support pressure remained constant. This enabled the analysis to be purely focused on the changes in friction resistance, which was 14.4 kN/m.

With tangential peak strength parameters for the sandstone are $c'_{t,p} = 50.8$ kPa and $\phi'_{t,p} = 47.8°$, the resulting σ_{EV} was -20.8 kN/m^2. This negative value of σ_{EV} was adjusted to zero using the previously mentioned back-analysis procedure (Terzaghi 1943). The resulting μ for this drive was 0.365. The back-analyzed average of 0.365 was somewhat greater than the recommended maximum limit of $\mu = 0.3$ for lubricated drives by Stein (2005). Earlier comparisons of lubricant injection volumes

3.5 Case Study 2: Pipe Jacking …

Table 3.10 Summary of outcomes from back-analysis of frictional pipe jacking from drives in highly weathered rocks

Geological information	Rock type	[–]	Sandstone (see Sect. 3.5.1.5.1)	Shale (see Sect. 3.5.1.5.2)	Phyllite (see Sect. 3.5.1.5.3)
Equivalent rock strength parameters	Peak tangential cohesion	$c'_{t,p}$ [kPa]	40.7	21.7	45.1
	Peak tangential friction angle	$\phi'_{t,p}$ [°]	49.8	38.7	46.2
Back-analysis of frictional jacking resistance	Effective overburden pressure	σ'_v [kN/m^2]	112	172	162
	Vertical stresses at pipe crown due to arching effect	σ_{EV} [kN/m^2]	−14.5	12.5	−14.4
	Pipe-rock coefficient of friction	μ [–]	0.365	0.356	0.080
Pipe jacking operational parameters	Microtunnel boring machine (MTBM) type	[–]	Herrenknecht AVN 1200TC	Iseki Unclemole Super TCS 1500	
	Tunnel diameter	D_e [m]	1.43	1.78	1.78
	Face support pressure	[kN/m^2]	104	Not measured	47
	Averaged frictional resistance from field measurements	F [kN/m]	14.4	37.2	4.8
	Volume of lubricant injected	[L/m]	47	1851	181
	Overcut volume	[L/m]	87	113	113
	Overcut utilization	[%]	55	1,638	160
	Jacking speed	[mm/min]	16	12	44

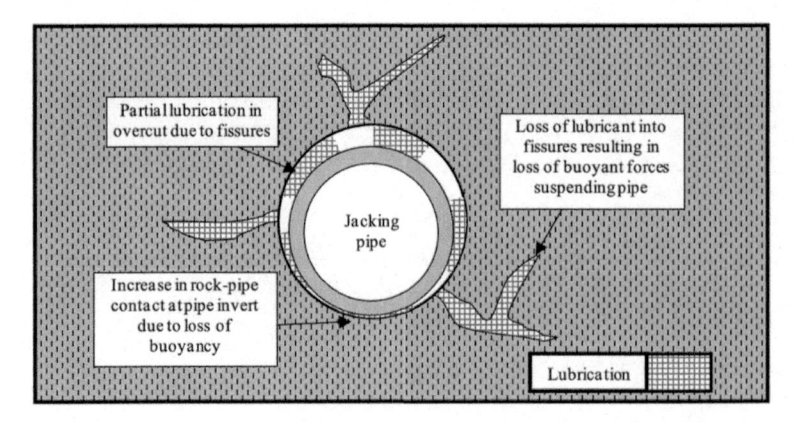

Fig. 3.33 Schematic diagram showing pipe behavior in shale, with loss of lubricant into rock fissures and its incurred effects

to theoretical overcut suggested that lubrication effects had not been completely utilized, with the overcut utilization being only 55%. Despite this lack of lubrication, the frictional resistance, F of 14.4 kN/m^2, may have been moderated due to the significant arching in this sandstone drive.

Drive in Highly Weathered Shale

For this drive in highly weathered shale, the Iseki Unclemole Super TCS 1500 was used to install a concrete pipeline extending 153 m at a depth of 21 m. An initial 73 m stretch of this drive traversed sand, followed by a 79 m section traversing shale. Jacking forces in the former sand section were not of interest, so only the latter 79 m in highly weathered shale was subject to back-analysis of jacking forces.

In the shale section of this drive, the volume of lubricant injected was 1851 L/m. This was 1638% of the anticipated overcut volume of 113 L/m, providing a clear indication of significant lubricant loss. The developed tangential peak MC parameters for shale spoil were $c'_{t,p} = 21.7$ kPa and $\phi'_{t,p} = 38.7°$, resulting in σ_{EV} of 12.5 kN/m^2 was obtained, showing reduced arching. From the measured jacking forces, the frictional resistance was 37.2 kN/m, with the back-analyzed μ being 0.356. This exceeded Stein's recommended upper bound for lubricated drives, i.e., $\mu = 0.3$. The increased lubricant use was attributed to lubricant leaking into the surrounding geology, as visualized in Fig. 3.33.

Thus, lubrication was only partially successful, as any buoyancy effects were reduced by lubricant seepage. The excessive lubricating effort, combined with reduced jacking speeds, was required to progress the drive through the shale. The large frictional resistance in shale can be attributed to the reduced arching, and the ineffective lubrication.

3.5 Case Study 2: Pipe Jacking …   87

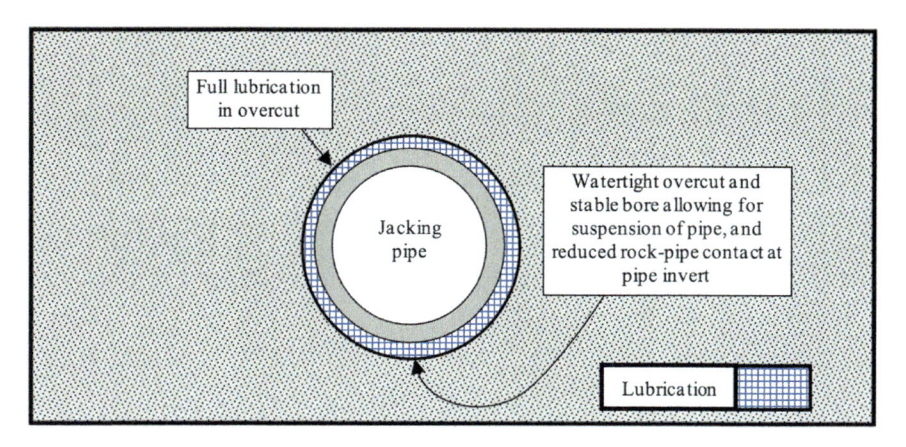

Fig. 3.34 Schematic diagram showing pipe behavior in phyllite, with minimal loss of lubricant in a stable bore

Drive in Highly Weathered Phyllite

The Iseki Unclemole Super TCS 1500 was used for this 120 m drive through highly weathered phyllite. There were no substantial delays in the pipe jacking operations, with the 120 m completed in 23 days. This was attributed to the phyllite of the Tuang Formation being bedded, firmly folded, and heavily sheared (Tan 1993). By plugging over existing joints in the phyllite, the tight and erratic folds generated a stable bore, preventing lubricant leakage into the tunnel surround.

The tangential peak MC values for phyllite spoil were $c'_{t,p}$ = 45.1 kPa and $\phi'_{t,p}$ = 46.2°. σ_{EV} was calculated to be -14.4 kN/m^2, indicating that substantial arching existed during the pipe jacking operations. The frictional jacking resistance, F, was 4.8 kN/m. The back-analyzed pipe-rock friction μ of 0.07 suggested that this drive in phyllite was well-lubricated. This can be corroborated by the field lubrication injection rate of 181 L/m, which only slightly exceeded the theoretical overcut of 113 L/m by approximately 60%.

The measured F of 4.8 kN/m was the lowest measured jacking force across the three drives. This was attributed to the existence of substantial arching and the tightly folded geology, which enabled the injected lubricant to be retained in the overcut with minimal loss to the surrounding geology. This concept is shown in Fig. 3.34.

3.5.1.6 Validation with Finite Element Analysis

To further substantiate the back-analysis, validation of the peak tangential strength parameters, $c'_{t,p}$ and $\phi'_{t,p}$, σ_{EV}, and the back-analyzed μ was performed via finite element analysis (FEA). A typical model geometry is shown in Fig. 3.35.

The peak tangential strength parameters were used as material properties for the modeled rock layers, while the back-analyzed μ was applied to the pipe-rock

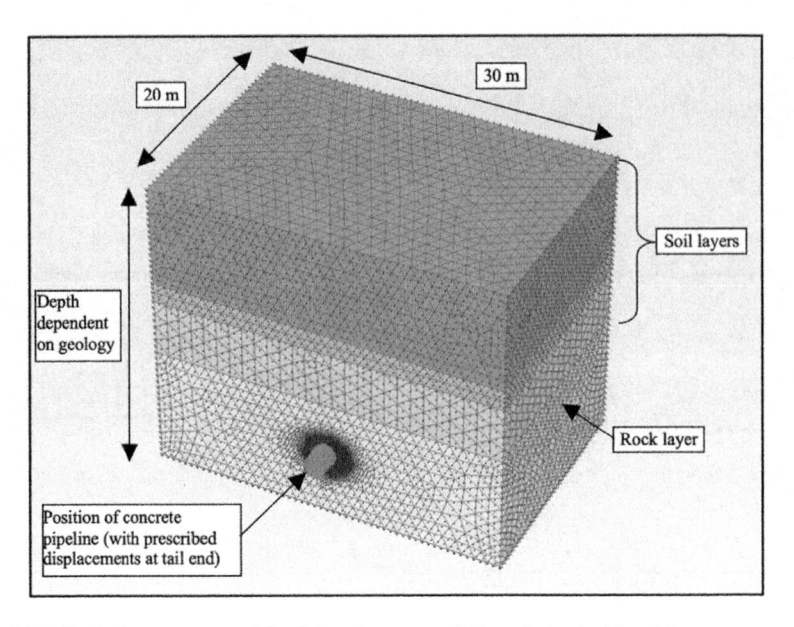

Fig. 3.35 Typical geometry used for finite element modeling of pipe jacking drives

interface elements as an interfacial friction angle, $\delta = \tan^{-1}\mu$. σ_{EV} was considered as part of a post-processing technique for consideration of arching. Table 3.11 shows the material sets used for the drive in sandstone.

The pipeline mesh for each modeled pipe jacking drive was produced by first constructing a circular polycurve on a vertical boundary parallel to the x–z plane. After that, the polycurve was extruded 20 m in the y-direction. The modeled pipes' front and back ends were both flushed with vertical model boundaries. A linear elastic model was used to model the pipeline material, i.e., reinforced concrete. The fixities at the front end of the modeled pipes were released (or set to 'Free') during the pipe

Table 3.11 Model material properties for the drive in highly weathered sandstone

Material	Sandstone	Rock-pipe interface	Pipe
Material model	MC	MC	Linear elastic
Cohesion, c' [kPa]	$c'_{t,p} = 40.7$	0	–
Friction angle, ϕ' [°]	$\phi'_{t,p} = 49.8$	$\tan^{-1}(\mu) = 20.05$	–
Elastic modulus, E' [kN/m^2]	7.1×10^6	1.000	27×10^9
Poisson's ratio, ν	0.35	0.35	0.15

3.5 Case Study 2: Pipe Jacking …

jacking simulation to allow for longitudinal displacement of the pipeline. This was accomplished by prescribing nonzero displacements at the modeled pipelines' tail ends.

The concrete pipelines were modeled as rigid elements. This was understandable because the stiffness of the concrete pipes was significantly higher than that of the surrounding rock pipe contact (Tian et al. 2014; Wang et al. 2009; Zhang et al. 2013). The elastic modulus of the modeled pipeline was artificially exaggerated to achieve this. Because of the rigidity of the modeled pipeline, axial stresses in the modeled pipeline were separated from shear stresses at the less rigid rock-pipe contact.

The simulation of the investigated pipe jacking drives required the selection of an adequate interface with respect to the stiffness of the modeled pipeline. Yin et al. (1995) proposed a very low interface tension modulus, E' of 1 kPa, for sliding failure based on numerical analysis of pull-out tests. The low value of E' was chosen to imitate soil and structure separation by allowing the interface to quickly flex when subjected to tensile loads. In the FEA validation of pipe jacking drives, low interface modulus values were chosen for the modeling of rock-pipe interface parts. The interface modulus, E_{inter} value of 1,000 kPa, was eventually allocated to the rock-pipe interface parts. The stiffness of the rock-pipe interface parts was still much lower than the stiffness of the modeled pipes and surrounding rock masses; therefore, this was justified.

The investigated drives were modeled in three sequential stages:

- Stage 1: Initial stress conditions are applied.
- Stage 2: Wish-in-place pipeline and rock-pipe interface elements activation.
- Stage 3: Displacements are applied at the pipeline's tail end.

To establish initial greenfield stress conditions, soil material qualities, ground water levels, K_0 condition, and model boundary conditions were assigned in Stage 1. In Stage 1, plate and interface elements for modeling the pipeline and the rock-pipe interface were present, but they were disabled.

The activation of plate elements was used to model the wish-in-place 20 m pipeline segment in Stage 2. The pipeline segment was modeled using the 'wish-in-place' technique, which meant that installation effects were neglected. The generated pipeline would simulate ongoing pipe jacking operations. At this point, the soil elements in the pipeline were deactivated and dried to eliminate the effects of hydrostatic pressure within the void pipeline. The rock-pipe interface parts were then activated.

At the tail end of the simulated pipeline, horizontal prescribed displacements were applied in the y-direction (longitudinal to the pipeline) in Stage 3. The displacements were uniformly prescribed along line elements generated from the pipeline's circumference at the tail end. Simultaneously, by introducing free prescribed displacements along the periphery of the plate elements at the front end of the modeled pipeline (which was flushed with a vertical model boundary), fixities at the front end of the pipeline (which was flushed with a vertical model boundary) were released. This aided the pipe's longitudinal displacement beyond the soil block's borders.

Table 3.12 Differences between results from FEA and field measurements

Rock type	Sandstone	Shale	Phyllite
Measured F [kN/m]	14.4	37.2	4.8
Numerical F from FEA [kN/m]	11.5	28.9	4.1
% difference between numerical and measured F	20.6	22.3	14.9

Table 3.12 shows the differences between the frictional resistances obtained from FEA against those obtained from field measurements.

Generally, the differences between numerical results and field measurements were not significant. This implied that the methodology of assessing frictional jacking resistance in highly weathered rocks was valid, encompassing the following key fundamental concepts:

- The direct shear testing of reconstituted tunneling rock spoil
- The development of equivalent design rock strength parameters ($c'_{t,p}$ and $\phi'_{t,p}$) using a generalized tangential approach with a power function
- Interpretation of σ_{EV} and μ based on field measurements of jacking forces and lubrication
- Modeling technique for FEA of pipe jacking drives.

3.5.1.7 Effect of Geology on Jacking Forces and Other Operational Parameters

One of the key findings arising from the assessment of jacking forces in highly weathered rocks is the influence of geology on pipe jacking operations. The rock types have significant effects on operational parameters (lubricant injected and jacking forces) as well as on jacking forces, including the pipe-rock interface. Table 3.13 shows a summary of these key findings, and how each rock type is favorable to the pipe jacking operational parameters and to pipe jacking in general.

The findings here summarily show that phyllite is most favorable to pipe jacking, with shale being the least favorable. This could be a useful tool for the planning of jacking spans, pipe jacking equipment and resources, as well as the siting of shafts.

Table 3.13 Favorability of each rock type to pipe jacking (Ong and Choo 2018)

	Pipe jacking operational parameters				Favorable for pipe jacking?
	Lubricant injected	Arching	Jacking forces	Pipe-rock interface	
Sandstone	M	G	M	M	Moderately favorable
Phyllite	G	G	G	G	Most favorable
Shale	P	P	P	P	Least favorable

*Note G = Good, M = Moderate, P = Poor

3.5.2 Estimating Geotechnical Strength Parameters Using In Situ Pressuremeter Test

Even though an in situ pressuremeter test (PMT) is often used to measure the in situ stiffness of stiff clay, (its operation is often limited by the maximum equipment pressure that can be applied) and not weathered rocks, this method is somewhat applicable to test the highly weathered and highly fractured phyllite of the Tuang Formation up until the limit pressure of the physical equipment is reached. Therefore, in order to mimic the complete PMT test, FEM is utilized to replicate the earlier behavior of PMT through back-analysis to enable the PMT test to be finished off after the yield pressure in the phyllite rock mass has been successfully reached in the FEM model. This two-phase calibration method has been discussed in detail in Chap. 2 of the book.

This contextualized method can work because weathering and fractures in a rock mass may reduce its strength to a point where it starts exhibiting soil-like behavior (Clarke and Smith 1992). In fact, highly fractured and weathered rock masses of Tuang Formation in Sarawak, Malaysia, has been successfully characterized using equivalent Mohr–Coulomb soil strength parameters for applications in pipe jacking (Choo and Ong 2015; Ong and Choo 2016, 2018). This opens an opportunity of characterizing highly fractured and weathered rocks using approaches or tests that are originally developed for soils, which includes pressuremeter test (PMT).

Pressuremeters can also be a viable alternative testing method for highly fractured and weathered rocks, as shown by studies conducted by Phangkawira et. al. (2016, 2017a, b). This is because PMTs are done in situ, which does not require rock sampling. This eliminates the challenges in extracting suitable rock core samples for conventional rock testing in laboratories. Furthermore, the results obtained from PMTs are also more representative of the in situ condition of the rock mass.

The above-mentioned points motivated the development of a methodology to characterize highly fractured and weathered rock mass using PMTs. This method was based on PMTs in highly fractured and weathered phyllite of Tuang Formation in Kuching, Malaysia. The results of characterization have been successfully applied for assessing pipe jacking forces in a local microtunneling project (Phangkawira 2018). In this book, two sets of PMT results are presented. The tests were done within the Sarawak River, next to a pipe jacking drive that will be used as a case study in this section. Figure 3.36 shows the location of the studied pipe jacking drive and PMTs. Figure 3.37 shows the ongoing PMT on a pontoon floating on the river. Figure 3.38 shows the borehole log at the PMT location. Results from PMT interpretation and the methodology to assess pipe jacking forces are explained, hereinafter.

It is recommended that readers revisit Chap. 2 of this book to familiarize themselves with the PMT concept first. Interpretation of PMT results was done using the semi-numerical method developed by Yu and Houlsby (1991), which is based on the fundamental theory of cavity expansion. This involved the use of several closed-form equations to generate a theoretical pressure-expansion curve, which was then fitted against the actual PMT results. The fitting process was done by adjusting the input

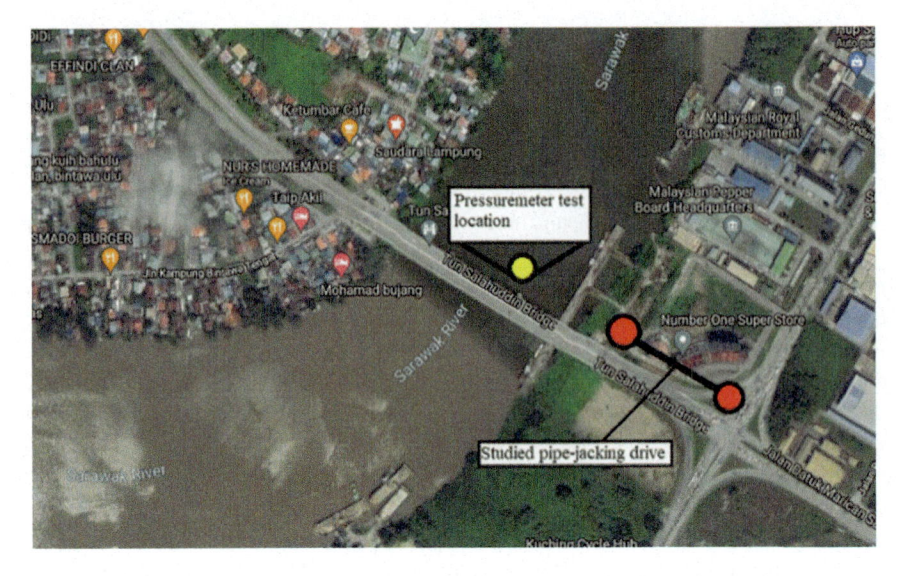

Fig. 3.36 Pressuremeter test (PMT) location with respect to pipe jacking drive

Fig. 3.37 Pressuremeter testing ongoing in the river

parameters that govern the theoretical pressure-expansion curves. The adjusted input parameters that can fit the actual PMT results were considered as the representative parameters that define the behavior of the rock mass. A separate finite element analysis was also performed to validate the reliability of these interpreted parameters. Further details regarding this method have also been included in Chap. 2.

The parameters used to generate the theoretical pressure-expansion curves are summarized in Table 3.14.

The results of comparison among the actual PMT results, the generated pressure-expansion curves, and the results extracted from the conducted finite element analysis

3.5 Case Study 2: Pipe Jacking … 93

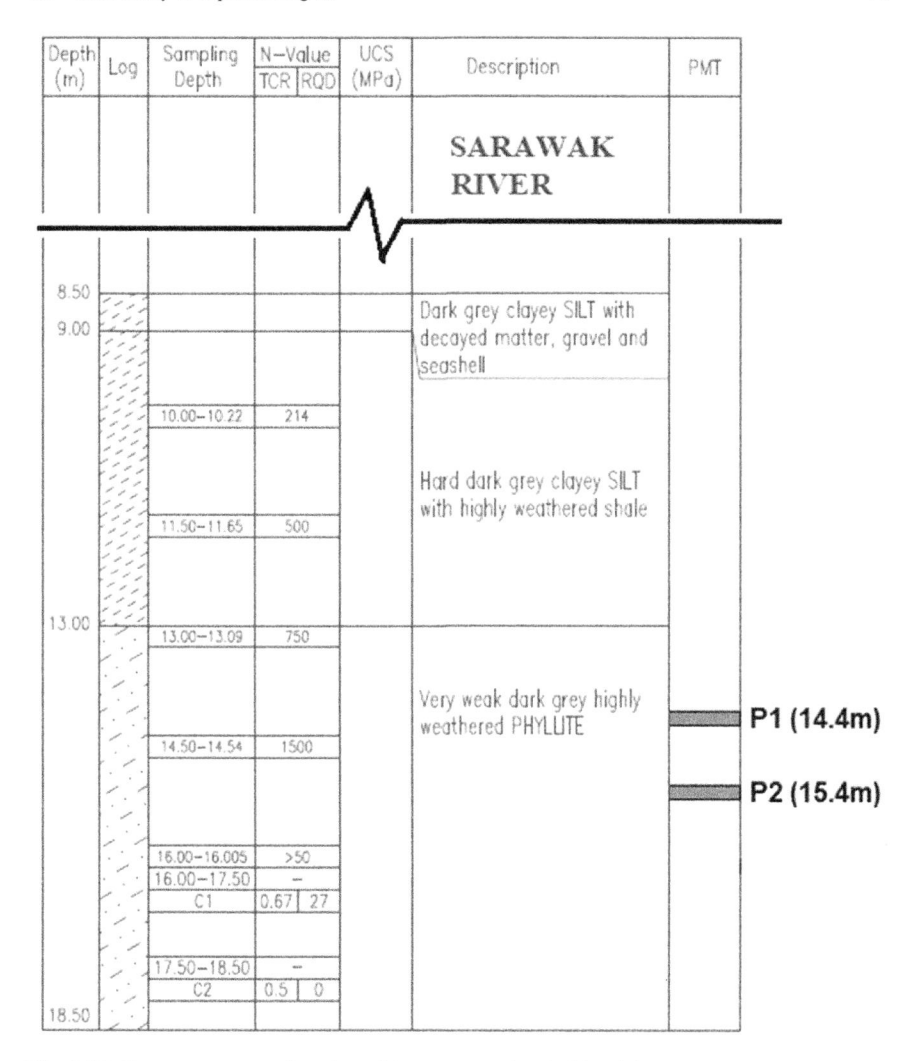

Fig. 3.38 Pressuremeter test location with respect to pipe jacking drive

Table 3.14 Parameters used to generate the theoretical pressure-expansion curves for tested phyllite rock masses

Pressuremeter Test	P1	P2
In situ stress, p_0 (kPa)	493	441
Poisson's ratio, v	0.3	0.3
Input Young's Modulus, E (MPa)	173	336
Cohesion, c' (MPa)	3.6	3.5
Friction angle, ϕ' (°)	53	52

Fig. 3.39 Comparison of pressuremeter test results against theoretical pressure-expansion curve and finite element analysis

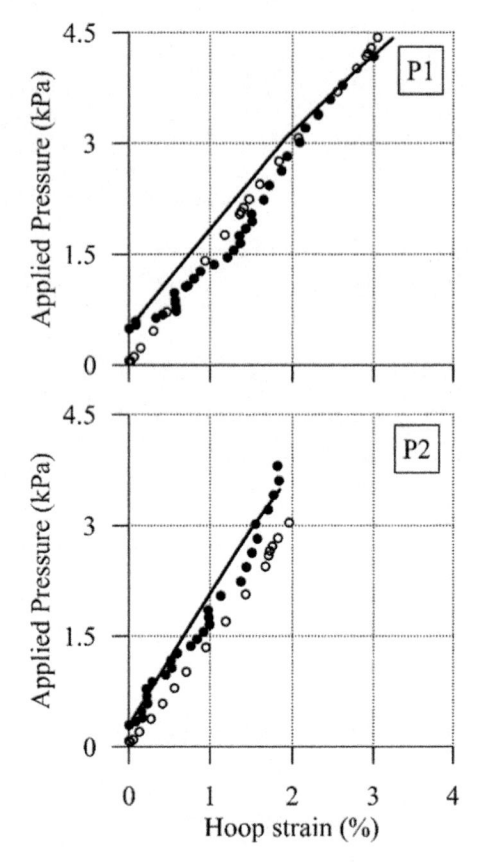

- • Results from pressuremeter test
- ○ Results from finite element analysis
- — Theoretical pressure-expansion curve

found in Chap. 2 are shown in Fig. 3.39, which shows acceptable agreement among the two plots. Therefore, this benchmarking exercise validated the reliability of the parameters. The optimized parameters were then used to assess jacking forces that were measured during the construction of the pipe jacking drive that traversed highly fractured and weathered phyllite lithology.

3.5.2.1 Back-Analysis of Jacking Forces Using Pressuremeter Results

The parameters shown in Table 3.14 were essentially the outcome of the PMT interpretation. These parameters were then used to back-analyze jacking forces. The previously presented pipe jacking works in phyllite (see Sect. 5.1.5.3) were used as a case study to assess the reliability of the interpreted parameters via PMT in the

3.5 Case Study 2: Pipe Jacking …

back-analysis of jacking forces. Figure 3.40 shows the compilation of pipe jacking measurements, which will be further described hereinafter.

Based on Fig. 3.40, the measured jacking force was averaged at 4.8 kN/m. Throughout the entire pipe jacking works, an average volume of injected lubricant of 181 L/m was observed. This value was compared against the theoretical overcut of 20 mm between the MTBM and the outer diameter of the jacked concrete pipes, which was calculated at 113 L/m. This shows that the actual volume of injected lubricant is comparable to the theoretical overcut region, which indicates that the injected lubricant managed to fill the theoretical overcut region without significant loss. This was possible since the phyllite lithology was tightly folded characteristics (Choo and Ong 2015). This would help to prevent the injected lubricant from excessively seeping into the surrounding rock mass. It is also worth noting that there was no extensive stoppage observed during the pipe jacking works. This was evidenced by

Fig. 3.40 Measured performance of pipe jacking works for the studied drive in highly fractured and weathered phyllite

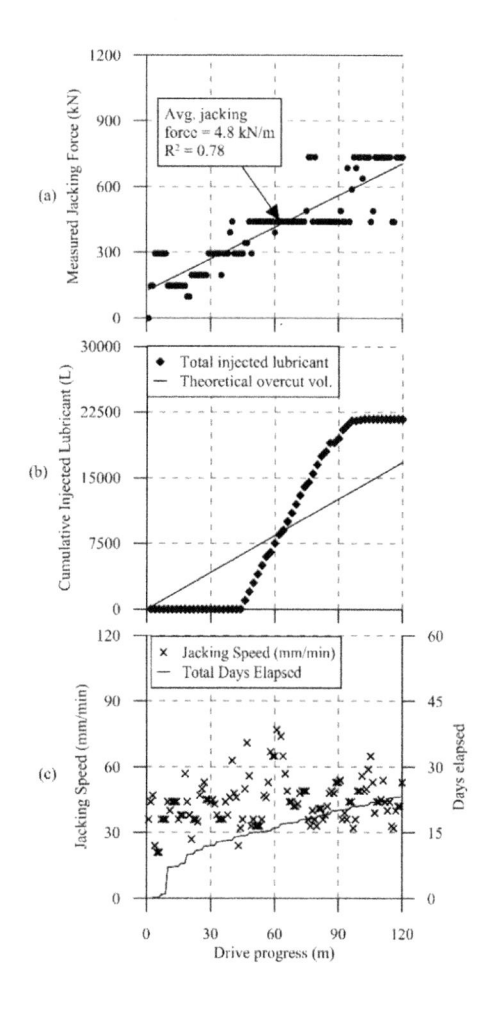

Fig. 3.41 Measured pipe jacking forces compared against represented forces based on recommended and back-analyzed μ values

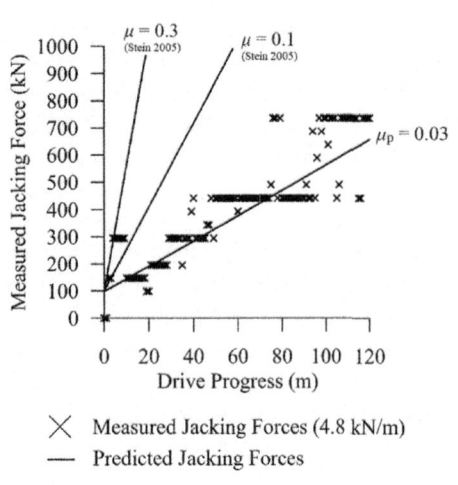

[Graph: Measured Jacking Force (kN) vs Drive Progress (m), showing lines for $\mu = 0.3$ (Stein 2005), $\mu = 0.1$ (Stein 2005), and $\mu_p = 0.03$]

× Measured Jacking Forces (4.8 kN/m)
— Predicted Jacking Forces

the lack of steep increments in the 'days elapsed' plot (see Fig. 3.39). This evidence suggests that a well-lubricated condition was achieved during the pipe jacking works.

With the interpreted parameters from the PMT, the back-analysis method for this pipe jacking drive was still based on the method developed by Pellet-Beaucour and Kastner (2002). The back-analyzed μ_p (based on PMT results) was found to be 0.03, which was lower than the lower limit value ($\mu = 0.1$) for a well-lubricated drive (Stein 2005). This suggests well-lubricated conditions during the pipe jacking works in highly fractured and weathered phyllite. The representative jacking forces based on the back-analyzed μ_p and the suggested limit values by Stein (2005) are compared against the actual jacking forces, as shown in Fig. 3.41.

3.5.2.2 Validation of Back-Analysis Using Finite Element Analysis

During the back-analysis of μ_p for this studied drive, results from the PMT interpretation were used to model the behavior of the highly fractured and weathered phyllite subjected to pipe jacking works. The modeled geological profile shown in Chap. 2 was used to represent the geological stratigraphy where the pipe jacking took place. A tunnel excavation in the highly fractured and weathered phyllite was then modeled to study the behavior of the modeled rock mass. Figure 3.42 shows the distribution of normal stresses and strains around the excavated tunnel.

The modeled tunnel excavation resulted in a maximum hoop strain, ε_θ of -0.2%. This was still within the range of elastic deformation as shown in the PMT results. The effective normal stress at the tunnel crown was calculated as 11 kPa, lower than the overburden pressure of 162 kPa. This shows that the arching phenomenon was captured in the modeled rock mass. It is worth noting that the effective normal stresses from the finite element analysis also corroborates with the back-analyzed μ_p.

3.5 Case Study 2: Pipe Jacking …

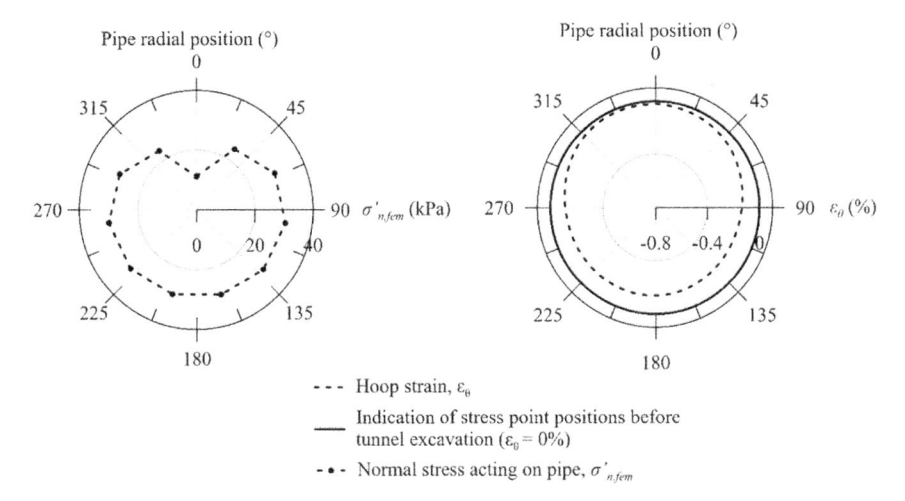

Fig. 3.42 Normal stress and strain distribution around the modeled tunnel cavity in the studied drive via finite element analysis

The findings from the back-analysis using the PMT results were comparable against the interpretation of the measured pipe jacking performance. Furthermore, these findings also arrived at similar conclusions with those reported by Ong and Choo (2018), who performed similar studies on highly fractured and weathered phyllite Tuang Formation. Therefore, the presented methodology has shown that PMT can be used to characterize highly weathered and weathered phyllite. Furthermore, the results of the interpretation can be used to assess pipe jacking performance.

3.6 Conclusions

The influence of geology on pipe jacking is multifaceted and complicated. Of particular interest currently and historically has been the assessment of jacking forces. Jacking forces (especially frictional resistance) are crucial to many aspects of pipe jacking, from planning, to design and construction. This chapter has curated the currently available knowledge, showing methods for assessing jacking forces for various soil types by considering soil arching, excavation stability, and ground convergence. Geotechnical engineering is a common approach for the prediction of jacking forces; however, operational pipe jacking parameters also influence the accrual of jacking forces during a drive. Some of these include observations made by the authors from various pipe jacking sites.

The appreciation of pipe jacking in soils is quite well-understood. However, the accounts of pipe jacking works being carried out in rocks continue to grow. Case studies have been presented for a drive in intact rocks, and for drives in highly weathered rocks. The case in the intact Altamura limestone demonstrated a methodology for assessing rock discontinuities, and their effects on pipe jacking forces through discontinuum modeling. Subsequently, cases were presented for drives in the Tuang Formation, showing novel methods for assessing jacking forces using laboratory tests (direct shear testing of reconstituted spoil) and in situ tests (pressuremeter testing). The principal challenges arose from difficulties in extracting intact rock specimens, thus necessitating the novel methods for developing equivalent rock strength parameters. One of the key findings was the effect of geology on jacking forces as well as on pipe jacking operational parameters. The structural features of the geology resulted in phyllite being most favorable for pipe jacking as compared to other rocks.

The findings in this chapter show that there is an extensive body of knowledge studying the complex nature between geology and pipe jacking, which continues to advance into new frontiers, some of which will be covered in the later chapters of this book.

References

AFTES (1982) Tunnels et ouvrages souterrains. Texte des réflexions sur les méthodes usuelles de calcul du revêtement des souterrains. Paris

Anyaegbunam AJ (2015) Nonlinear power-type failure laws for geomaterials: synthesis from triaxial data, properties, and applications. Int J Geomech 15(1). https://doi.org/10.1061/(ASCE)GM.1943-5622.0000348

Atkinson JH, Mair RJ (1981) Soil mechanics aspects of soft ground tunnelling. Ground Eng 14(5):20–28

ATV-A (German Association for Water Environment) (1990) ATV-A 161, Structural calculation of driven pipes. German Association for the Water Environment

Barla M, Camusso M, Aiassa S (2006) Analysis of jacking forces during microtunnelling in limestone. Tunn Undergr Sp Tech 21(6):668–683. https://doi.org/10.1016/j.tust.2006.01.002

Bergeson W (2014) Review of long drive microtunneling technology for use on large scale projects. Tunn Undergr Sp Tech 39:74–80. https://doi.org/10.1016/j.tust.2013.02.001

Broms BB, Bennermark H (1967) Stability of clay at vertical opening. J Soil Mech Found Div 93(1):71–94. https://doi.org/10.1061/JSFEAQ.0000946

Chapman DN, Ichioka Y (1999) Prediction of jacking forces for microtunnelling operations. Tunn Undergr Sp Tech 14:31–41. https://doi.org/10.1016/S0886-7798(99)00019-X

Charles JA, Watts KS (1980) The influence of confining pressure on the shear strength of compacted rockfill. Géotechnique 30(4):353–367. https://doi.org/10.1680/geot.1980.30.4.353

Charles JA, Soares MM (1984) Stability of compacted rockfill slopes. Géotechnique 34(1):61–70. https://doi.org/10.1680/geot.1984.34.1.61

Chen H, Xiao C, Yao Z et al (2019). Prediction of TBM tunneling parameters through an LSTM neural network. In: Proceeding of the IEEE international conference on robotics and biomimetics

Cheng WC, Ni JC, Shen JSL et al (2017) Investigation into factors affecting jacking force: a case study. Proc Insti Civ Eng—Geotech Eng 170:322–334. https://doi.org/10.1680/jgeen.16.00117

Cheng WC, Bai XD, Sheil BB et al (2020) Identifying characteristics of pipejacking parameters to assess geological conditions using optimisation algorithm-based support vector machines. Tunn Undergr Sp Tech 106. https://doi.org/10.1016/j.tust.2020.103592

Choo CS, Ong DEL (2012) Back-analysis of frictional jacking forces based on shear box testing of excavated spoils. In: Proceedings of 2nd international conference on geotechnique, construction materials and environment, GEOMATE International Society, Kuala Lumpur, Malaysia

References

Choo CS, Ong DEL (2015) Evaluation of pipejacking forces based on direct shear testing of reconstituted tunneling rock spoils. J Geotech Geoenviron Eng 141(10). https://doi.org/10.1061/(ASCE)GT.1943-5606.0001348

Choo CS, Ong DEL (2016) The Development of a frictional jacking force model for nonlinear soil behavior. In: Chan SH, Ooi TA, Ting WH, Chan SF, Ong DEL (eds) Deep excavation and ground improvement—proceedings of the 19th SEAGC and 2nd AGSSEAC, 31 May–3 June 2016, Subang Jaya, Malaysia. ISBN 978-983-40616-4-7

Choo CS, Ong DEL (2017) Impact of highly weathered geology on pipe-jacking forces. Geotech Res 4(2):94–106. https://doi.org/10.1680/jgere.16.00022

Clarke BG, Smith A (1992) Self-boring pressuremeter tests in weak rocks. Constr Build Mater 6:91–96

Collers P, Staheli K, Bennet RD et al (1996) A review of jacking foces by both theoretical and empirical methods as compared with 20 years of practical experience. In: Proceeding of international NO-DIG'96, New Orleans, pp 126–150

Collins IF, Gunn CIM, Pender MJ et al (1988) Slope stability analyses for materials with a non-linear failure envelope. Int J Numer Anal Meth Geomech 12(5):533–550. https://doi.org/10.1002/nag.1610120507

De Mello VFB (1977) Reflections on design decisions of practical significance to embankment dams. Géotechnique 27(3):281–354. https://doi.org/10.1680/geot.1977.27.3.281

Drescher A, Christopoulos C (1988) Limit analysis slope stability with nonlinear yield condition. Int J Numer Anal Meth Geomech 12(3):341–345. https://doi.org/10.1002/nag.1610120307

FSTT (French Society for Trenchless Technology) (2004) Microtunnelling and horizontal drilling: recommendations. Hermes Science Publishing Ltd., 335 pp

Geocomp Corp. (2010) ShearTrac-II user's manual. Geocomp Corp., Massachusetts, USA

Hadri MSAM, Mohammad H (2020) Case Study of sewerage pipe installation using pipe jacking and micro-tunnelling boring machine (MTBM) in Ipoh. IOP Conf Ser: Mater Sci Eng 932. https://doi.org/10.1088/1757-899X/932/1/012047

Haslem RF (1986) Pipe-jacking forces: from practice to theory. In: Conference on infrastructure renovation and waste control: international centenary, Manchester

Hunt SW, Del Nero DE (2009) Rock microtunneling—an industry review. International No-Dig 2009, Toronto, Ontario, Canada, 29 March–3 April 2009

Iscimen M (2004) Shearing behavior of curved interfaces. Master of science thesis, Georgia Institute of Technology

Ji X, Zhao W, Ni P et al (2019) A method to estimate the jacking force for pipe jacking in sandy soils. Tunn Undergr Sp Tech 90:119–130. https://doi.org/10.1016/j.tust.2019.04.002

Kastner R, Pellet-Beaucour AL, Guilloux A (1997) Microtunnelling: comparison between friction forces prediction and in situ measurements. In: Proceeding of International NO-DIG'97, Taipei, Paper A-3, 16 pp

Kimura T, Mair RJ (1981) Centrifugal testing of model tunnels in soft clay. In: Proceeding 10th international conference on soil mechanics and foundation engineering, Stockholm, vol 1, pp 319–322

Kirsch G (1898) Die Theorie der Elastizität und die Beduerfnisse der Festigkkeitslehre. Zeitshrift Des Vereines Deutscher Ingenieure 42:797–807

Lade PV (2010) The mechanics of surficial failure in soil slopes. Eng Geol 114(1–2):57–64. https://doi.org/10.1016/j.enggeo.2010.04.003

Lamé MG (1852) Leçons sur la Théorie Mathématique de l'Elasticité des Corps Solides. Bachelier, Paris

Lauritzen, R, Sande, O, Slatten A (1994) Europipe landfall tunnel. In: Proceeding of international NO-DIG'94, Copenaghen, article G2, 10 pp

Leroueil S, Tavenas F, Le Bihan JP (1983) Propriétés caractéristiques des argiles de l'est du Canada. Can Geotech J 20(4):681–705. https://doi.org/10.1139/t83-076

Mair RJ (1979) Centrifugal modelling of tunnel construction in soft clay. PhD Thesis, Cambridge University, Cambridge, UK

Milligan GWE, Marshall MA (1995) Ground movements due to construction of pipe jacked tunnels. In: 11th European conference on soil mechanics and foundation engineering, Copenhagen, Denmark

Milligan GWE, Norris P (1994) Pipe jacking: research results and recommendations. Pipe Jacking Association, London, UK, 55 pp

Milligan GWE, Norris P (1999) Pipe–soil interaction during pipe jacking. Proc Inst Civ Eng—Geotech Eng 137(1):27–44. https://doi.org/10.1680/gt.1999.370104

Norris PM, Milligan GWE (1991) Field instrumentation for monitoring the performance of jacked concrete pipes. In: 3rd international symposium on field measurements in geomechanics, Oslo, Norway, 9–11 Sept 1991

Norris PM, Milligan GWE (1992a) Frictional resistance of jacked concrete pipes at full scale. In: Henry J-P, Mermet M (eds) No-Dig 1992, Paris, France, pp 121–8

Norris PM, Milligan GWE (1992b) Pipe end load transfer mechanisms during pipe jacking. No-Dig 92, Washington DC, USA

Ong DEL, Choo CS (2011) Sustainable construction of bored pile foundation system in erratic phyllite. In: ASEAN Australian engineering congress 2011, 25–27 Jul 2011, Kuching, Sarawak, Malaysia

Ong DEL, Choo CS (2016) Back-analysis and finite element modelling of jacking forces in weathered rocks. Tunn Undergr Sp Tech 51:1–10. https://doi.org/10.1016/j.tust.2015.10.014

Ong DEL, Choo CS (2018) Assessment of non-linear rock strength parameters for the estimation of pipejacking forces. Part 1. Direct shear testing and backanalysis. Eng Geol 244:159–172. https://doi.org/10.1016/j.enggeo.2018.07.013

O'Reilly MP, Rogers C (1987) Pipe jacking forces. In: Proceeding of international conference on foundations and tunnels, Edinburgh

Osumi T (2000) Calculating jacking forces for pipe jacking methods. No-Dig International Research, October, 2000, 40–42

Panet M (1995) Le calcul des tunnels par la méthode convergence-confinement. Presses de l'Ecole Nationale des Ponts et Chaussées, Paris

Peerun I, Ong DEL, Choo CS (2015) Behaviour of reconstituted sand-sized particles in direct shear tests using PIV technology. Jpn Geotech Soc Spec Publ 2:354–359. https://doi.org/10.3208/jgssp.MYS-06

Pellet AL (1997) Analyse des interactions sol-canalisation et solmachine pour la pose de conduites par microtunnelage. Thesé Doct, INSA de Lyon, December 97, p 248

Pellet-Beaucour AL, Kastner R (2002) Experimental and analytical study of friction forces during microtunneling operations. Tunn Undergr Sp Tech 14:83–97. https://doi.org/10.1016/S0886-7798(01)00044-X

Phangkawira F (2018) Characterisation of highly weathered phyllite via in-situ pressuremeter test for the assessment of pipejacking forces. PhD thesis, Swinburne University of Technology

Phangkawira F, Ong DEL, Choo CS (2016) Characterisation of highly fractured and weathered rock mass using pressuremeter test. In: Chan SH, Ooi TA, Ting WH, Chan SF, Ong DEL (eds) Deep excavation and ground improvement—proceedings of the 19th SEAGC and 2nd AGSSEAC, 31 May–3 June 2016, Subang Jaya, Malaysia. ISBN 978-83-40616-4-7

Phangkawira F, Ong, DEL, Choo CS (2017a) Numerical verification of tensile stress development in highly fractured phyllite characterised via pressuremeter test. In: Proceedings of the 6th international young geotechnical engineers' conference (iYGEC6), Seoul, South Korea

Phangkawira F, Ong DEL, Choo CS (2017b) Numerical prediction of plastic behavior of highly fractured and weathered phyllite subjected to pressuremeter testing. In: Proceedings of the 19th international conference on soil mechanics and geotechnical engineering, Seoul, South Korea

Phelipot A (2000) Interaction sol-structure lors d'opérations de microtunnelage. Thèse Doctorat, INSA de Lyon

Ribacchi R, Riccioni R (1977) Stato di sforzo e deformazione intorno ad una galleria circolare. Gallerie e Grandi Opere Sotterranee 4:7–17

References

Ripley KJ (1989) The performance of jacked pipes. PhD thesis, Magdalen College, Oxford University, viewed 4 Feb 2010

Rogers CDF, Yonan SJ (1992) Experimental study of a jacked pipeline in sand. Tunn Tunnelling 24(6):35–38

Schofield AN (2005) Disturbed soil properties and geotechnical design. Thomas Telford, Heron Quay, London

Shao B, Ma B, Shi L (2009) A sewer pipeline installation using pipejacking in Lang Fang. Am Soc Civ Eng 1413–1424. https://doi.org/10.1061/41073(361)148

Shou K, Yen J, Liu M (2010) On the frictional property of lubricants and its impact on jacking force and soil–pipe interaction of pipejacking. Tunn Undergr Sp Tech 25:469–477. https://doi.org/10.1016/j.tust.2010.02.009

Soon SC, Drescher A (2007) Nonlinear failure criterion and passive thrust on retaining walls. Int J Geomech 7(4):318–322. https://doi.org/10.1061/(ASCE)1532-3641(2007)7:4(318)

Staheli K (2006) Jacking force prediction: an interface friction approach based on pipe surface roughness. PhD thesis, Georgia Institute of Technology

Stein D (2005) Trenchless technology for installation of cables and pipelines. Prof. Dr.-Ing. Stein & Partner GmbH, Bochum, Germany

Stein D, Möllers, K, Bielechi R (1989) Microtunnelling: installation and renewal of nonman-size supply and sewage lines by trenchless construction method. Ernst, Berlin, Germany

Sterling RL (2020) Developments and research directions in pipe jacking and microtunnelling. Tunn Undergr Sp Tech 5:1–19. https://doi.org/10.1016/j.undsp.2018.09.001

Szechy K (1970) Traité de Construction des Tunnels. Dunod, Paris, pp 157–702

Tan DNK (1993), Geology of the Kuching Area, West Sarawak, Malaysia. Geological Survey of Malaysia, Kuching, Malaysia

Terzaghi K (1936) Stress distribution in dry and in saturated sand above a yielding trap-door'. In: Proceedings of 1st international conference on soil mechanics and foundation engineering, Graduate School of Engineering, Harvard University, Cambridge, Massachusetts, USA

Terzaghi K (1943) Theoretical soil mechanics. Wiley, New York, 528 pp

Thomson J (1993) Pipejacking and microtunnelling. Blackie Academic & Professional, London, UK, 273 pp

Tian Y, Cassidy MJ, Randolph MF et al (2014) A simple implementation of RITSS and its application in large deformation analysis. Comput Geotech 56:160–167. https://doi.org/10.1016/j.compgeo.2013.12.001

Wang D, Hu Y, Randolph MF (2009) Three-dimensional large deformation finite-element analysis of plate anchors in uniform clay. J Geotech Geoenviron Eng 136(2):355–365. https://doi.org/10.1061/(ASCE)GT.1943-5606.0000210

Wei XJ, Wang X, Wei G et al (2021) Prediction of jacking force in vertical tunneling projects based on neuro-genetic models. J Mar Sci Eng 9. https://doi.org/10.3390/jmse9010071

Working Group No. 3 (1994) Microtunnelling jacking force. Japan Society of Trenchless Technology, Japan, 96 pp

Yang XL, Yin JH (2004) Slope stability analysis with nonlinear failure criterion. J Eng Mech 130(3):267–273. https://doi.org/10.1061/(ASCE)0733-9399(2004)130:3(267)

Yin ZZ, Zhu H, Xu GH (1995) A study of deformation in the interface between soil and concrete. Comput Geotech 17(1):75–92. https://doi.org/10.1016/0266-352X(95)91303-L

Yu HS, Houlsby GT (1991) Finite cavity expansion in dilatant soils: loading analysis. Geotechnique 41(2):173–183. https://doi.org/10.1680/geot.1991.41.2.173

Zhang X, Krabbenhoft K, Pedroso DM et al (2013) Particle finite element analysis of large deformation and granular flow problems. Comput Geotech 54:133–142. https://doi.org/10.1016/j.compgeo.2013.07.001

Chapter 4
Deep Shaft Excavation: Design, Construction, and Their Challenges

4.1 Introduction

Temporary shafts are constructed to facilitate the jacking and receiving of microtunnel boring machines (mTBM) used in pipe jacking works, hence the terminology 'jacking' or 'launching' shafts and 'receiving' or 'retrieval' shafts. Shafts vary in diameters and depths due to the varying lengths and diameters of the jacking pipes used to create the sewer network. The design of shafts is very much aligned to the design of a typical deep excavation.

After mTBM retrievals, permanent prefabricated cylindrical manholes are built bottom-up in these temporary shafts. The surrounding spaces between the larger temporary shafts and smaller permanent manholes (typically only about 1 m in diameter) are backfilled and compacted with suitable soils.

4.2 Types of Shafts

4.2.1 *Circular Concrete Caisson*

In constructing a shaft using circular concrete caisson where the upper soil layers consist of soft and/or loose soils, sheet piles are initially driven around the footprint of the shaft. For soft clays, the installed sheet piles will reduce the risk of basal heave due to the increasing differences in overburden pressure inside and outside the shaft that is being progressively excavated. For loose sands, this action will lengthen the ground water flow path to reduce the risk of hydraulic uplift at the excavation floor. Cast in situ circular concrete caisson rings (typically each measuring 1.3 m in height, 300 mm thick with 100 mm overlap) are then progressively installed top down to provide structural support to the surrounding soils so that excavation inside the shaft could be carried out. The mobilized caisson wall friction is expected to carry the self-weight of the concrete caisson rings.

Fig. 4.1 Typical shaft constructed using circular concrete caisson rings with sheet pile curtain

In this method, the concrete caisson rings serve as temporary works and the design is mainly based on hoop stresses generated by the confining lateral soil pressures. As such, the steel reinforcement is kept at nominal by using steel mesh since the hoop stresses are designed to be resisted by the adequate provision of the concrete wall thickness. Therefore, it is important that the concrete caisson rings are cast as circular as possible in each excavation lift to ensure that the hoop stresses can be resisted safely. Such construction method, otherwise, also known as hand-dug caisson, has been documented by Allenby et al. (2009), Yee et al. (2001), and Yee (2006) and is usually effective when the founding shaft level is in weathered rocks. The main advantage of this construction method is that the completed shaft remains strut-free thus easing construction works in the shaft such as removing excavation spoil, launching or removing mTBMs.

Figure 4.1 shows a typical shaft constructed using circular concrete caisson rings with sheet pile curtain. The shaft has a diameter of 7 m and is 16 m deep. This shaft was constructed in a layer of 8.5-m-thick, soft soils, followed by 3.4 m of medium dense sand. Highly to moderately weathered rocks were found below the sand layer. The shaft base was constructed at approximately 16 m depth. The ground water was 1.5 m below ground level.

4.2.2 Steel Sheet Pile

Temporary steel sheet pile shaft is typically considered when the surrounding soils have SPT N values of less than about 25–30 right down to the proposed sheet pile toe, to ensure sufficient penetration depth. When observed in plan view, they are typically circular or rectangular in nature. If circular sheet pile shaft design is adopted, circular concrete waler beams are typically used to resist the hoop stresses generated by the lateral soil pressures.

4.2 Types of Shafts

Fig. 4.2 Typical rectangular shaft constructed using steel sheet piles and shoring system

If rectangular sheet pile shaft is to be designed, then conventional strut–waler bracing system is adopted to ensure safe transfers of lateral soil pressures to the sheet pile wall. The walers are then designed to resist the soil-wall pressures, while the struts ensure that the walers are adequately propped at suitable spacing in the plane strain direction.

In ensuring stability of the sheet pile shaft constructed in fine-grained soils, basal stability check (see Fig. 4.6) is critical to prevent excessive basal heave beyond serviceability condition which could lead to failure. If the shaft is being constructed in medium dense sand with a relatively high water table, then hydraulic uplift check at the excavation floor is necessary. Figure 4.2 shows a typical rectangular shaft constructed using steel sheet piles with its corner strut–waler system.

4.2.3 Diaphragm Wall

'Circular' shafts constructed with cast in situ diaphragm wall panels are not as common simply due to costs, panel geometry (since panels are cast in rectangular pieces owing to the shape of the clamshell bucket), and time. Diaphragm walling entails a laborious construction process; consistent circulation of bentonite slurry for trench stability, construction of guide wall templates for verticality control, need for end-stops and waterstops to create secure water-tight joints, relatively heavy reinforcement for crack control, use of specialized machinery, and the need for highly skilled machine operators.

Only in very exceptional cases that diaphragm walls are utilized in shaft construction, such as when the surrounding ground consists of very dense sand with high surrounding water table when all other types of shafts could not be practically designed to achieve the necessary factors of safety. Figure 4.3 shows a shaft being constructed using diaphragm wall panels. Circular concrete walers are usually still required to work together with the diaphragm wall panels to effectively resist the lateral soil pressures from the surrounding soils.

Fig. 4.3 Typical 'circular' shaft constructed using diaphragm wall panels

4.2.4 Sinking Circular Concrete Caisson

Sinking circular concrete caisson, which is another method of shaft construction, is typically adopted in soft soil and loose sandy soil conditions. Part of the concrete shaft is constructed above ground (can consists of either precast panels or cast in situ) and then progressively sunken into the ground up to the required depth when the soils within the shaft footprint are removed or when the panels are vertically jacked into the ground. The disturbances caused by the excavation process remolds the surrounding soils and with the concrete self-weight allow the cast circular concrete caisson to be sunken into the ground when the skin friction between the concrete wall and the soil is overcome. Under exceptional circumstances, concrete blocks which function as additional self-weight are stacked and applied to allow the shaft to sink to greater depths.

One of the major advantages of sinking circular concrete caisson is the wet excavation process, which allows the whole excavation and sinking process to be carried out without pumping of ground water from the shaft. With the presence of the stabilizing ground water inside and outside of the excavation, any potential basal stability (if construction is in relatively softer fine-grained soils) and liquefaction (if construction is in sandy soils with a relatively high water table) issues are eliminated. Upon reaching the required excavation depth, the concrete base is then cast using tremie method to seal the shaft.

The sinking circular concrete caisson method requires highly skilled machine operators to ensure smooth sinking process. Unfavorable sinking procedure causing tilt will lead to damaged concrete walls or even misalignment of the shaft. Figure 4.4 shows the process of constructing a typical circular concrete caisson shaft via the sinking method.

4.2.5 Cased or Oscillation-Drilled Shaft

Cased or oscillation-drilled shaft is usually considered for shallow shaft with excavation depths not exceeding 9 m, provided that the surrounding soils have SPT N values

4.2 Types of Shafts

Fig. 4.4 Typical circular shaft constructed using sinking method

Fig. 4.5 Typical circular cased or oscillation-drilled shaft being constructed

of less than 15. This construction method utilizes a circular steel casing (typically 3 m diameter), which is oscillated into the ground using an oscillator machine, while the soil inside the shaft is gradually removed. Upon reaching the required depth, a concrete base slab is cast to seal the shaft. Under exceptional circumstances where basal heave or sand liquefaction is of concern, wet excavation method can be adopted.

The steel oscillator casing method is simple and fast in terms of construction. However, this method is limited by the excavation depth, surrounding soil conditions and shaft size. Figure 4.5 shows a typical steel circular cased or oscillation-drilled shaft being constructed.

4.3 Design Philosophy and Selection Guide for Shafts

This section is intended for readers who have some prior experience in the process of excavation. It is written focusing on the salient points regarding the excavation process in the context of shaft construction and thus should not be used solely as a complete design guideline for a general deep excavation process.

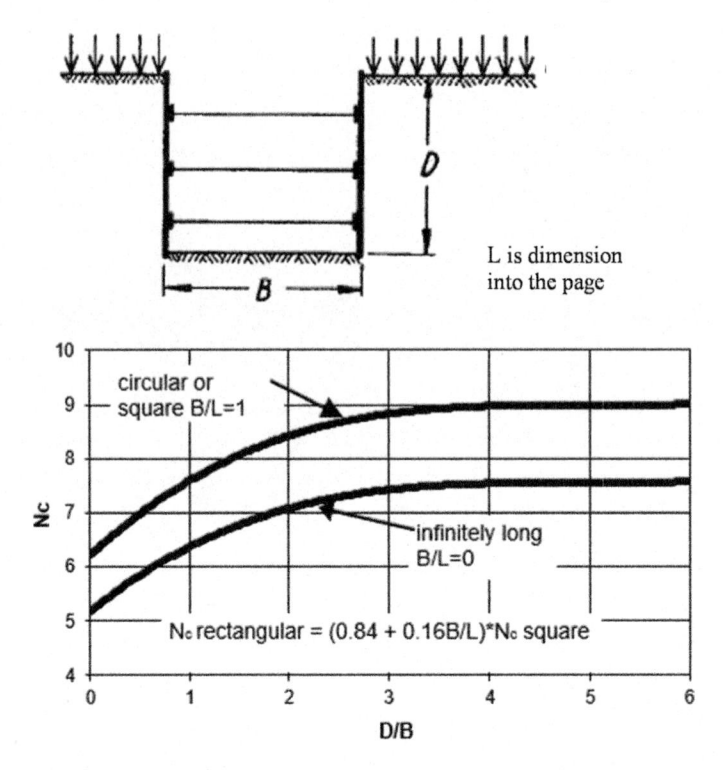

Fig. 4.6 Basal stability check for a typical excavation process (after Bjerrum and Eide 1956)

4.3.1 Bottom-Up or Top-Down Construction

An excavation can progress either in a bottom-up or top-down manner. In a typical bottom-up excavation, the permanent structure is constructed starting with its base slab first at the bottom of the excavation after the excavation has reached its deepest level. This is carried out after the temporary retaining structure has been fully and successfully completed. This is the usual method used for shaft construction.

On the other hand, top-down excavation involves construction of the upper-most slab or the structural restraint of the underground portion of a permanent structure first before the next stage of excavation is performed. The permanent slab or restraint is connected to the cast in situ retaining wall, which usually forms part of the permanent structure. In the context of shaft construction for mTBM access, this excavation method is not common due to the much smaller shaft footprint as compared to a general excavation say for an underground station box.

4.3.2 Jacking and Receiving Shafts

Shafts constructed for mTBM access can be categorized as either a 'jacking' or 'receiving' shaft. The terms 'jacking' and 'receiving' are often used interchangeably with 'launching' and 'retrieval,' respectively.

4.3.2.1 Position of Jacking and Receiving Shafts

Temporary jacking and receiving shafts are usually strategically positioned along wider road medians, car parks, garden beds, or foot paths to minimize traffic disturbance. As pipe jacking works are usually carried out in urban areas, it is often that the jacking lengths vary as a result of locating such open spaces.

4.3.2.2 Size Requirement of Shafts and Work Areas

A jacking shaft is typically larger than a receiving shaft for the same sewer pipe or mTBM diameter due to the need to install a thrust block and the jacking frame system in the former. For example, for a sewer pipe or mTBM of 2.2 m diameter, a jacking and receiving shaft can measure some 7.0 m and 5.0 m in diameter, respectively. For the same shaft sizes, barricaded work areas measure some 200 m^2 and 100 m^2, respectively.

4.3.2.3 Safety Aspects

Since shafts are often constructed in urban areas, traffic management, or diversion plans must be enforced to minimize traffic disturbance. At night, warning or hazard lights are to be installed to prevent road accidents. Gates and fences are also compulsory to be installed to safely barricade off the work areas to prevent wandering or curious public from entering the work areas. It is common to hear and see bizarre accidents such as vehicles veering and dropping into shafts, or a drunk person tripping and falling into one!

4.3.3 Shaft Excavation in Soils

4.3.3.1 Sands

Being granular, sandy soils or sands are often seen as competent bearing materials. However, in the case of an excavation, typical sandy ground usually has greater permeability often between 10^{-4} and 10^{-7} m/s as compared to clayey ground and

Table 4.1 Seepage condition at different exit gradients

Exit gradient, i	Seepage condition
0.0–0.5	Light/no seepage
0.2–0.6	Medium seepage
0.4–0.7	Heavy seepage
0.5–0.8	Sand boils

Source US Army Corps of Engineers Engineering Technical Letter No. 1110-2-569 Design Guidance for Levee Underseepage (May 2005)

thus will encourage water seepage through its voids under a positive hydraulic gradient that usually exists between the surrounding higher ground water table and the progressing excavation depth.

In any shaft design or construction, sand particles under the influence of increasing hydrostatic water pressures should be properly accounted for during its excavation to prevent the occurrence of liquefaction or boiling. Under such circumstance, the presence of water pressure (hydrostatic or seepage) will reduce the soil effective stress (or confining pressure), thus reducing its shear strength. If the surrounding ground water pressures overcome the interparticle frictional resistance, liquefaction will occur whereby the sand particles will flow like water into the shaft from the progressing excavation floor, thus undermining construction safety. Liquefaction is also known as boiling (vertical-dominant direction) or piping (horizontal-dominant direction).

Hydraulic uplift check is often a necessity to counter liquefaction or boiling that may occur at the excavation floor. This check simply involves ensuring that the saturated soil overburden pressure under the excavation floor is at least 1.5 times larger than the uplift water pressure as a result of the excavation. The uplift water pressure can be measured from the water standpipe elevation head reading (then multiplied by the unit weight of water, i.e., 10 kN/m^3). The tip of the water standpipe in this case would have to be installed in the corresponding sand layer beneath the excavation floor.

It is also the authors' preference to perform (effective stress) seepage analyses to determine the potential exit gradient during excavation as an extra layer of design check against boiling. Table 4.1 provides reliable magnitudes of exit gradients that are often used as a safety benchmark. GeoStudio SEEP/W package is an example of a seepage analysis software that is often used to develop flow nets to check against boiling.

4.3.3.2 Silts and Clays

As opposed to sandy soils, silty and clayey soils contain proportionately more of the fine-grained soil particles than sand particles. As such, it is expected that the permeability of such soils often varies between the order of 10^{-8} and 10^{-11} m/s. In

4.3 Design Philosophy and Selection Guide for Shafts

this circumstance, total stress or undrained basal stability check is often carried out to determine its factor of safety against upheaval of excavation floor (see Fig. 4.6). It is often that temporary wall (e.g., most basic is sheet pile flexible wall) penetration depth governs the design factor of safety, which can be conservatively estimated using Eq. 4.1 by Bjerrum and Eide (1956):

$$\text{FS} = \frac{C_u N_c}{\gamma H + q} \tag{4.1}$$

where c_u is the undrained shear strength underneath the excavation, N_c is the coefficient depending on the dimensions of the excavation, γ is the unit weight of saturated soil, H is the excavation depth, and q is the surface surcharge.

Typical coupled-consolidation analysis via finite element modeling (FEM) using Biot's theory (1941) addressing simultaneous soil stress and deformation responses as a direct result of time-dependent pore water pressure changes in excavation has been documented and observed by Ong (2008, 2019).

If shaft excavation is to be carried out in plastic, soft clays where basal stability is usually of concern, then some form of ground improvement method should be executed to improve the factor of safety before excavation commences. One of the more common ground improvement methods in this case is the use of jet grouted piles (JGP) to form stiffer and stronger soil–cement mix columns at the depths where design analyses have shown that basal instability could be of concern. For a typical cylindrical shaft, the guiding principle is to have a relatively thick JGP layer formed below the final excavation depth. Since the JGP layer is cast in situ with no provision of steel reinforcement, it is recommended that the minimum thickness is at least half the shaft outer diameter to ensure that the JGP block remains completely in compression throughout the entire excavation process. The at-depth JGP layers are normally installed first prior to commencement of excavation.

4.3.3.3 Organic Soils

It is not uncommon to have the need to perform shaft excavation in organic soils. The main challenge in performing excavation in organic soils is the much wider extent of ground subsidence (if triggered) which may detrimentally affect the serviceability performance of adjacent buildings and infrastructure.

When performing field permeability tests be it falling or constant head test method in soils, often at times, the vertical soil permeability is recorded due to the localized small footprint of the testing point. It is often misconstrued that field falling or constant head tests would provide a true representation of the soil actual permeability, as the resultant of x–y flow directions. While it may be closer to the truth for weathered residual soils, it is not really the case for soils deposited through the sedimentation process, e.g., marine clays, whose horizontal permeability is often established as being twice more permeable than its vertical permeability. Similarly, from the research outcomes of Mesri and Ajlouni (2007), Malinowska and Szymanski (2015), and CIDB (2019), it is demonstrated that the permeability of organic soils, often formed via the sedimentation process as well, is usually anisotropic in flow, i.e., being 5–10 times more permeable in the horizontal direction than in the vertical direction, thus presenting a challenge in interpreting standard falling or constant head permeability test results.

In an established case study by Ong (2019) (also presented as Case Study 1 in Sect. 4.6.1), a 16-m-deep shaft was excavated through 8.5-m-thick over-consolidated organic soils (initial moisture content 120%) and 3.4 m of dense sand before a fracture was observed in the highly to moderately weathered bedrock of sandstone–mudstone interbeds. The rock fracture was somehow connected to the much permeable upper organic soil and sand layers thus depressing the ground water table as much as 6.5 m, triggering ground subsidence of more than 120 mm and up to 200 m radius (recorded as at least 15 times the shaft excavation depth). This highlights the detrimental effect of excavation in organic soils where minor to moderate building and infrastructure damages have been reported in the case study.

4.3.4 Shaft Excavation in Rocks

Deep shaft excavation in rocks can be common if the design philosophy of self-cleansing, gravity flow deep sewerage network is adopted to channel wastewater from dwellings to a centralized wastewater treatment plant. For example, the Deep Tunnel Sewerage System (DTSS) of Singapore, Kuching Wastewater Management System of Malaysia, and the Bulimba Creek Sewer Upgrade of Brisbane, Australia, are just some projects where some of the main trunk sewers were installed via pipe jacking method to amazing depths of more than 30 m.

Normally at those depths, rocks are encountered. In the South-East Asian countries of Malaysia and Singapore, rocks encountered are usually highly weathered to relatively greater depths as these countries are located on the Equator and hence the terminology of 'soft rocks' is often used in this region. One of the main challenges faced working with soft rocks is that they are often friable (as opposed to being weak) when extracted, thus making engineering properties difficult to be accurately quantified. Being neither a typical soil nor a typical rock, excavation in soft rocks is often problematic and can lead to unexpected construction challenges. These points are briefly discussed hereinafter.

4.3.4.1 Sedimentary Rocks

Sedimentary rocks are formed near or on the surface of the earth. The processes that often lead to the formation of 'new' sedimentary rocks are associated to the breaking down of the 'old' larger rocks or historical landscapes through erosion, weathering, dissolution, precipitation, and lithification. While erosion, weathering, and dissolution physically disintegrate larger rock masses over time into finer sediments, precipitation and lithification build new rocks or minerals through (horizontal) accumulation of sediments and (natural) compaction under the influence of large overburden pressures. As such, sedimentary rock masses are often described in practice as being striped or laminated and are often more easily identifiable.

Some examples of more commonly found sedimentary rocks are sandstone, siltstone, mudstone, shale, and limestone. In practice, quick and simple identification of sandstone, siltstone, and mudstone is often done crudely on site by rubbing one's fingers on the cored rock samples; sandstone having the most grit (imagine coarse sandpaper) followed by siltstone (less grit) then mudstone (smooth). Shale often appears flaky and will be slakable (having high degree of disintegration) upon overexposure to the elements. Simple and quick identification methods are often useful to appreciate the greater geology present at a specific site that could be broadly related to 'pipe jackability.' The jackability of pipes in various geological formations has been discussed earlier in Chap. 3, so the discussion points presented herein are focused on excavation of shafts (not tunnels), where jack hammers or hammer-mounted mini excavators are used without much issue. Shaft excavation into sedimentary rocks (except for karst limestone) are often water-tight, thus making this type of geology less problematic usually.

Limestone Cavities

Karst limestone is often regarded as problematic especially in underground construction due to possible presence of cavities or large underground voids such as streams and caverns. It is also often associated to being connected to a network of natural underground water sources, thus the dreaded possibility of perpetual recharge of ground water. If a shaft or a (micro)tunnel was to be constructed in karst, often a time a contingency or mitigation plan is accompanied, for example:

- Tube-a-Manchette (TAM) or compaction grouting to fill fully or partially the cavities to bridge the rock masses
- Recharge well system to minimize ground water loss so as to mitigate the detrimental effects of ground subsidence on nearby buildings and infrastructure
- Increased number in soil investigation borehole points or probes to locate cavities
- Nondestructive testing (e.g., seismic refraction, electrical resistivity, ground penetrating radar or microgravity survey) to map cavity size or distribution. However, these methods are often limited by the eventual size or distribution of the cavity that is to be detected. The bigger or more widespread the cavities are, the easier they are to be contrasted against the surrounding 'solid' grounds.

Fig. 4.7 Occurrence of a sinkhole near a deep shaft being excavated

Sinkhole

The formation of a sinkhole on the ground surface near to a shaft being excavated is often associated to ground loss where liquefied sands or plastic clays squeeze its way through gaps, crevices, or weakest flow paths, connecting the outside soils to the inside of the shaft during the excavation process. Usually if the karst cavity roof or wall collapses, wider and deeper sinkholes are formed as shown in Fig. 4.7. In dealing with sinkholes, prevention is always much safer and more economically viable than rectifying a collapsed one which may detrimentally affect the structural integrity of the nearby infrastructure.

4.3.4.2 Metamorphic Rocks

Metamorphic rocks are formed when the original parent rocks (be it sedimentary or igneous) are subject to high heat, high pressure, and exposed to hot, mineral-rich liquid but not to the extent of being melted in the deep earth's crust or at moving tectonic plate boundaries. Ancient geological uplifts often raised metamorphic rocks from the earth's deep crust to the surface. Metamorphic rocks will also 'appear' after the upper soil covers are eroded via surface erosion or weathering. Examples of challenging metamorphic rocks that have been experienced in shaft excavation include metagraywacke and phyllite.

Fissures in Metagraywacke

Loosely speaking, metagraywacke consists of poorly sorted angular, coarse sand grains set in compact clay matrix as a result of high pressure and high temperature effects in the metamorphism process. Tunnel excavation in such material may induce uncontrolled fissures within the affected rock mass leading to potential loss of fluids such as grout or bentonite when used in pipe jacking activities. In shaft excavation, such phenomenon may cause ground water leakage into the shaft being excavated thus contributing to detrimental ground subsidence within the excavation influence zone.

Slickenside and Sub-vertical Discontinuities in Phyllite

Shale as mentioned before often appears flaky due to the presence of mica minerals. Upon experiencing low-grade regional metamorphism, shale will turn into slate. On the other hand, if slightly higher-grade regional metamorphism is experienced, shale will turn into phyllite. Both slate and phyllite are foliated, but the distinguishable difference is that the latter often has somewhat a reflective sheen or slickenside. Since metamorphic rocks are subject to directional tectonic plate movements accompanied by high pressures in ancient times, they tend to develop compressed 'slaty cleavage' features often loosely described as sub-vertical joints (naturally occurring) and fractures (movement induced) or broadly termed as discontinuities.

Direction-Dependent Strength

With the continuous presence of discontinuities, the rock strength is direction-dependent. This means that resistance against shear is expected to be lesser than resistance under compression load, i.e., when the rock mass is confined under overburden stress. Phyllite is one such example.

RQD Versus Pressuremeter Test

As discontinuities in phyllite often occur sub-vertically, when cored using a core barrel in a vertical coring direction, it is often that the rock quality designation (RQD) recorded is 0% or near 0% because the core barrel mechanically cores through the sub-vertical discontinuities leading to disintegration when collected in the barrel. This often gives the false impression that phyllite is 'weak' due to low RQD values (Choo and Ong 2015, 2017; Ong and Choo 2011, 2018).

In fact, if the more appropriate in situ pressuremeter test (see Fig. 4.8 or Chaps. 2 and 3 for more technical details) is used instead of mechanical coring, one will find that the in situ confined compressive strength is much higher and thus being more representative of the phyllite rock mass (Phangkawira et al. 2017 and Phangkawira 2019). This knowledge is important in the selection of suitable excavation tools and cutter bits for the microtunneling boring machine (mTBM) since these activities involve understanding the confined compressive strength (Shi et al. 2015) as

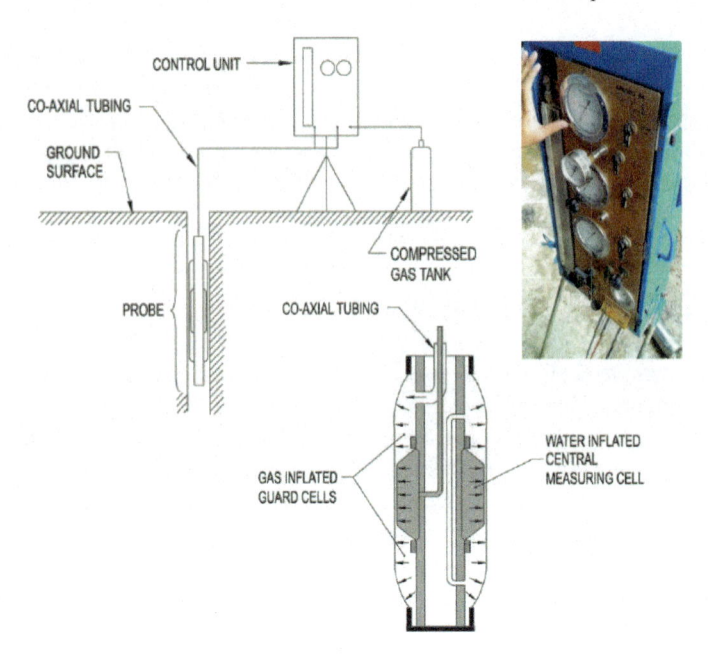

Fig. 4.8 Schematic and setup of the Menard pressuremeter system and its inflatable packer

opposed to the unconfined compressive strength (after core extraction), which is usually governed by shear failure.

4.3.4.3 Igneous Rocks

Igneous rocks are formed when magma or earth's molten rock cools and crystallizes. Igneous rocks are called 'extrusive' if lava quickly solidifies after it surfaces from the crust of the earth. Due to rapid cooling when in contact with the atmosphere, extrusive igneous rocks often have smooth, glassy appearance with no clear crystals. Examples of such rocks are obsidian and basalt. On the other hand, 'intrusive' igneous rocks cool very slowly without leaving the earth's crust thus having much more time for crystallization to occur. Granite is a good example where sizeable, shiny crystals can be observed, hence the nickname 'dalmatian' rock. As igneous rocks are formed through solidification process, they do not readily display any bedding planes nor any folds unlike sedimentary rocks. Neither have they got slickenside features unlike metamorphic rocks. Thus, igneous rock can be easily recognized by its solid, homogeneous appearance with glassy surface (extrusive) or sparkling crystals (intrusive). Such appearances also explain why igneous rocks tend to be the strongest compared to sedimentary and metamorphic rocks at similar weathering grade.

4.3 Design Philosophy and Selection Guide for Shafts   117

High Rock Strength

Excavation in igneous rock mass is less problematic except for its higher strength, when contrasted against sedimentary and metamorphic rocks. In this case, appropriate mTBM disk cutters should be aptly chosen to ensure that the tunneling works progress with minimal disruption. In shaft excavation, often a high-powered mechanical breaker mounted on a mini excavator is used to break down the rock mass. In this case, noise pollution is usually a problem in densely built-up areas.

4.3.5 Strut-Less or Strutted Shaft Design

A shaft can be designed to be either a strut-less or a strutted structure depending on its usage, location, geology, and costs. The former usually takes the shape of a circle and the latter a rectangle (in plan view).

4.3.5.1 Strut-Less Shaft Design and Circumferential Stress

For jacking and receiving shafts, a strut-less design is usually preferred since it provides obstruction-free access during the lowering or raising of the mTBM and its ancillary components. Often in a strut-less shaft design, circular (in plan view) shape shafts are adopted to rely on the development of internal compressive circumferential stresses to counteract the external lateral earth pressures acting perpendicularly to the shaft cylindrical wall. As concrete is very competent in resisting compressive stresses, top-down construction of shaft using concrete caisson (sometimes segmental) rings are often the natural and cost-effective choice. In such construction, steel wire meshes are used to 'bridge over' the next cast sequence to ensure continuity and connectivity, hence the eventual monolithic structure. A typical design of a concrete caisson shaft is shown in Fig. 4.9.

4.3.5.2 Strutted Shaft Design and Plane Strain Condition

If circular shafts cannot be constructed usually due to space constraint, then strutted rectangular shafts are the next best option. Such shafts may take the form of sheet piles (see Fig. 4.2), secant-pile wall or even diaphragm wall (see Fig. 4.3) depending on the type of geology as well as strength and serviceability requirements. Rectangular shafts are usually designed based on a 2D plane strain condition in which the wall deforms in a single dominant plane, in this case, toward the excavation and the other perpendicular, longitudinal plane runs 'infinitely long into the page,' hence often the design notation of 'per meter run' forces induced in the wall running into the page.

Strutting members over certain intervals are then required to be determined in the (plane strain) horizontal and vertical directions to ensure that the bending moment

DETAIL FOR 1200mm HIGH RING (FULL RING)

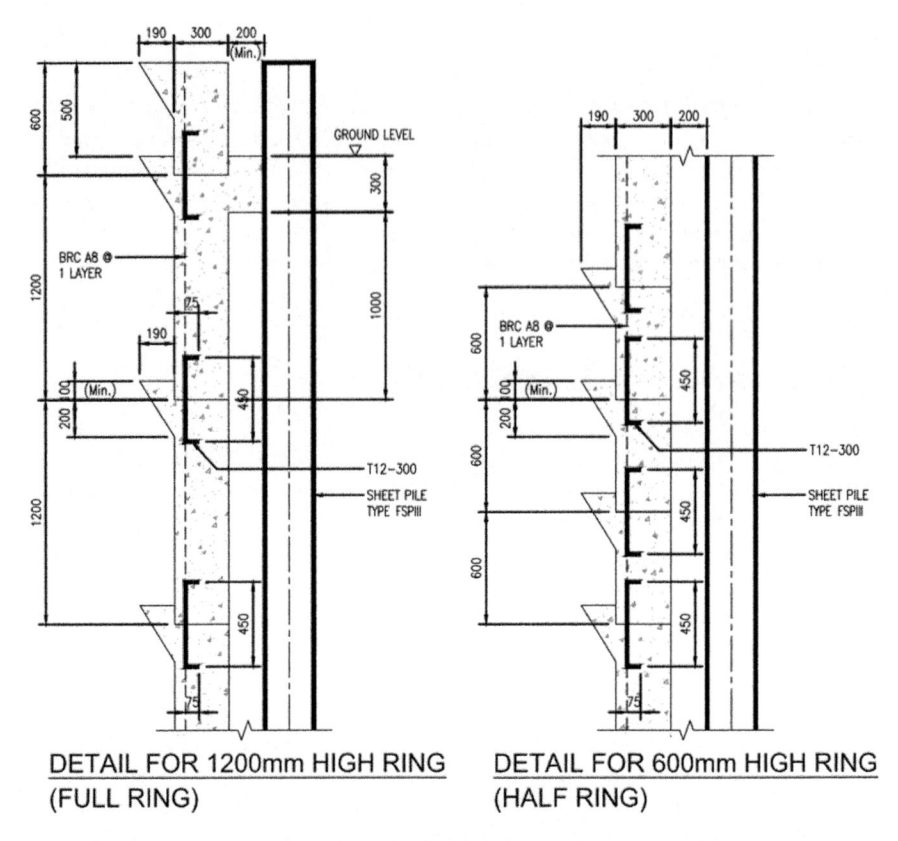

DETAIL FOR 600mm HIGH RING (HALF RING)

Fig. 4.9 Schematic of a full ring and half ring concrete caisson wall design

and bending stress induced in the retaining wall could be resisted. Struts are usually designed steel members (e.g., universal columns, beams, or pipes) and thus would be able to resist both compression and tension forces to prop open the excavation retaining walls being squeezed sideways by the lateral earth pressures. In deeper excavation, struts may be preloaded to efficiently minimize retaining wall deflections. Readers are advised to continue developing the knowledge of deep excavation shoring system in greater detail by reading Ou (2006) and Puller (2003).

4.3.5.3 Concrete Thrust Block

The concrete caisson rings, which function as a temporary works, are designed to withstand lateral pressures from the surrounding soils. The advantage of casting the concrete ring in a circular shape, which is naturally strong in compression due to the confining hoop stresses, allows the design of the caisson ring to have nominal concrete thickness and minimal steel reinforcement. Throughout casting of each

Fig. 4.10 Concrete thrust block as reaction wall to facilitate jacking process

caisson ring downwards, sufficient overlapping of concrete from each cast minimizes water ingress from the construction joint.

To facilitate subsequent pipe jacking works, a thick concrete thrust block is cast between the setup jacks and the concrete caisson wall. The thrust block enables the hydraulic jacks to be jacked against it, in order to propel the pipe string forward. The concrete thrust block is designed with sufficient thickness and steel reinforcement to allow the pipe jacking forces to be distributed evenly on the caisson wall before mobilizing the passive soil resistances behind the caisson wall. Figure 4.10 shows an example of a concrete thrust block.

4.3.6 Nearby Infrastructure Risk Assessment Documentation and Analyses

4.3.6.1 Dilapidation or Baseline Survey

Typically, before the commencement of physical shaft and tunneling excavation work, a dilapidation survey or a pre-construction condition baseline survey is to be commissioned. This survey is typically executed by an insurance loss adjuster or a qualified building surveyor. Typical report contains a written and photographic record of building or infrastructure locations and broad observations of the existing exterior conditions as observed from public easements and sidewalks within reasonable distances from the construction site. The dilapidation survey provides an independent photographic and documentary report evidencing existing defects and distressed building elements prior to commencement of construction activities. It is often helpful that the photos taken are provided with timestamps for easy reference. It is not uncommon today that the photos can also be taken to include global positioning system (GPS) coordinates and bearing directions to ease identification and documentation processes. From a contractor's perspective, the dilapidation report is primarily a loss control and claims defense mechanism against property owners' allegations of damages due to the current constructor's on-site activities. Usually, at this stage of work, only externally visible damages are documented since accesses to individual properties are often limited.

The dilapidation survey often includes photos showing pre-existing damages on external facades such as visible cracks, distortions, separations of the following items:

- Buildings including houses, shophouses; components such as walls, roofs, columns, beams, and floors
- Above-ground structures; abutments, approach structures, underpass ramps, footpaths, kerbs, retaining walls, foot bridges and ancillary structures
- Existing pavements; road carriageway, roundabouts and slip roads including road shoulders
- (exposed part) Underground structures; underpasses, pump sumps and below-grade road structures
- Drainage culverts, catch basins, open drain structures and wing-walls
- Road furniture, public fire hydrants and signages

4.3.6.2 Methodology and Analysis for Ground Movement

Effect of Tunneling

To assess soil–structure interactions that consider ground movements due to stage excavation, normally finite element analysis (FEA) (e.g., PLAXIS, FLAC, ABAQUS, GeoStudio's SIGMA/W etc.) is carried out prior. Effective stress or drained analysis is usually carried out to estimate the long-term ground movements. The prediction of ground settlement profile for pipe jacking works is usually taken from the concept of larger bored tunneling works and is based on the empirical method suggested by Mair et al. (1996).

The methodology begins by estimating the potential transverse ground settlement profile. In greenfield area where there is no prior tunneling experience, volume loss is often estimated from the TBM overcut (in percentage) with reasonable allowance to err on the conservative side. Otherwise, volume loss is based on the past experience of tunneling works performed in similar ground conditions and similar tunneling method. For example, in the established Singaporean experience, Shirlaw (2002) and Shirlaw et al. (1995, 2001) reported the following observed total volume losses (upper bound) due to large-diameter TBM driving in the different types of soils as:

- <1% for weathered rocks of the Jurong (Sedimentary) Formation without presence of Kallang (Quaternary) Formation above tunnel
- <1% for mixed rocks and residual soil grades of the Bukit Timah (Igneous) Granite
- <1% for the Old Alluvium (Over-consolidated deposits comprising clay–sand mix)
- <2% for residual soils resulting from weathered of the Bukit Timah (Igneous) Granite

- <2% for a mixed face of Kallang (Quaternary) Formation soils and Old Alluvium (Over-consolidated deposits comprising clay–sand mix)
- <3% for weathered rocks of the Jurong (Sedimentary) Formation with presence of Kallang (Quaternary) Formation above tunnel
- <4% for a mixed face of Kallang (Quaternary) Formation and weathered rocks of the Jurong (Sedimentary) Formation
- <4% for soils (fluvial sand, estuarine clays) of Kallang (Quaternary) Formation
- <4% for the Marine Clay of Kallang (Quaternary) Formation

The ground settlement profile from tunnel boring is obtained from the Gaussian distribution curve by specifying the appropriate volume loss using the equations shown from Eqs. 4.2–4.4.

For a single tunnel, immediate surface settlement, S_v, shall be determined from the relationship:

$$S_v = S_{max} \exp[-y^2/2i^2] \qquad (4.2)$$

where S_{max} is the maximum settlement above the tunnel, y is the horizontal distance to the tunnel, and i is the horizontal distance from the tunnel center line to the point of inflexion on the settlement trough.

$$i = K z_o \qquad (4.3)$$

where K is a parameter, which varies between 0.5 (for clay) and 0.25 (for sand) and z_o is the depth to the center of the tunnel.

$$S_{max} = [0.0031 V D^2]/i \qquad (4.4)$$

where D is the excavated diameter of the tunnel, and V is the volume loss expressed as a percentage of the excavated tunnel face area.

Typical coefficient, K and volume loss, V parameters for the Singaporean experience can be found in LTA (2019).

The tunnel settlement contours are usually prepared in chainages of every (approximately) 50 m cross-sections (including corresponding FEA runs) and plotted on the topographical map to further assess the impact on adjacent buildings, structures, and underground utilities as described hereinafter.

Effect of Deep Excavation

The transverse ground settlement profiles obtained from the FEM analysis are usually inclusive of consolidation settlements. Then the corresponding ground settlement contours (plan view) are prepared based on the estimated ground settlement profiles as shown in Fig. 4.11. The extent of ground surface settlement usually taken as

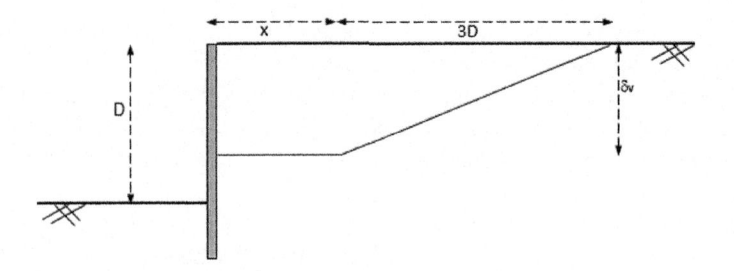

Fig. 4.11 Definition of the influences of a deep excavation process

3D (D = depth of excavation) from the location of maximum ground settlement is adopted based on the recommendations of Clough and O'Rourke (1990) to develop the idealized settlement contour plan. The estimated maximum settlement value, δ_v and the distant from the excavation face to the maximum settlement, x are obtained from the FEA.

Combined Effects of Tunneling and Deep Excavation

Finally, the settlement contours are prepared and superimposed from the settlement profile plan obtained from both the deep excavation and the bored tunneling.

The settlement profiles are fitted using a sine curve to further assess the damage potential of the buildings and structures, while the settlement contours are used to assess the damage potential of the angular rotation of the utility network.

4.3.6.3 Methodology and Analysis for Building Damage

Sensitive structures around the shaft and within the zone of influence are identified and located on the settlement contour plan. For each affected structure, horizontal, bending, and diagonal strains in both hogging and sagging zones are calculated based on the estimated ground movements in both vertical and horizontal directions (extracted from FEA results) as well as the configuration of the structure's footprint projected onto the contour plan.

The degree of severity of damage to the affected structure is then assessed based on the estimated tensile strains, since tensile strains cause the occurrences of cracks. The classification of the degree of severity of damage for each structure is based on the relationship of limiting tensile strain as outlined in the work of Boscardin and Cording (1989) is presented below in Table 4.2.

Buildings and structures are considered low risk if the predicted degree of damage falls into the first three categories 0–2, i.e., 'negligible' to 'slight.' The division between damage categories 2 and 3 represents an important threshold and is therefore of particular significance in the process of risk assessment.

4.3 Design Philosophy and Selection Guide for Shafts

Table 4.2 Relationship of limiting tensile strain with category of damage

Limiting tensile strain (%)	Normal degree of severity	Category of damage
0.000–0.050	Negligible	0
0.050–0.075	Very slight	1
0.075–0.150	Slight	2
0.150–0.300	Moderate	3
Greater than 0.300	Severe to very severe	4–5

Angular rotations for structures with shallow foundations have also to be checked. This additional check is to ensure that no building experiences a slope exceeding 1:300 (Skempton and MacDonald 1956).

For buildings and structures supported on pile foundations and where pile details are known, the foundation piles will be checked for the induced bending moment as a result of lateral ground movements due to impact of the nearby excavation and/or tunneling (Ong et al. 2003a, b, 2006, 2015; Chong and Ong 2020; Leung et al. 2006).

The abovementioned calculations are often intense, so the development and use of spreadsheet and purpose-written sub-routines (VBA macros) are quite common.

4.4 Soil–Structure Interactions

Soil–structure interactions (SSI) is a broad definition that describes the active interactions between the (infra)structure and the corresponding ground that supports the (infra)structure. In civil engineering, structures are built on or in the ground. The large contact areas between the structural elements and the surrounding ground ensures that the effect felt by one would create a corresponding response in the other. In this regard, SSI is a common feature in today's civil engineering design philosophy where structures are no longer only designed to its limit state for strength but also to a conforming performance-based specification, where movements (settlements or deflections) are expected to be within the predetermined set limits over the entire service lives of the structures to ensure occupational comfort and public safety.

In this regard, the following design philosophies related to SSI are briefly discussed in the following subsections and can be used to assess the challenges in shaft construction as discussed in Sect. 4.3.

4.4.1 Analytical Method

Analytical method is regarded as the application of scientific techniques, tools or software to problem-solve a typical problem systematically. Broadly, engineers rely on this concept to solve day-to-day challenges by contextualizing the larger problem and then breaking it down into idealized situations for easy manipulation. For example, Figs. 4.6 and 4.11 demonstrate a contextualized technique of solving complex engineering problem.

4.4.2 Empirical or Closed-Form Method

Empirical or closed-form method is considered a subset of an analytical method and uses established equations for problem-solving. However, this method is often restrictive in the sense that the specific values of the coefficients used are usually based on 'local experience' for validation purpose. Even though the overall analytical method is consistent, authors from various locations may adopt slightly different coefficients due to, for example, variations in soil conditions or rainfall intensity from one region or country to another.

For instance, Eqs. 4.2–4.4 are widely accepted Gaussian equations for ground subsidence due to tunneling, and with much research that has been carried out on the Singapore soils, locally derived volume loss parameters have been confidently adopted in the country.

4.4.3 Field Observational Method

This method employs the use of evidence-based technique via the reliable geotechnical instrumentation system that includes a continuous data monitoring and collation program over the entire construction period and sometimes post-construction as well. Reliable field performance data is often used for back-analysis or for sensitivity analysis in order to shed light on some ambiguous problems that require further understanding. For example, Case Study 1 that is discussed in detail in Sect. 4.6.1 is a back-analysis method carried out to ascertain the existence of flow anisotropy in organic soils in an excavation dewatering process (Ong 2019).

4.4.4 Numerical Method and Finite Element Method (FEM)

Numerical method uses mathematical techniques to solve engineering analyses that cannot be readily or possibly solved by analytical methods. Numerical method is a systematic and efficient 'trial-and-error' procedure. Typically, an initial solution with a selected increment of the variable is first input into the simulation with the selected increment which the intended solution is required to cover. Unstable numerical increment of the variable will subsequently result in 'oscillation' or becoming unbounded in the trend of the projected values. Stable numerical outcome is the result of convergence of the range of the values analyzed.

Finite element method (FEM) is born as the natural extension of the efficiency of the numerical method adopted in solving partial differential equations arising in complex engineering or mathematical modeling. To efficiently solve a problem in hand, this advanced technique subdivides a large problem into discretized, smaller parts called finite elements via the construction of interconnecting sets of meshes consisting of nodes and elements. PLAXIS is a widely used geotechnical stress–strain finite element code. FEM method is also utilized by GeoStudio's SEEP/W software in analyzing steady-state and transient seepage problems, where an example of this can be found in Sect. 4.5.1. Basic explanation and comparison of different types of FEM software performance and practical applications have been reported by Ong et al. (2006b, 2008, 2018) and Choo and Ong (2020). To model soil–structure interfaces and to appreciate soil arching around shaft or tunnel in both granular and cohesive soils, readers are recommended to follow the work of Peerun et al. (2019, 2020) and Cheng et al. (2020), respectively.

4.4.5 Artificial Intelligence

The emergence of artificial intelligence (AI) techniques is envisaged to have an enormous potential in addressing geotechnical problems that involve complex soil–structure interactions. Jong et al. (2021) found that applications of AI techniques in studying underground soil–structure interactions are often focused on aspects such as characterization of soils and rocks, pile foundations, deep excavations, and tunneling. The increase of AI applications in underground construction indicates that they are in great demand in underground works, due to their ability to analyze complex relationships among various variables and produce meaningful output for interpretation.

Examples of AI techniques used in underground applications include support vector machines (SVM), logistics regression (LR), decision tree (DT), random forest (RF), and Bayesian network (BN). Genetic programming (GP) and artificial neural network (ANN) have been used to rank the importance of various parameters affecting geopolymer strength development (Leong et al. 2015, 2018b), whose concept could be adapted to represent soil stabilization parameters.

4.4.6 Consistency of Methods

Typically, in any benchmarking exercise in design where confidence level is to be improved, broad consistency achieved in the various methods as discussed in the preceding sections is encouraged. At least two methods executed from Sects. 4.4.1–4.4.4 should be used as standard verification.

4.5 Typical Construction Challenges

This section is written focusing on the salient points regarding the possible challenges encountered during the construction of deep shafts and should not be used solely as a complete design guideline for general excavation.

4.5.1 Soil Conditions

The soil condition surrounding the excavation plays a key role throughout the overall excavation process. This is a determining factor in selecting the appropriate shafts or construction method to be adopted.

4.5.1.1 Sand Liquefaction or Boiling

During deep excavation in relatively high ground water table condition, the extraction of ground water from the shaft creates a natural seepage flow, whereby water travels through the voids within the soils from higher hydraulic head to lower hydraulic head.

As the water flows, seepage pressure is exerted onto the soil particles. Especially in high permeable sandy soil, the relatively high permeability of sand allows the surrounding ground water to flow freely into the shaft, resulting in the built-up of seepage pressure. Seepage pressure increases proportionally with the height of hydraulic gradient or as the excavation gets deeper into the ground. Sand liquefaction or also known as sand boiling, is the flow of soil particles with water and this occurs when the seepage pressure exceeds the hydrostatic pressure and the overburden pressure (see Mehdizadeh et al. (2016, 2017) for further understanding).

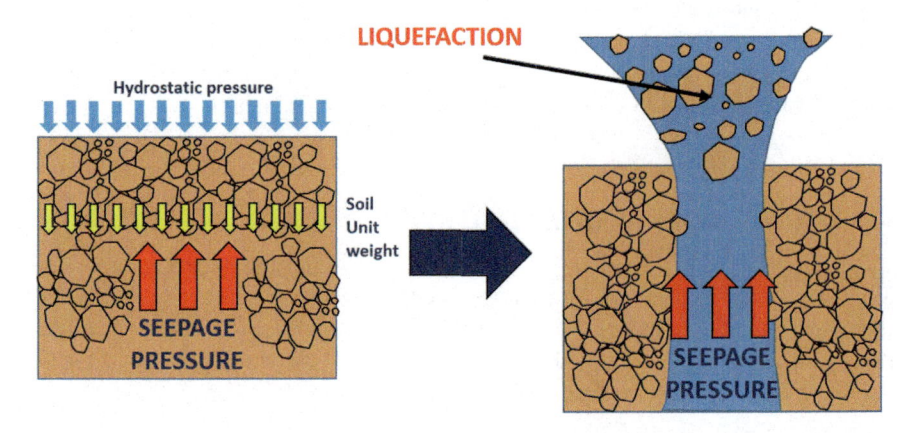

Fig. 4.12 Schematic showing the occurrence of liquefaction (boiling) under high seepage pressure

4.5 Typical Construction Challenges

Figure 4.12 illustrates the phenomenon of boiling. Under normal circumstances, sand boiling can be mitigated by lengthening the seepage flow path (e.g., extending the cut-off wall toe) or by mixing the soils with cement (Liu et al. 2020, 2021) or sustainable binder (Leong et al. 2016a, 2016b, 2018a; Ngu et al. 2019; Omoregie et al. 2016, 2017, 2019a, b, c, 2020) in the form of ground improvement work.

4.5.1.2 Soft Soil Bottom Heave

Basal failure to an excavation by means of soil heaving from the excavation floor is usually a concern when excavating into plastic, soft soils. The stress relief due to the removal of soils within the shaft results in a nonequilibrium of effective overburden pressures inside and outside of the shafts. Once the nonequilibrium stresses cause the soil shear strength to be fully mobilized or when limiting condition is reached, bearing capacity failure will be resulted at the base of the excavation, thus the occurrence of basal heave failure. Figure 4.13 illustrates the case of basal heave failure within an excavation.

Basal stability of an excavation especially in soft soils can be improved by having deeper wall toe penetration into the underlying more competent ground, where wall friction could be readily mobilized as additional resistance. Besides, ground improvement such as deep soil mix (DSM) pile or jet grouted pile (JGP) techniques can be used to enhance the shear strength of soils around the excavation floor.

4.5.2 Seepage Pressure

Hydraulic uplift occurs when the underground structure such as a concrete slab is uplifted as the artesian pressure exceeds the dead weight of the structure. This situation is particularly prevalent at sites having sandy soils with consistent and

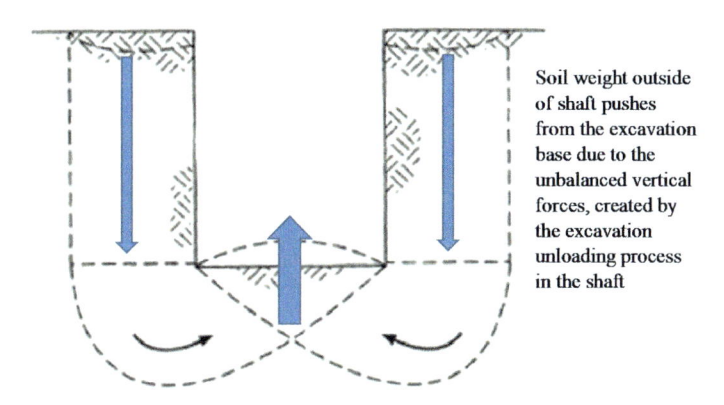

Fig. 4.13 Schematic showing basal instability of an excavation process

high water table. During the excavation process, the difference in hydraulic gradient inside and outside the shaft results in the surrounding ground water flowing into the shaft, creating a zone of low effective stress within the sandy soil mass. When the sand particle-to-particle interactions are diminished under the increasing hydrostatic pressure, the sand particles would then get 'washed' into the bottom of the shaft under the seepage pressure. This phenomenon is akin to sand liquefaction or boiling.

In terms of design, a flow net analysis is required to be conducted first. If the exit gradient (see Table 4.2) at the excavation floor is found to be threatening, methods to lengthen the seepage flow must be implemented to reduce the exit gradient. Some of these methods include installing deeper cut-off wall, performing TAM grouting around the base of the shaft, or simply having a series of controlled water pressure relief outlets at the excavation floor.

4.5.2.1 Water Pressure Relief Outlet

A water pressure relief outlet allows the dissipation of built-up hydrostatic water pressure under the concrete base slab, thus preventing the sudden occurrence of hydraulic uplift. The simple system consists of having a 350-mm-thick layer of clean aggregates sandwiched (preferably throughout) between the excavation floor and the base slab. Each outlet has a PVC tube (perforated at embedded bottom and nonperforated above ground) controlled using an individual gate on/off valve. Figure 4.14 shows a typical detail of the water pressure relief outlet.

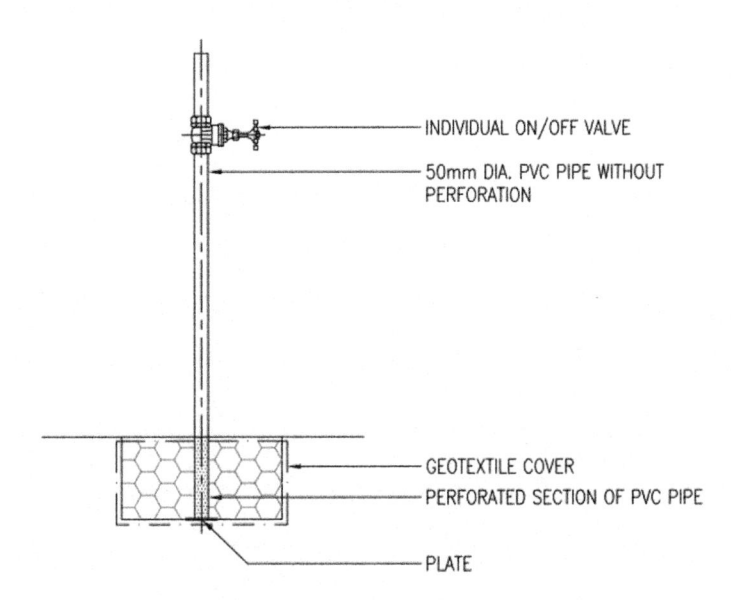

Fig. 4.14 Schematic showing the design of a typical water pressure relief outlet placed at the floor of the excavation

4.5.3 Ground Water Intrusion

Ground water intrusion is one of the most commonly faced challenges during shaft excavation in ground with high water table. Excessive water intrusion into the shaft being excavated will result in the depression of the surrounding ground water table, which consequently leads to ground subsidence. The effects are damaging especially in built-up areas congested with buildings and network of infrastructure.

4.5.3.1 Limitation of Sheet Pile Penetration into Stronger Underlying Material

Under normal circumstances, temporary works using steel sheet piles are one of the most commonly used methods in minimizing water intrusion. Steel sheet piles which are installed in a continuous interlocking manner are preinstalled outside the excavation footprint before the commencement of excavation works. Steel sheet piles can be installed down to socket into competent, relatively impermeable soils, to cut off any possible flow regime within the more permeable upper soil layers.

Nevertheless, the typical installation method of steel sheet piles is often limited especially into the underlying stronger or denser material with SPT N values greater than 25–30. However, due to technological advancement in recent decades, sheet pile installation rigs which could perform simultaneous pre-boring and water jetting to loosen the denser soils have become available in the market. These technologically advanced rigs are efficient in advancing sheet pile penetration into denser ground or soft rock, thus effectively cutting-off water intrusion. Figure 4.15 shows a GIKEN Silent Piler rig used to socket sheet pile toe into the underlying stiffer soils or weathered soft rocks.

Fig. 4.15 Installation of steel sheet piles into weathered rock using GIKEN Silent Piler

4.5.3.2 Ground Improvement Works

Ground improvement works can be deployed to minimize ground water intrusion into the shaft. For example, permeation or chemical (sodium silicate) void grouting using Tube-a-Manchette (TAM) pipes would infill the soil voids thus reducing the permeability of the treated soils.

Rock or fissure grouting is often carried out to reduce the rock mass permeability especially along discontinuities in an aquifer. Nonshrink microfine grout is normally injected in isolated stages from pre-drilled boreholes. In order to target specific grouting depths or sections, either a single-packer or double-packer method can be deployed to seal the end(s) of the section to be grouted in order to minimize grout loss.

Ground improvement works via grouting method requires specialized machineries and highly skilled machine operators to ensure that the grouting works are carried out efficiently with minimal wastage. Some grouting operators strategize their grouting procedure by having alternating grouting patterns, i.e., primary and secondary grout holes or by allowing sufficient rest periods before commencement of subsequent grouting operations. Nonetheless, it is often that the specific grout intake volumes are recorded to demonstrate effectiveness. Figure 4.16 illustrates a typical grouting strategy, i.e., by having a series of alternating primary and secondary grouting holes along the outer perimeter of the shaft footprint to minimize water intrusion.

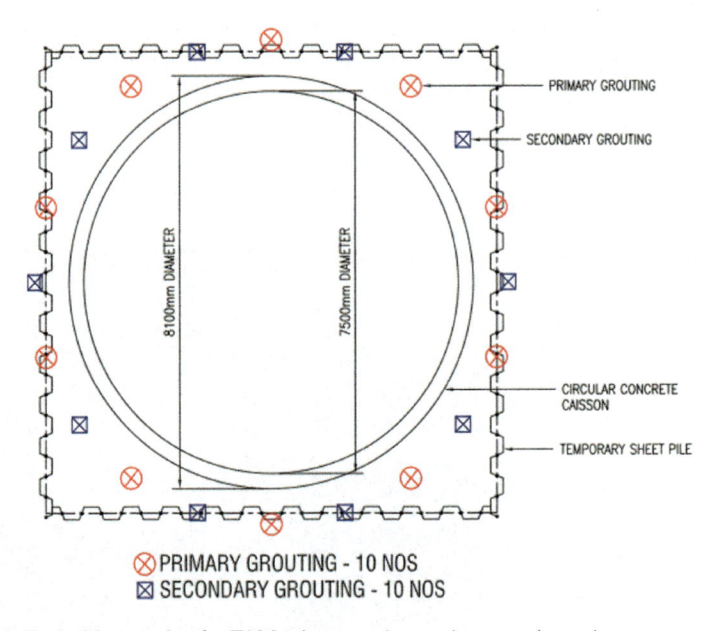

Fig. 4.16 Typical layout plan for TAM primary and secondary grouting points

4.5.4 Space Constraints

Excavation works carried out within built-up areas call for the need of proper feasibility study of the area before deciding on the actual position of the shaft. A certain footprint of construction site is required for the physical shaft itself, placement of heavy machineries as well as a loading/unloading area for trucks and for pipe stockpile.

Considering an excavation period which may take up to weeks or months, the suggested areas should be properly planned and established in order to minimize any inconvenience to nearby property owners or road users. Under normal circumstances where excavation works are carried out along an existing road or road reserves, a proper traffic management plan is required to be submitted to the local authority for approval before commencement of work to ensure the safety of road users. Grass verge, pedestrian walkway, public carpark, and garden beds are typical examples of areas that can be temporarily used as construction spaces for pipe jacking works and be made good after completion of the works.

4.5.5 Utility Detection in Urban Areas

To prevent any unwanted damages or destruction to existing underground utilities or pipelines, locations of the utilities are required to be verified first before commencement of excavation or installation of temporary works (e.g., sheet piles).

4.5.5.1 Nondestructive Testing

Nondestructive testing refers to a technologically advanced soil investigation technique where investigative work is carried out on a precise operation to minimize inconveniences and potential damages to underground structures or utilities.

Vacuum Excavation

As the name implies, vacuum excavation is an advanced excavation process whereby soils are loosened by high-pressured air or water and then vacuumed away into a holding tank for subsequent reuse or disposal, thus leaving no mess behind and with no or minimal damage to the underground utilities. It is an advancement of recent nonevasive investigation technique. By performing vacuum excavation, access to underground utilities is made safer and easier like never before, besides being environment friendly and sustainable.

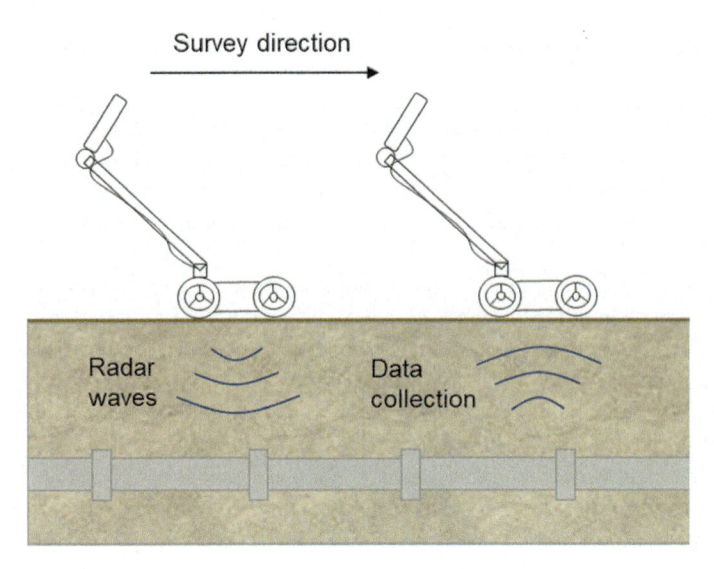

Fig. 4.17 Ground penetrating radar used for utility detection

Ground Penetrating Radar

The application of ground penetrating or probing radar (GPR) allows the detection of underground utilities via the emission and subsequent re-detection of the electromagnetic radiation waves reflected by any existing underground objects. Figure 4.17 illustrates an example of underground utility detection using GPR.

The advantage of using GPR is that it is nondestructive, where physical excavation such as trial trenching is not required throughout the whole process of attempting to detect utilities. This can provide users some reliable preliminary findings and information on the presence of underground utilities, within a short period of time. However, the accuracy of underground utilities detection using GPR is highly dependent on the electrical conductivity of the subsurface condition and the material that the utilities are made of. For example, GPR is more effective in drier and denser sands or gravels than wetter clays since the electrical conductivity of soils increases with increasing water, clay and soluble salt contents. Increased electrical conductivity means increased rapid attenuation of radar energy that subsequently restricts penetrating depths. Generally, GPR is particularly effective at depths between 3 and 4 m. Further discussion can be found in Chap. 2.

4.5.5.2 Conventional Method

Conventional method of utility detection by means of manual excavation is labor intensive and time consuming, but it has the advantage of being able to see the actual condition of the exposed utility.

Fig. 4.18 Laborious trenching work in progress

Trial Trench

Test pitting or trial trenching may be carried out prior to the commencement of shaft excavation, by either hand-augering or trench-excavation. These methods allow contractors to verify and confirm the types and actual positions of underground utilities.

Nevertheless, these physically destructive methods may be inefficient and time consuming as compared to the GPR nondestructive method. Multiple trial trenches may be required to map out the network of underground utilities within the excavation site and temporary shoring measures may also be required to locate deeper buried underground utilities. For trial trenching to be carried out on road pavements, this will undoubtedly result in inconveniences to road users and costly reinstatement. Nevertheless, the advantage of performing trial trenching is that this method is evidence-based, and the outcomes are often irrefutable. This method is particularly useful when there is an absence of as-built drawings or when it is carried out to locally verify the outcomes of GPR assessment. Figure 4.18 shows a trial trenching method in progress to verify the locations of underground utilities.

4.6 Case Studies

4.6.1 Case Study 1: Shaft X in Organic Soils

The geology of the area is characterized by deltaic deposits consisting of peat, clay, silt and sand layers overlying Kuching's Tuang Formation. Figure 4.19 shows the

Fig. 4.19 Excavated spoils showing organic soil (brownish) mixed with clayey soil (grayish)

texture of organic soils found at the project site. The underlying Tuang Formation rock has been described as 'erratic' by Ong and Choo (2011, 2012) in view of its highly sheared and weathered rock mass.

4.6.1.1 Details of Proposed Deep Shaft

Figure 4.20 shows the details of the proposed 7 m diameter, 16-m-deep shaft. Sheet piles were first driven around the shaft in an attempt to lengthen the ground water seepage paths in order to reduce the risk of basal instability or hydraulic uplift due to the difference in hydrostatic pressure inside and outside the shaft that was being excavated. Advancing cast in situ circular concrete caisson rings (each measuring 1.3 m in height, with 100 mm overlap) were then installed to provide structural support to the surrounding soils so that excavation inside the shaft could be executed.

In this method, the concrete caisson rings served as temporary works. The design was mainly based on hoop stresses generated on the rings by the lateral soil pressures. As such, the steel reinforcement used was nominal since the hoop stresses were designed to be taken entirely by the concrete.

4.6.1.2 Effect of Leakage at Underlying Fractured Rock

At one of the many challenging excavation sites, when the 7 m diameter, 16-m-deep shaft that was being excavated reached the moderately to highly weathered sandstone mass at depth of about 13.3 m, major leakage occurred as shown in Fig. 4.21. Hence, the deep excavation works had to be suspended. Consequently, and inevitably, ground water drawdown of about 5 m occurred, triggering settlement of 125 mm to nearby infrastructure, mainly due to the relatively high permeability of the 8.5 m upper soft, organic soils with SPT $N = 0$.

4.6 Case Studies

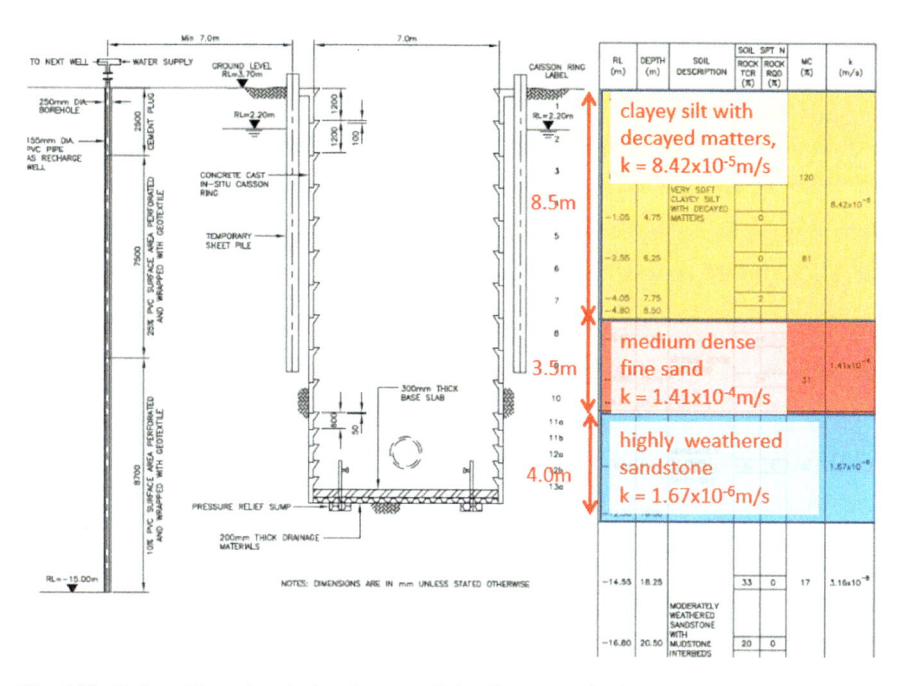

Fig. 4.20 Soil profile and typical recharge well details near to the location of the proposed deep shaft

Fig. 4.21 Major leakage occurred at the weathered sandstone interface

4.6.1.3 Ground Water Responses

The quantity and rate of the ground water that seeped into the leaking shaft could be estimated on site by simply calculating upwards from the base of the shaft the number of caisson rings flooded in the shaft after it was successfully pumped dry. The amount of water that rose in the shaft overnight (no pumping activity) would then give a reliable estimate of the water inflow rate into the shaft. In this case, the 7.0 m diameter shaft simply acted like a large water standpipe. The volumes of water and flow rates vary on different days because the exposed rock faces had different characteristics and permeability values as the contractor excavated into the rock mass over time.

4.6.1.4 Finite Element Model

On average, the measured seepage was 13.6 m^3/h, which was reasonably comparable to the estimated ground water seepage rate of 17.4 m^3/h based on the finite element seepage analyses conducted during the design stage (see Fig. 4.22). The analyzed seepage rate is important because it was subsequently used to design the recharge well system to replenish the depressed ground water table caused by the leakage.

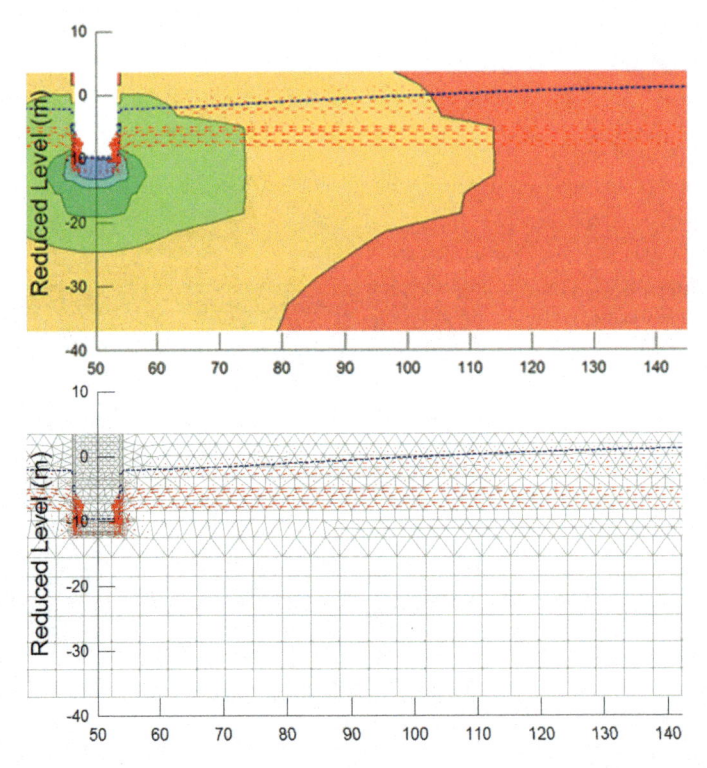

Fig. 4.22 Finite element-based transient seepage analysis showing ground water table being drawn down due to the major leakage in the shaft

4.6 Case Studies

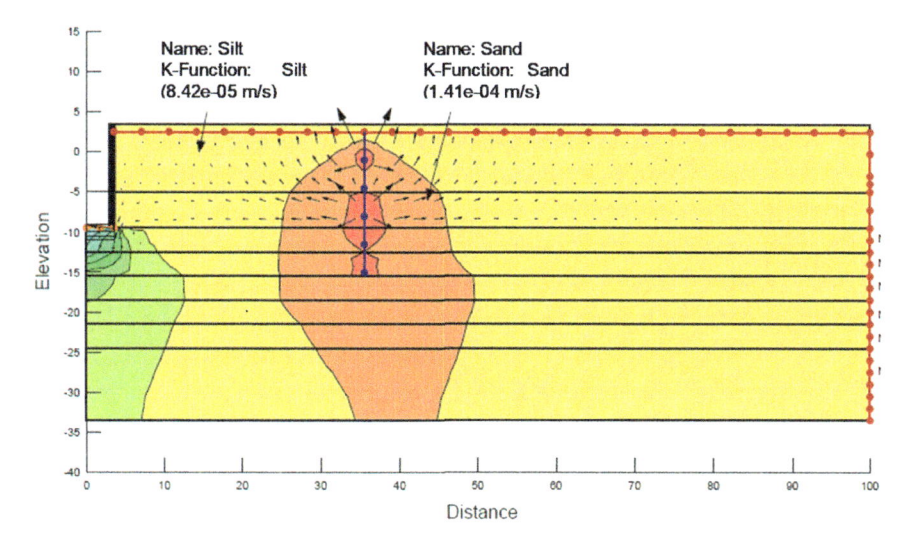

Fig. 4.23 Typical transient seepage analysis carried out to ascertain the recharge capability of the proposed recharge well system

4.6.1.5 Ground Treatment Works

As a form of ground treatment works, large-diameter recharge wells were designed (see Fig. 4.23) and installed to recharge the depressed ground water table at site. Field monitoring showed that ground settlements of not more than 7.5 mm were observed when the excavation works resumed. It took a further 17 days to complete the remaining excavation works, which consisted merely a further excavation of 1.8 m and casting of the base slab. Pressure relief sumps, control valves as well as lined drainage materials were constructed at the shaft floor to control water seepage coming into the shaft as the base slab was deliberately designed not to cater for the full uplift forces. Readings from the nearby water standpipes also showed that ground water did not drop more than 2.5 m during further construction works, due to constant water recharging via the recharge wells. Field observations showed that particles wash out into the shaft due to recharging did not occur.

4.6.1.6 Ground Responses

Finite element analysis was used to predict the ground settlement. Figure 4.24 shows the predicted and measured ground surface settlement profiles at various stages of the excavation. It was found that the ground settlement magnitudes and profiles are sensitive toward the ratio of $k_h:k_v$ of the top organic soil layer, where k_h is the horizontal soil permeability and k_v the vertical. The predicted and measured values show good agreement. Based on back-analyses and with the outcomes shown in Fig. 4.24, the predicted ground surface settlement profiles using $k_h:k_v$ ratios of 1, 3, 5,

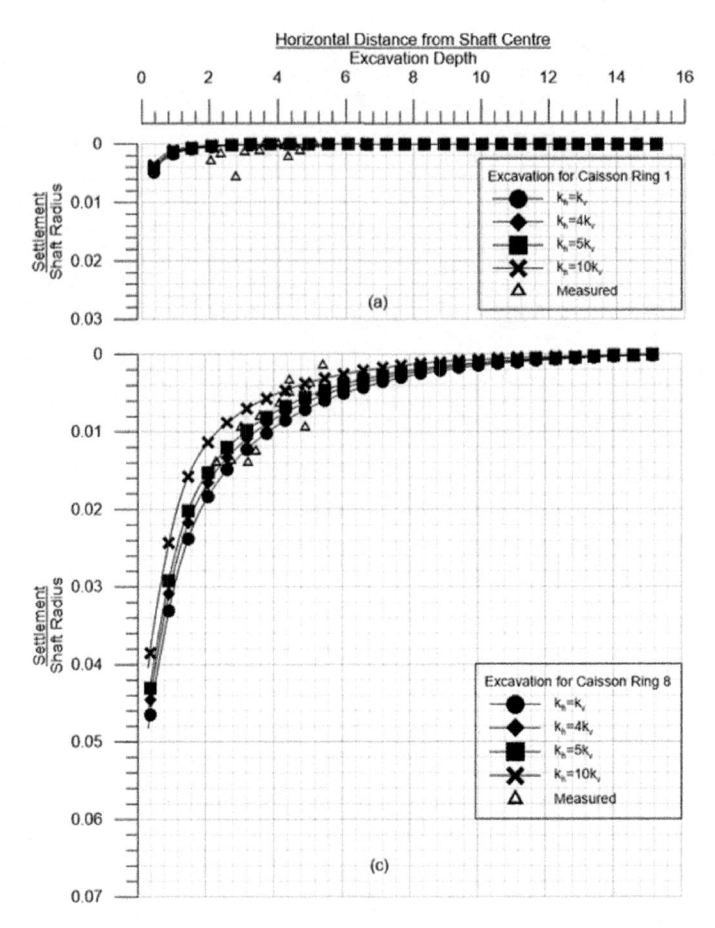

Fig. 4.24 Predicted and measured ground settlement profiles when excavation depth approaches concrete caisson rings **a** no. 1 and **b** no. 8

and 10 consistently fall within the range of measured settlement values, even though ratio of 5 seems to provide relatively better fit to the measured values in general. Similar anisotropy behavior in organic soil permeability had also been reported by Mesri et al. (1997), Younger et al. (1997), and Sobhan et al. (2007). This observation shows that that the horizontal permeability is definitely much greater than vertical within the soft silt layer with organic materials and most likely could be due to the natural sedimentation process (horizontal-dominant direction).

4.6 Case Studies

4.6.2 *Case Study 2: Shaft Y in Karst*

Excavation for a shaft in karst (limestone) formation is a big challenge due to the highly variable rockhead profiles coupled with the unpredictable nature of cavities (size and network of water-bearing channels) within the formation. Such a situation is experienced in one of the excavation works during the construction of a temporary deep shaft in the Kuching Wastewater Management System project.

4.6.2.1 Details of Proposed Deep Shaft

Shaft Y was constructed using a 6 m diameter circular concrete caisson ring method with steel sheet pile surround.

The excavation depth was up to 23 m. Based on the borehole information, limestone rockhead was found at 9.6 m depth whereby cavities were recorded at (i) 16.8–17.45 m depth and (ii) 18.6–20.1 m depth. Figure 4.25 illustrates the shaft excavation scheme with the anticipated soil conditions.

4.6.2.2 Water Intrusion and Remedial Works

During excavation up to 14.4 m depth (or 4.8 m excavation into karst but not a cavity level yet) within a greenfield site, the ground water intrusion rate increased from a few cubic meters per hour to 24.4 m^3/h within a short period of time. Immediately, excavation was suspended, and the shaft was backfilled.

Compaction grouting via Tube-a-Manchette (TAM) grouting method was then proposed to fill and seal the cavities in order to minimize potential water intrusion through the cavity channel. The grouting works were first carried out through the 250 mm diameter primary grouting holes (denoted by series A) which were spaced approximately 6 m apart and subsequently via the alternating secondary grouting holes (denoted by series B). Throughout the grouting process, the grout volume intake was monitored, and it was observed that the primary grout holes received relatively higher grout-take as compared to the secondary holes. The cavities were successfully plugged during the grouting process. Figure 4.26 illustrates the primary and secondary TAM grouting hole layout plan.

Upon completion of the primary and secondary TAM grouting, the excavation continued, and water intrusion rate reduced from 24.4 to 4.2 m^3/h, which is less than the work suspension value of 5.0 m^3/h. This case study shows that TAM grouting has successfully minimized water ingress into the shaft. Figure 4.27 illustrates the rate of water ingress into the shaft before and after the completion of the primary and secondary TAM grouting of karst found at the site. It is also noted that after the grouting exercise, the ground water table re-established itself and maintained way above the base of the shaft signifying the most of water-bearing channels in the karst geology had been successfully plugged.

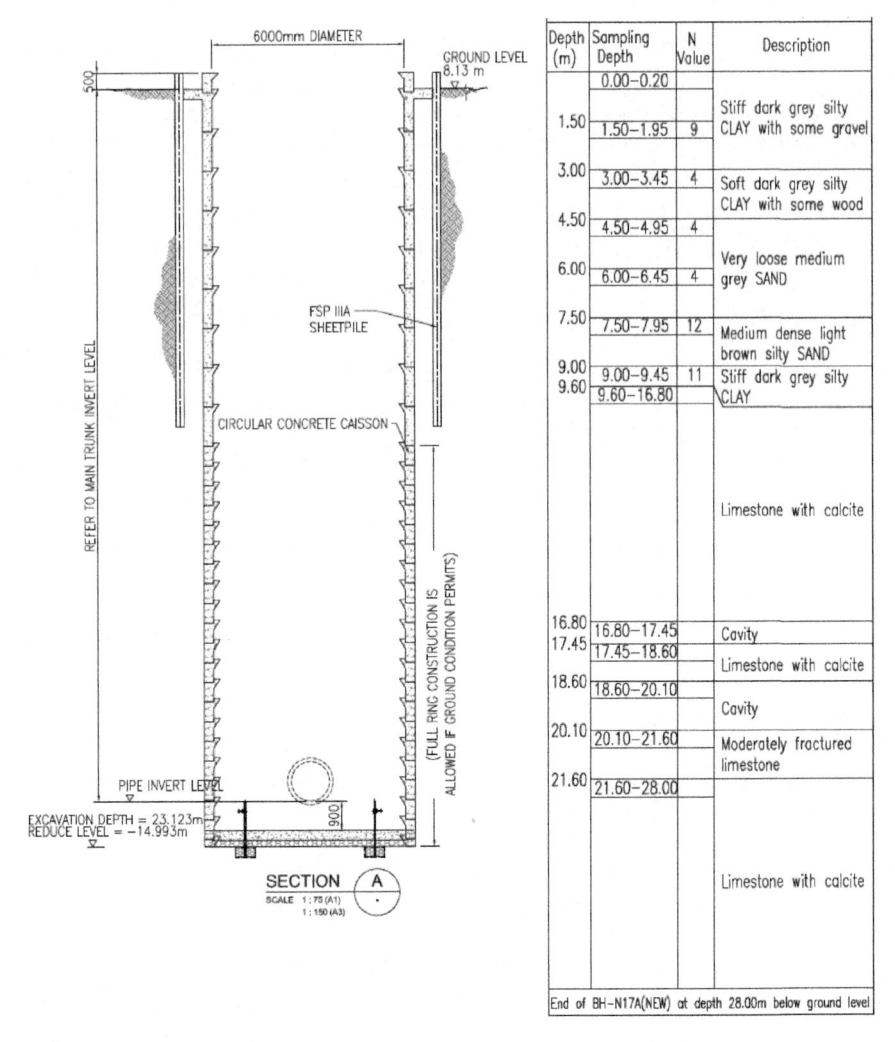

Fig. 4.25 Typical design of a 23-m-deep shaft in relation to available borehole information

4.7 Conclusions

It is important to note that the design of each shaft will most likely be different due to the variability of the subsurface profiles. From shaft selection to design philosophy and from strutting system to soil–structure interactions, indeed there are many design and construction aspects that geotechnical engineers have to consider and be well-versed in. The ability to comprehend, interpret and adopt reliable field and laboratory test results for implementation in design and for practical construction are key aspects in a successful deep shaft construction especially in challenging geological settings. It is imperative that adequate geotechnical investigations, up-to-date analytical tools,

4.7 Conclusions

Fig. 4.26 Example of grout-take volume in a karst geology during the construction of a 23-m-deep shaft

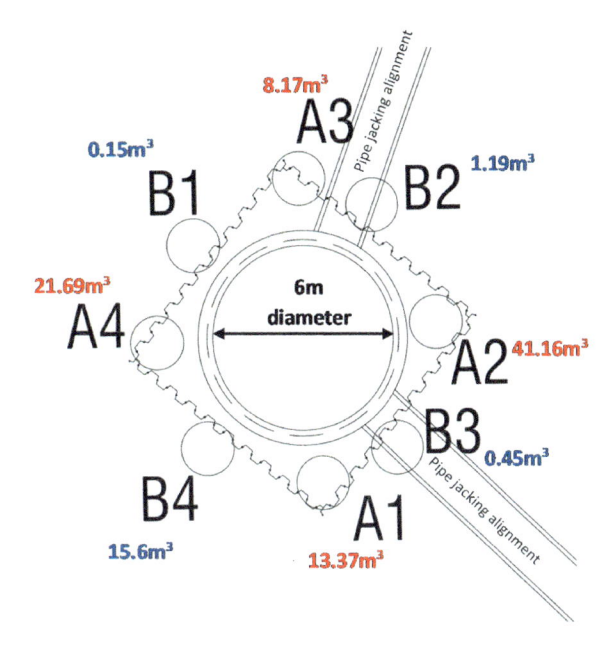

and real-time monitoring program are made available to help the decision-making process and to minimize construction risks. In the recent decades, many ground improvement techniques have entered the market and have since become available at the disposable of geotechnical engineers to help mitigate risks and losses. Forensic engineering, back-analysis, and parametric studies of real-life challenges such as those presented in this chapter and the book must be encouraged to provide learning and research opportunities to keep on improving our own design.

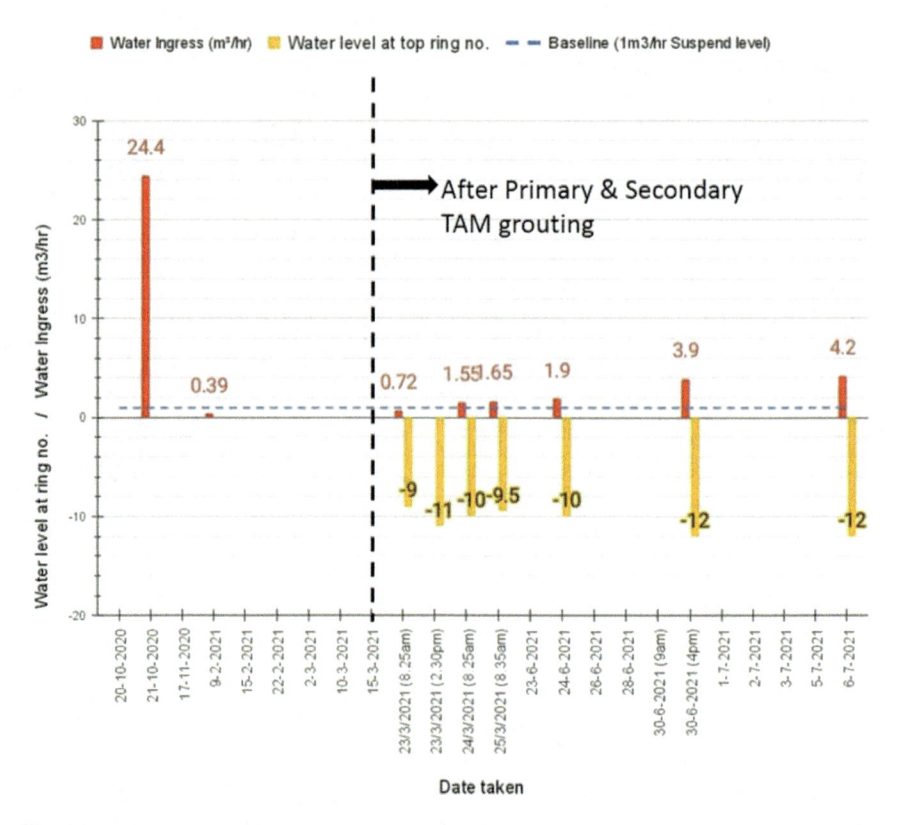

Fig. 4.27 Reduced water ingress rates into shaft and re-establishment of ground water level after the successful compaction grouting operation in karst environment

References

Allenby D, Waley G, Kilburn D (2009) Examples of open caisson sinking in Scotland. Proc Inst Civ Eng: Geotech Eng 162(GE1):59–70. https://doi.org/10.1680/geng.2009.162.1.59

Biot MA (1941) General theory of three-dimensional consolidation. J Appl Phys 12:155–164

Bjerrum L, Eide O (1956) Stability of strutted excavations in clay. Géotechnique 6(1):32–47. https://doi.org/10.1680/geot.1956.6.1.32

Boscardin MD, Cording EG (1989) Building response to excavation induced settlement. J Geo Eng 115(1):1–21

Cheng WC, Li G, Ong DEL, Chen SL, Ni JC (2020) Modelling liner forces response to very close-proximity tunnelling in soft alluvial deposits. Tunn Undergr Space Technol 103:103455. https://doi.org/10.1016/j.tust.2020.103455

Chong EEM, Ong DEL (2020) Data-driven field observational method of a contiguous bored pile wall system affected by accidental groundwater drawdown. Geosciences 10(7):268. https://doi.org/10.3390/geosciences10070268

Choo CS, Ong DEL (2015) Evaluation of pipe-jacking forces based on direct shear testing of reconstituted tunneling rock spoils. J Geotech Geoenviron Eng 141(10). https://doi.org/10.1061/(ASCE)GT.1943-5606.0001348

Choo CS, Ong DEL (2017) Impact of highly weathered geology on pipe-jacking forces. Geotech Res 4(2):94–106. https://doi.org/10.1680/jgere.16.00022

Choo CS, Ong DEL (2020) Assessment of non-linear rock strength parameters for the estimation of pipe-jacking forces. Part 2. Numerical modeling. Eng Geol. https://doi.org/10.1016/j.enggeo.2019.105405

CIDB (2019) Guidelines for construction on peat and organic soils in Malaysia. Construction Research Institute, Construction Industry Development Board, Ministry of Works, Malaysia

Clough GW, O'Rourke TD (1990) Construction induced movements of in-situ walls. In: Proceedings of a specialty conference on design and performance of earth retaining structures, Cornell University. ASCE, New York, pp 439–470

Jong SC, Ong DEL, Oh E (2021) State-of-the-art review of geotechnical-driven artificial intelligence techniques in underground soil-structure interaction. Tunn Undergr Space Technol 113:103946. https://doi.org/10.1016/j.tust.2021.103946

Leong HY, Ong DEL, Sanjayan JG, Nazari A (2015) A genetic programming predictive model for parametric study of factors affecting strength of geopolymers. RSC Adv 5:85630–85639. https://doi.org/10.1039/C5RA16286F

Leong HY, Ong DEL, Sanjayan JG, Nazari A (2016a) The effect of different Na_2O and K_2O ratios of alkali activator on compressive strength of fly ash based-geopolymer. Constr Build Mater 106:500–511. https://doi.org/10.1016/j.conbuildmat.2015.12.141

Leong HY, Ong DEL, Sanjayan JG, Nazari A (2016b) Suitability of Sarawak and Gladstone fly ash to produce geopolymers: a physical, chemical, mechanical, mineralogical and microstructural analysis. Ceram Int 42:9613–9620. https://doi.org/10.1016/j.ceramint.2016.03.046

Leong HY, Ong DEL, Sanjayan JG, Nazari A (2018) Strength development of soil-fly ash geopolymer: assessment of soil, fly ash, alkali activators, and water. J Mater Civ Eng 30(8):1–15. https://doi.org/10.1061/(ASCE)MT.1943-5533.0002363

Leong HY, Ong DEL, Sanjayan JG, Nazari A, Kueh SM (2018) Effects of significant variables on compressive strength of soil-fly ash geopolymer: variable analytical approach based on neural networks and genetic programming. J Mater Civ Eng 30(7):1–17. https://doi.org/10.1061/(ASCE)MT.1943-5533.0002246

Leung CF, Ong DEL, Chow YK (2006) Pile behavior due to excavation-induced soil movement in clay. I: Stable wall. J Geotech Geoenviron Eng 132(1):36–44. https://doi.org/10.1061/(ASCE)1090-0241(2006)132:1(36)

Liu Z, Liu Y, Bolton M, Ong DEL, Oh E (2020) Effect of cement and bentonite mixture on the consolidation behavior of soft estuarine soils. Int J GEOMATE 18(65):49–54

Liu Y, Liu Z, Oh E, Ong DEL (2021) Strength and microstructural assessment of reconstituted and stabilised soft soils with varying silt contents. Geosciences 11(8):302. https://doi.org/10.3390/geosciences11080302

LTA (2019) Civil design criteria for road and rail transit systems. Engineering Group Document, E/GD/09/106/A2, Land Transport Authority

Mair RJ, Taylor RN, Burland JB (1996) Prediction of ground movement and assessment of risk of building damage due to bored tunnelling. In: Proceedings of the 9th international symposium on geotechnical aspects of underground construction, Balkema Rotterdam, pp 713–718

Malinowska EE, Szymanski A (2015) Vertical and horizontal permeability measurements in organic soils. Ann Warsaw Univ Life Sci, Land Reclamation 47(2):153–161

Mehdizadeh A, Disfani MM, Evans R, Arulrajah A, Ong DEL (2016) Discussion of 'development of an internal camera-based volume determination system for triaxial testing' by S. E. Salazar, A. Barnes, and R. A. Coffman. Geotech Test J 39(1):165–168. https://doi.org/10.1520/GTJ20150153

Mehdizadeh A, Disfani MM, Evans R, Arulrajah A, Ong DEL (2017) Mechanical consequences of suffusion on undrained behaviour of a gap-graded cohesionless soil—an experimental approach. Geotech Test J 40(6):1026–1042. https://doi.org/10.1520/GTJ20160145

Mesri G, Ajlouni M (2007) Engineering properties of fibrous peats. J Geotech Geoenviron Eng 133(7):850–866. https://doi.org/10.1061/(asce)1090-0241(2007)133:7(850)

Mesri G, Stark TD, Ajlouni MA, Chen CS (1997) Secondary compression of peat with or without surcharging. J Geotech Geoenviron Eng 123:411–421. https://doi.org/10.1061/(ASCE)1090-0241(1997)123:5(411)

Ngu LH, Song JW, Hashim SS, Ong DEL (2019) Lab-scale atmospheric CO_2 absorption for calcium carbonate precipitation in sand. Greenh Gases 9(3):519–528. https://doi.org/10.1002/ghg.1869

Omoregie AI, Senian N, Li PY, Hei NL, Ong DEL, Ginjom IRH, Nissom PM (2016) Ureolytic bacteria isolated from Sarawak limestone caves show high urease enzyme activity comparable to that of *Sporosarcina pasteurii* (DSM33). Malays J Microbiol 12(6):463–470

Omoregie AI, Khoshdelnezamiha G, Senian N, Ong DEL, Nissom PM (2017) Experimental optimisation of various cultural conditions on urease activity for isolated *Sporosarcina pasteurii* strains and evaluation of their biocement potentials. Ecol Eng 109:65–75. https://doi.org/10.1016/j.ecoleng.2017.09.012

Omoregie AI, Ngu LH, Ong DEL, Nissom PM (2019) Low-cost cultivation of *Sporosarcina pasteurii* strain in food-grade yeast extract medium for microbially induced carbonate precipitation (MICP) application. Biocatal Agric Biotechnol 17:247–255. https://doi.org/10.1016/j.bcab.2018.11.030

Omoregie AI, Ong DEL, Nissom PM (2019) Assessing ureolytic bacteria with calcifying abilities isolated from limestone caves for biocalcification. Lett Appl Microbiol 68(2):173–181

Omoregie AI, Palombo EA, Ong DEL, Nissom PM (2019c) Biocementation of sand by *Sporosarcina pasteurii* strain and technical-grade cementation reagents through surface percolation treatment method. Constr Build Mater 228:116828. https://doi.org/10.1016/j.conbuildmat.2019.116828

Omoregie AI, Palombo EA, Ong DEL, Nissom PM (2020) A feasible scale-up production of *Sporosarcina pasteurii* using custom-built stirred tank reactor for in-situ soil biocementation. Biocatal Agric Biotechnol 24:101544. https://doi.org/10.1016/j.bcab.2020.101544

Ong DEL (2008) Benchmarking of FEM technique involving deep excavation, pile-soil interaction and embankment construction. In: Proceedings of the 12th international conference of international association for computer methods and advanced in geomechanics (IACMAG), Goa, India, 1–6 Oct 2008, pp 1–6

Ong DEL (2019) Impact of anisotropy in permeability of peaty soil on deep excavation. In: Geotechnical design and practice. Developments in geotechnical engineering, pp 149–158. https://doi.org/10.1007/978-981-13-0505-4_13

Ong DEL, Choo CS (2011) Sustainable bored pile construction in erratic phyllite. In: ASEAN-Australian engineering congress, Kuching, Malaysia, July 2011, pp 30–45

Ong DEL, Choo CS (2012) Bored pile socket in erratic phyllite of tuang formation. In: GEOMATE 2012, 2nd international conference on geotechnique, construction materials and environment, Kuala Lumpur, Malaysia, pp 167–171

Ong DEL, Choo CS (2018) Assessment of non-linear rock strength parameters for the estimation of pipe-jacking forces. Part 1. Direct shear testing and backanalysis. Eng Geol 244:159–172. https://doi.org/10.1016/j.enggeo.2018.07.013

Ong DEL, Leung CF, Chow YK (2003a) Piles subject to excavation-induced soil movement in clay. In: Proceedings of the 13th European conference on soil mechanics and geotechnical engineering, Prague, Czech Republic, vol 2, pp 777–782

Ong DEL, Leung CF, Chow YK (2003b) Time-dependent pile behaviour due to excavation-induced soil movement in clay. In: Proceedings of the 12th Pan-American conference on soil mechanics and geotechnical engineering, Boston, U.S.A., Massachusetts Institute of Technology, vol 2, pp 2035–2040

Ong DEL, Leung CF, Chow YK (2006a) Pile behavior due to excavation-induced soil movement in clay. II: collapsed wall. J Geotech Geoenviron Eng 132(1):45–53. https://doi.org/10.1061/(ASCE)1090-0241(2006)132:1(36)

Ong DEL, Yang DQ, Phang SK (2006b) Comparison of finite element modelling of a deep excavation using SAGECRISP and PLAXIS. In: Proceedings of the 2006 international conference on deep excavations, pp 50–63

References

Ong DEL, Leung CF, Chow YK, Ng TG (2015) Severe damage of a pile group due to slope failure. J Geotech Geoenviron Eng 141(5):04015014. https://doi.org/10.1061/(ASCE)GT.1943-5606.0001294

Ong DEL, Sim YS, Leung CF (2018) Performance of field and numerical back-analysis of floating stone columns in soft slay considering the influence of dilatancy. Int J Geomech 18(10):1–16. https://doi.org/10.1061/(ASCE)GM.1943-5622.0001261

Ou CY (2006) Deep excavation—theory and practice. CRC Press. ISBN 9780415403306

Peerun MI, Ong DEL, Choo CS (2019) Interpretation of geomaterial behavior during shearing aided by PIV Technology. J Mater Civ Eng 31(9):04019195. https://doi.org/10.1061/(ASCE)MT.1943-5533.0002834

Peerun MI, Ong DEL, Choo CS, Cheng WC (2020) Effect of interparticle behavior on the development of soil arching in soil-structure interaction. Tunn Undergr Space Technol 106:103610. https://doi.org/10.1016/j.tust.2020.103610

Phangkawira F (2019) Characterisation of highly weathered phyllite via in-situ pressuremeter test for the assessment of pipe-jacking forces. PhD thesis, Swinburne University of Technology Sarawak Campus, Malaysia

Phangkawira F, Ong DEL, Choo CS (2017) Numerical prediction of plastic behavior of highly fractured and weathered phyllite subjected to pressuremeter testing. In: Proceedings of the 19th international conference on soil mechanics and geotechnical engineering, Seoul, South Korea, 17–21 Sept 2017

Puller M (2003) Deep excavations: a practical manual, 2nd edn. ICE Publishing, ISBN 9780727734594

Shi X, Meng Y, Li G, Li J, Tao Z, Wei S (2015) Confined compressive strength model of rock for drilling optimization. Petroleum 1:40–45. https://doi.org/10.1016/j.petlm.2015.03.002

Shilaw NJ, Busbridge J, Yi X (1995) Consolidation settlements over tunnels: a review. Can Tunn 253–265

Shirlaw JN (2002) Controlling the risk of excessive settlement during EPB tunnelling. In: Proceeding of IES conference on case studies in geotechnical engineering, Nanyang Technological University, Singapore, 4–5 July 2002

Shirlaw JN, Ong JCW, Rosser HB, Osborne NH, Tan CG, Heslop PJE (2001) Immediate settlements due to tunneling for the North-East line. In: Proceedings of underground, Singapore

Skempton AW, Macdonald DH (1956) The allowable settlement of buildings. Proc Inst Civ Eng 3(3):168–178

Sobhan K, Ali H, Riedy K, Huynh H (2007) Field and laboratory compressibility characteristics of soft organic soils in Florida. In: Geo-Denver, ASCE, New peaks in geotechnics: advances in measurement and modeling of soil behaviour

Yee YW (2006) Caisson and well foundation. In: Huat BBK, Faisal HA, Husaini O, Singh H (eds) Foundation engineering: design and construction in tropical soils. Taylor & Francis, London, pp 129–138

Yee YW, Pan JKL, Guo YY (2001) Construction and design of hand dug caisson foundation in Malaysia. In: Ho KKS, Li KS (eds) Proceedings of the fourteenth southeast asian geotechnical conference. Swets & Zeitlinger, Lisse, pp 1071–1074

Younger JS, Barry AJ, Harianti S, Hardy RP (1997) Construction of roads over soft and peaty ground. In: Proceedings of the conference on recent advances in soft soil engineering, Kuching, Malaysia

Chapter 5
Irregular-Shaped Pipe Jacking Technique

5.1 Introduction

The rectangular or noncircular pipe jacking technology is defined as irregular shape of pipe jacking in the book. It has been developed and practiced worldwide particularly in Japan and China. In 1970, rectangular pipe jacking was developed and successfully applied for the first time in a Tokyo underground passage (Tunnel 2017). In 1995, the first rectangular pipe jacking machine was designed, followed by the trial project in Shanghai, China (Yu et al. 2018). In 2021, the world's first super-large cross-section rectangular hard rock pipe jacking machine was designed and manufactured in China (CREG 2021).

The technology of rectangular pipe jacking and other noncircular pipe jacking is widely applied to the construction of common utility tunnels, pedestrian crossings, and subway entrance/exit passageways. Table 5.1 shows a summary of rectangular pipe jacking projects in some parts of China since 2006. Figure 5.1a and b show the dimensions of the cross-section of a 105 m long tunnel with 10.12 m in width and 7.27 m in height, constructed by rectangular pipe jacking in Henan, China (Li 2017).

Projects No. 6 and No. 7 in Table 5.1 are the two case studies that will be discussed in detail in this chapter. It is to be noted that the dimensions of the cross-sections of the tunnels for Projects No. 8 and No. 9 in the Table 5.1 are 9.8 m in width and 5.0 m in height, 7.2 m in width and 6.5 m in height, respectively, while the maximum dimensions of the cross-section of the tunnels for the remaining projects in Table 5.1 are within 6.9 m in width and 4.9 m in height.

The rectangular and noncircular pipe jacking technology has been improved over the years due to increased market demand. Valuable project experience has been accumulated from industry practice in terms of the manufacturing of pipe jacking machines and the applications of construction method. This chapter will first present the technological development of the jacking machines, followed by further detail study and discussion of the major issues to be considered in the selection of the machine. Two case studies are then presented for further discussion in the last section.

Table 5.1 Summary of rectangular pipe jacking projects (2006–2017) in China

No	Projects	Details*	Jacking schedule
1	Pedestrian at Kaiming St and Medicine St, Ningbo, Zhejiang	2 × 50	Jun 2006–Sept 2006
2	Entrance No. 5 and No. 6 at Qibao Station, Shanghai	2 × 56	Jan 2007–Mar 2007
3	Exit No. 5, No. 7 of Interchange station, Xinjianwan city, Shanghai	1 × 56.5 1 × 43	Aug 2008–Dec 2008
4	Pedestrian at Simon St Memorial of Nanjing Water	1 × 60	Dec 2009–Jan 2010
5	Link between two existing buildings at Zhongshan hospital, Shanghai	1 × 78.5	Sept 2010–Feb 2011
6	Foshan, underground space	4 × 60.5	Dec 2011–May 2012
7	Baotou pedestrian	1 × 65	Nov 2012–April 2013
8	Exit for MRT project at Wuhan	1 × 70	Mar 2017–July 2017
9	Pedestrian link to airport and MRT at Guangzhou Baiyun Airport	1 × 101	July 2017–Oct 2017

*Indicates the number of tunnels × jacking length (m)

Fig. 5.1 **a** Breakthrough of the cutterheads; **b** Internal view of the tunnel during jacking (Yu et al. 2018)

5.2 Types of Irregular-Shaped Pipe Jacking Machines

Irregular-shaped pipe jacking machines mean that the jacking pipes are noncircular compared with conventional pipe jacking. The most common types of irregular-shaped pipe jacking in the industry are square, rectangular, horseshoe-shaped pipe jacking. The common practice for irregular-shaped pipe jacking is to adjust the conventional jacking machines to fit the irregular shape of the pipes. The stability of the pipe sections needs to be addressed as the jacking process is dynamic. Therefore, most irregular-shaped pipes adopt either rectangular, square, or arch door shapes. Rectangular pipes are the most basic and widely used shape, which will be the main focus in this chapter.

5.2 Types of Irregular-Shaped Pipe Jacking Machines

Three types of rectangular pipe jacking machines will be discussed, including rectangular jacking machines with combined multi-cutterheads, eccentric multi-axis pipe jacking machines, and square planetary gear-type jacking machines (Yu et al. 2018).

5.2.1 Rectangular Jacking Machines with Combined Multi-cutterheads

5.2.1.1 Rectangular Pipe Jacking Machines with Different Sizes of Cutterheads

The main structure of the combined cutterheads with large and small cutters rotated by motor(s) for a rectangular pipe jacking machine is shown in Fig. 5.2. As the path for rotation and excavation of a cutterhead is circular, while the cross-section of the pipe jacking machine is rectangular, the cutterheads cannot cover 100% of the rectangular cross-section of the pipe jacking machine during excavation. However, to maximize the area of excavation and mixing, the present invention adopted combined cutterheads. A large cutterhead with an excavation diameter slightly larger than the height of the pipe jacking machine is set up in the center of the pipe jacking machine. A

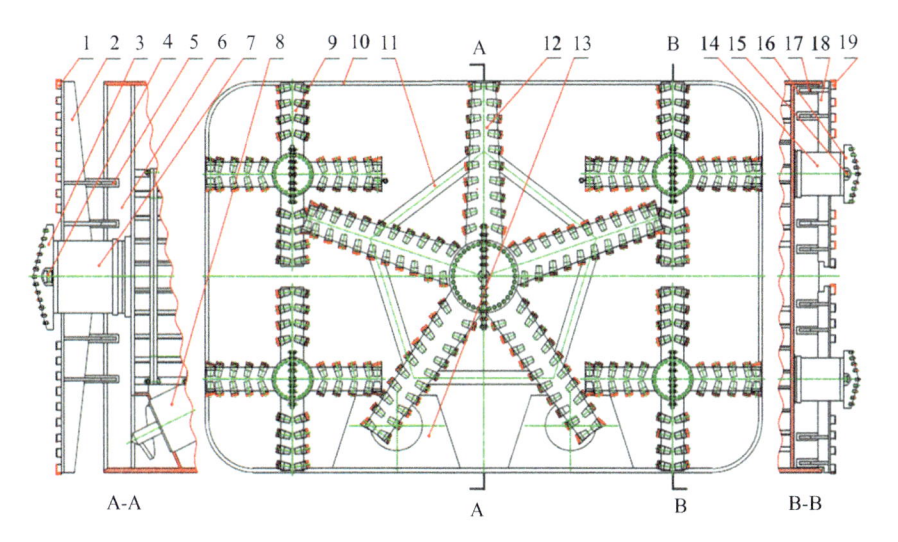

Fig. 5.2 Combined large and small cutterheads of a rectangular pipe jacking machine. 1—Large cutter; 2—Stiffener for the large cutterhead; 3—Center cutter; 4—Large grouting hole; 5—Large mixing rod; 6—Excavation chamber; 7—Base of large cutterhead; 8—Screw conveyor; 9—Panel of small cutterhead; 10—Shell; 11—Link; 12—Panel of large cutterhead; 13—Inlet; 14—Base of small cutterhead; 15—Small grouting hole; 16—Small cutter; 17—Small mixing rod; 18—Panel of small cutterhead; 19—Small cutter

small cutterhead is also designed to be at each of the four corners of the pipe jacking machine. The large cutterhead extends forward a little bit, while the small corner cutterheads are indented back a little bit. The excavation ranges of the large and small cutterheads have a certain amount of overlap to reduce the cutter blind areas. With such arrangements, the excavation area of the five cutterheads can reach more than 90% of the rectangular cross-section of the pipe jacking machine, resulting in a very limited blind area. The major blind areas are located in the middle of both sides of the rectangular pipe jacking machine and are fitted with some scraping devices. Shovel teeth are also designed on the edge of the cutterheads at blind areas. In practical engineering projects, the working and receiving shafts are strengthened with various forms of reinforcement, such as jet grout piles and mixing pipes, with compressive strength ranging from 3 to 8 MPa. Under these conditions, the rectangular pipe jacking machines with multi-cutterheads satisfy the strength and stability requirements.

Figure 5.3 is the rear view of the combined cutterheads of a rectangular jacking machine. The large cutterhead consists of large cutters, stiffener and link, panel, and base, while the small cutterheads are composed of small cutters, panel, stiffener, and base. The large cutterhead uses multiple motors to form its driving device through the planetary reducer and the gearbox. Similarly, each of the four small cutterheads employs two electric motors to form its driving device through a planetary reducer and gearboxes. There are two personnel accesses on the left and right sides of the driving device of the large cutterhead. In the event of a large obstacle, appropriate soil reinforcement measures must be applied, and the personnel accesses are opened to remove the obstacle under the conditions that no surge of soil will occur.

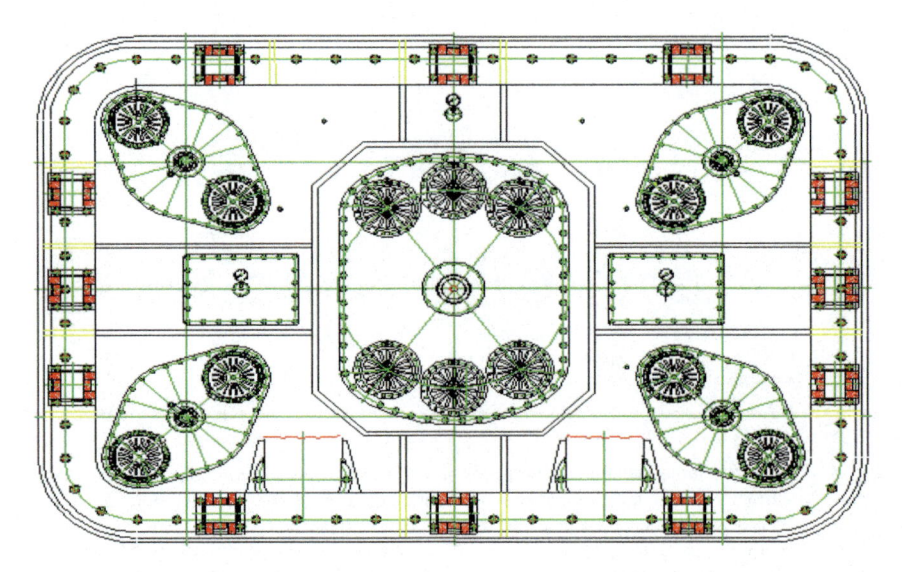

Fig. 5.3 Rear view of the combined cutterheads of the rectangular pipe jacking machine

5.2 Types of Irregular-Shaped Pipe Jacking Machines 151

Fig. 5.4 Control panel of the rectangular pipe jacking machine with combined large and small cutterheads

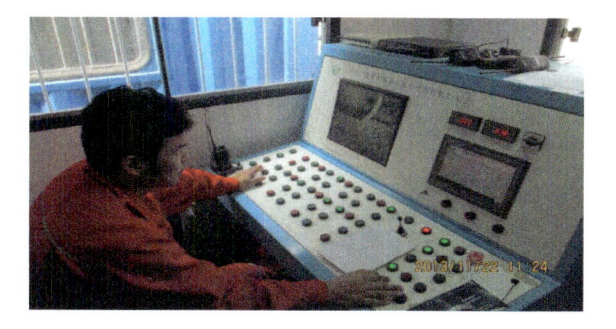

The rectangular pipe jacking machine with combined cutterheads works based on the earth pressure balancing principle. The large cutterhead and four small cutterheads work together to excavate the soil and mix it up at the same time, which makes the excavated soil having good plasticity, mobility, and a certain impermeability. When the soil in the excavation chamber has a large sand content, grouting can be conducted by injecting a certain proportion of mud into the excavation chamber through the large and small grouting holes in the large and small cutterheads, respectively, to improve its impermeability.

In addition, two screw conveyors are installed on the compartment board on the left and right sides of the driving device of the large cutterhead. The front and rear shells are connected using 12 cylinders for deviation rectification.

Figure 5.4 shows the control panel for the rectangular pipe jacking machine with combined large and small cutterheads. It includes a 17-inch monitor to display three different views at the same time, with a relatively large one as the main view and two other relatively small views. The main view and two small views can be switched as necessary. The three views come from three cameras mounted in the pipe jacking machine, and the cameras can choose their capture ranges according to the operational requirements. These views usually display the instrument panel and the discharge state of two screw conveyors.

5.2.1.2 Assembled Rectangular Pipe Jacking Machine with Multi-cutterheads

Traditional rectangular pipe jacking machines are single pieces. This single piece works fine when the size is relatively small. When the rectangular pipe jacking machine is large, the shell of the rectangular pipe jacking machine with multi-cutterheads (Fig. 5.5) is assembled with the upper and lower parts to address the problem of lifting and transportation during construction. From Fig. 5.5, it can be seen that the cutterheads of the rectangular pipe jacking machine consist of three front cutterheads and three rear cutterheads.

The shell of the assembled rectangular pipe jacking machine with multi-cutterheads (Fig. 5.6) is composed of front upper and lower half-shells and rear

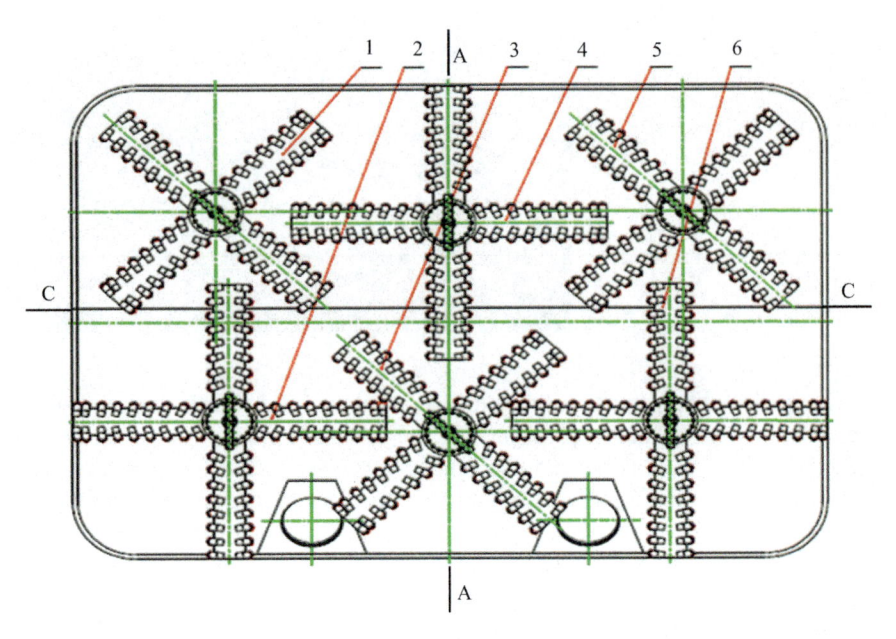

Fig. 5.5 Arrangement of the multi-cutterheads. 1, 3, and 5 are the three front cutterheads; 2, 4, and 6 are the three rear cutterheads

Fig. 5.6 Shell of the machine. 7—Front upper part; 8—Front lower part; 9—Rear upper part; 10—Rear lower part

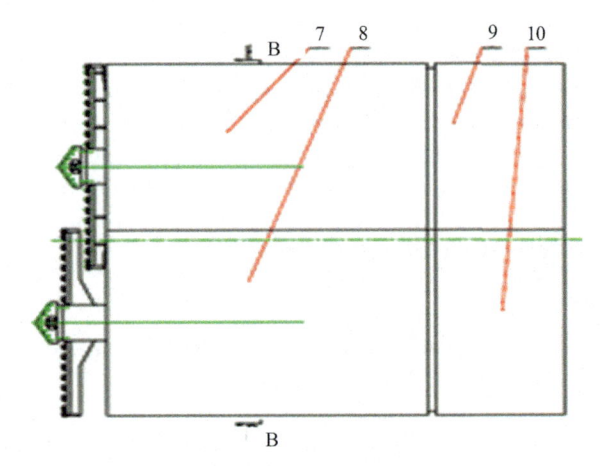

upper and lower half-shells. The bottom cutterhead shown in Fig. 5.6 is one of the so-called front-stretched cutterheads, which is at a certain distance between the upper and lower cutterheads in the axial direction to avoid interference with each other.

As shown in Fig. 5.7, bolts are used to connect the upper and lower half-shells in the front and rear, while pins are applied to prevent the displacement between upper and lower parts produced during jacking. Seal ring is placed between the front and

5.2 Types of Irregular-Shaped Pipe Jacking Machines

Fig. 5.7 Shell structure of the machine (Section A-A). 11—Seal ring; 12—Shell connection bolts; 13—Pin; 14—Cylinder for deviation rectification; 15—Motor; 16—Planetary decelerator; 17—Spindle gearbox; 18—Mixing rod; 19—Small cutter; 20—Stiffeners; 21—Bits; 22—Shell seal

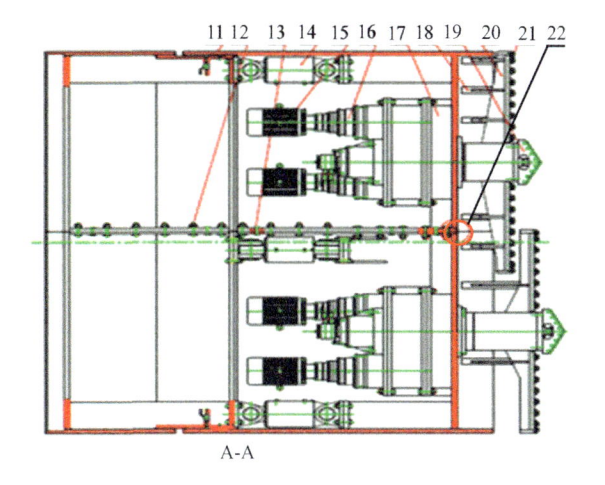

rear shells to prevent mud infiltration outside the pipe jacking machine, while the cylinder is used for deviation rectification.

The cutterhead is driven by a number of motors which may range from 1 to 8, depending on the size of the cutterhead, through the planetary reducer and the spindle gearbox. The cutterhead consists of a mixing rod, a center cutter, stiffeners, and auxiliary cutters.

Figure 5.8 shows a close-up view of the seal between the upper and lower half-shells. As shown in the rear view of the pipe jacking machine in Fig. 5.9, each cutterhead is driven by three electric motors, and the muck will be discharged after

Fig. 5.8 Close-up view of the shell seal in Fig. 5.7

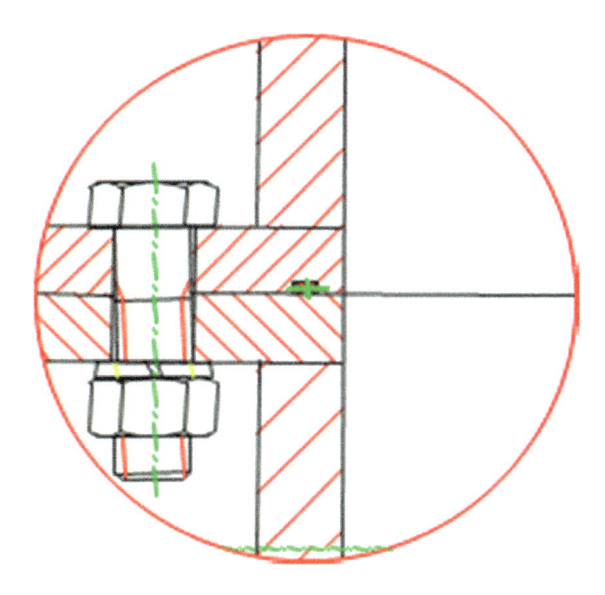

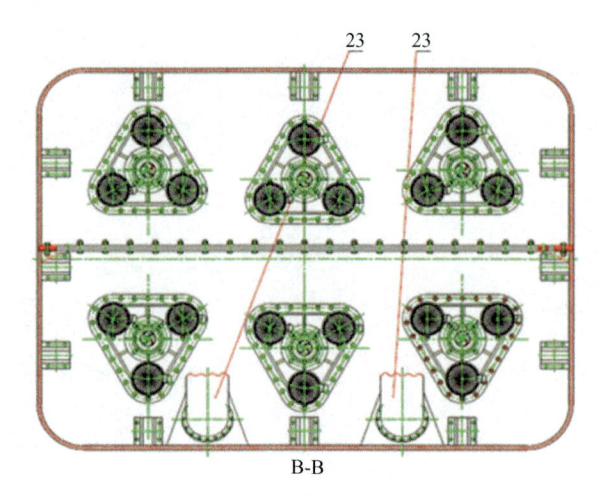

Fig. 5.9 Rear view of the pipe jacking machine (Section B-B). 23—Screw conveyors

being excavated and mixed by the cutterheads via the two screw conveyors located in the lower part of the pipe jacking machine.

Figure 5.10 shows the interface between the upper and lower half-shells and the seals. The interface between the upper and lower half-shells is a straight line

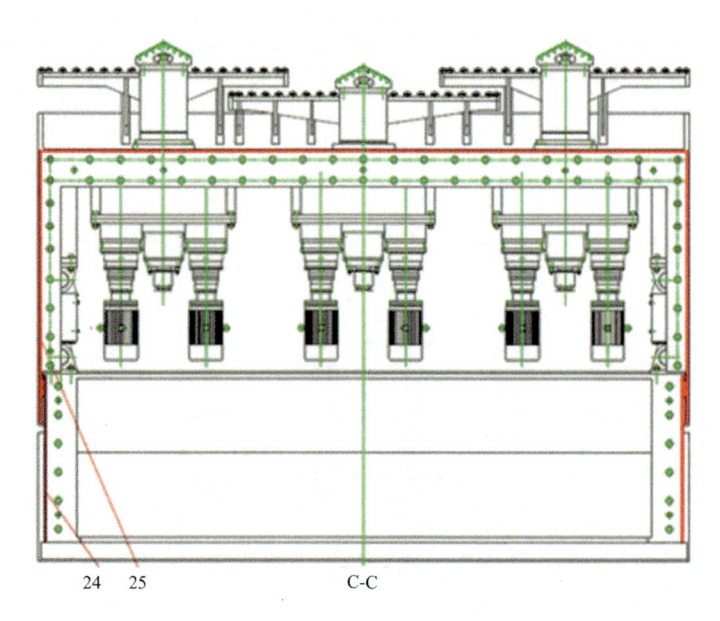

Fig. 5.10 Shell seal (Section C-C). 24—Interface between the upper and lower half-shells; 25—Seals

5.2 Types of Irregular-Shaped Pipe Jacking Machines

Fig. 5.11 Configuration of the cutterheads

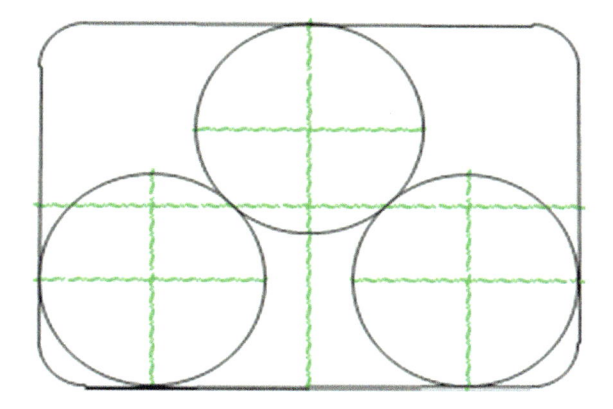

and symmetrical at the center axis of the pipe jacking machine. The seals seal the compartment board and the left and right area of the shell.

For arrangement of the cutterheads, regardless of the ratio of the height to the width of the cross-section of the rectangular pipe jacking machine as shown in Fig. 5.11, three circles of the same size are drawn within the rectangular shape, with two circles at the bottom and the other one on the top. The top circle is tangent to the edge of top rectangular shell. The bottom left circle on both sides is tangent to the left and bottom edges of the rectangular shell, respectively. The bottom right circle is tangent to the right and bottom edges of the rectangular shell, respectively. Then, the diameters of the three circles are rounded down to integers and reduced appropriately so that they do not interfere with one another. Finally, the top, bottom, left, and right ends of the adjusted circles are made tangent to the four edges of the rectangular shell, as described earlier. In this way, the adjusted circles become the three rear-retracted cutterheads.

The arrangement of the other three front cutterheads as shown in Fig. 5.5 is similar to the three rear cutterheads. Meanwhile, the blind areas shaded in red in Fig. 5.12

Fig. 5.12 Blind area distribution diagram

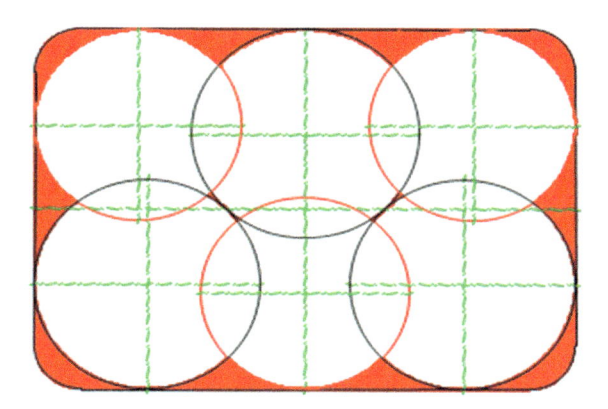

should be minimized, and the interference among the various cutterheads should be avoided.

5.2.2 Eccentric Multi-axis Pipe Jacking Machines

5.2.2.1 Principle and Structure of Eccentric Multi-axis Cutter Drive

The principle and structure of the eccentric multi-axis pipe jacking machine cutterheads are shown in Fig. 5.13. On the eccentric block connected with each of the four rotating axes, there is an eccentric axis in the same direction and with the same eccentricity. Each of the four corners of the rectangular cutterhead frame is connected to a corresponding eccentric axis. Thus, when the four rotating axes rotate in the same direction, the excavating trajectory of the cutterhead is a rectangle with two eccentric margins larger than the width and height of the cutterhead frame, and the radii of the arcs at the four corners equal to the eccentricity.

Sometimes, the eccentric multi-axis cutterhead is only used as an auxiliary cutterhead, and under this condition, it only needs to be driven by a single electric motor. Figure 5.14 shows a single motor drive mechanism for an auxiliary eccentric multi-axis cutterhead. The motor is decelerated by the planetary reducer to drive a driving gear in gearbox to rotate. This driving gear at the same time drives the rotation of the other three driven gears in the gearbox, which also drives the eccentric axis located on the three driven gears to rotate together, thus driving the rotation of the cutterhead frame.

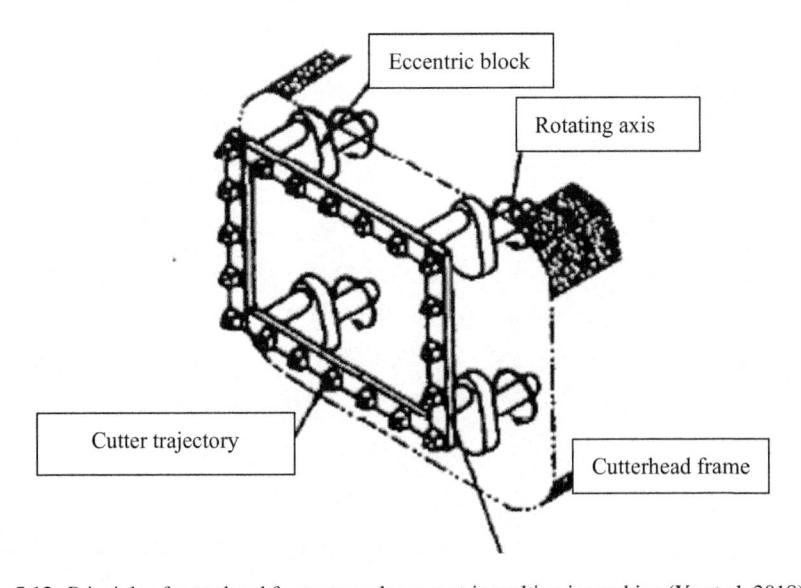

Fig. 5.13 Principle of cutterhead for rectangular eccentric multi-axis machine (Yu et al. 2018)

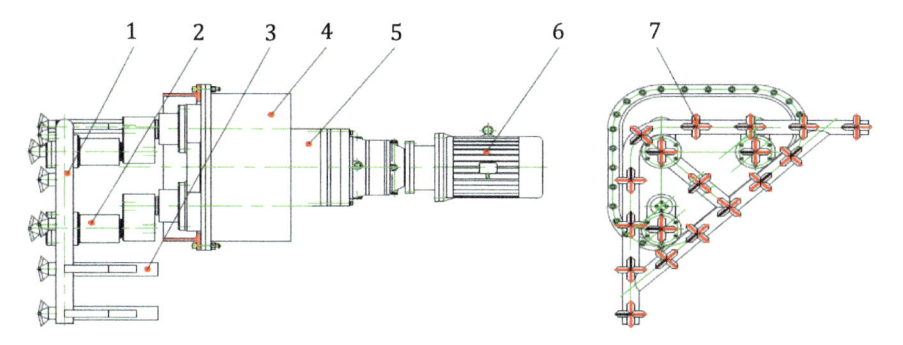

5.2 Types of Irregular-Shaped Pipe Jacking Machines

Fig. 5.14 Single motor drive of eccentric multi-axis cutterheads. 1—Frame of the cutterhead; 2—Eccentric axis; 3—Mixing rod; 4—Gearbox; 5—Planetary reducer; 6—Motor; 7—Cutter

There are four driving axes in the eccentric multi-axis cutterhead as shown in Fig. 5.13. However, the number of the driving axes can also be two with a diagonal arrangement, and the other two diagonal eccentric axes can be driven axes. Each cutterhead must have at least three eccentric axis supports as shown in Fig. 5.14.

5.2.2.2 Design and Calculation of Eccentric Multi-axis Cutter Drive

As mentioned earlier, the reaction force of the eccentric multi-axis cutterhead has a considerable effect on the stability of the shell of the pipe jacking machine. This is particularly evident when the cutterhead is pressured downward. The shell of the machine will be lifted, and the amount of lifting is proportional to the amount of over-excavation by the cutterhead. A larger amount of over-excavation will cause the loose soil around the shell to form a large space, and the range of tolerance of the shell will also be large. Therefore, the amount of over-excavation should be minimized during the machine design. In addition, it is beneficial if the amount of extension of the cutterheads out of the shell can be reduced appropriately during the design.

Meanwhile, as the eccentricity is generally limited, the mixing rod on the soil is not sufficient. The eccentricity is determined based on soil conditions. If the soil is soft clay, the requirement for mixing is not high. A small eccentricity can be used, and the driving power can also be small. If it is sandy soil, there will be a high requirement for mixing. Therefore, a large eccentricity should be used, and the driving power and the strength of the components should be large. However, the maximum eccentricity should not exceed 400 mm.

When designing the rotational speed of the cutterhead, the linear speed of the cutterhead motion should be controlled at about 5 m/min. With this requirement and the eccentricity, the speed of the cutterhead can be determined.

In designing the eccentric multi-axis cutterhead, a coefficient β is used to determine the torque of the cutterhead. The torque T (kNm) of the eccentric multi-axis cutterhead is formulated as follows (Yu et al. 2018):

$$T = \beta \cdot r \cdot D^2 \tag{5.1}$$

$$D = \sqrt{\frac{4A}{\pi}} \tag{5.2}$$

where β is the coefficient of torque (kN/m²), r is the eccentricity (m), and D is the diameter of the equivalent circle (m). The coefficient β in Eq. 5.1 is normally between 45 and 100. It is small for soft soil condition and large for hard soil condition. A is the cross-sectional area of the excavating of the pipe jacking machine (m²).

5.2.2.3 Characteristics of the Eccentric Multi-axis Pipe Jacking Machine and Its Suitable Soil Conditions

As the cutterhead of the eccentric multi-axis pipe jacking machine has a very large opening rate and cross-section, it is very difficult to balance the soil pressure and the ground water pressure on the excavation face even though the grouting with a high mud-to-water ratio is used. Therefore, the earth pressure balance principle as shown in Fig. 5.15 has been widely adopted for the rectangular jacking machine.

The characteristics of the eccentric multi-axis pipe jacking machine are described as follows:

- The most important feature of the eccentric multi-axis pipe jacking machine is the flexibility in the types of the cross-section, which can be circular, oval, rectangular, horseshoe-shaped, arch-shaped, and other arbitrary shapes

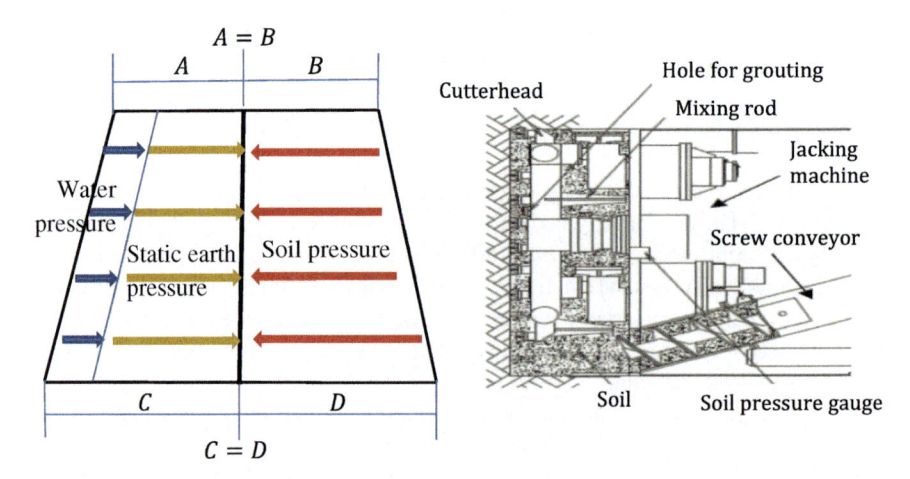

Fig. 5.15 Schematic of an eccentric multi-axle pipe jacking machine based on the soil pressure balance principle (Yu et al. 2018)

5.2 Types of Irregular-Shaped Pipe Jacking Machines

Table 5.2 Soil conditions for eccentric multi-axis pipe jacking machines

No	Soil conditions	N values	Suitability
1	Clay	0	Relatively applicable
2	Silty clay	0–2	Applicable
3	Sandy silty clay	0–5	Applicable
4	Sandy silty clay	5–10	Applicable
5	Silt	10–20	Applicable
6	Sandy silt	15–25	Applicable
7	Sandy silt	Greater than 25	Applicable
8	Mudstone	Greater than 50	Relatively applicable
9	Sandy silty clay	10–15	Applicable
10	Loose sandy soil	10–30	Applicable
11	Dense sandy soil	Greater than 30	Applicable
12	Gravel	10–40	Applicable
13	Consolidated gravel	Greater than 40	Applicable
14	Sand and cobbles	0–2	Applicable
15	Cobbles	0–2	Relatively applicable

- Eccentric multi-axis pipe jacking machines can be applied to a wide range of soil conditions from soft to hard soil, from sand to gravel
- The structures of the eccentric multi-axis pipe jacking machines are very simple
- As the rotational radius of the cutterheads of the eccentric multi-axis pipe jacking machine is small, its driving power is also small. The cutters are not easy to wear, suitable for long-distance jacking
- It is easy to carry out soil improvement if needed as the soil improvement mud can be injected from the front of the cutterhead to increase the efficiency
- It is easy to carry out maintenance and disassembly as the parts are not big

The applicable soil conditions for eccentric multi-axis pipe jacking machines are summarized in Table 5.2. The N values are determined from the standard penetration tests (e.g., ASTM D1586/D1586M 2018).

5.2.2.4 Eccentric Multi-axis Rectangular Pipe Jacking Machine

Figure 5.16 shows the schematic of an eccentric multi-axis rectangular pipe jacking machine. It is powered by four motors to rotate the driving axis around the rotary center. The two ends of the eccentric block are connected to the driving axis and the eccentric axis. The eccentric axis is connected to the cutterhead frame through the

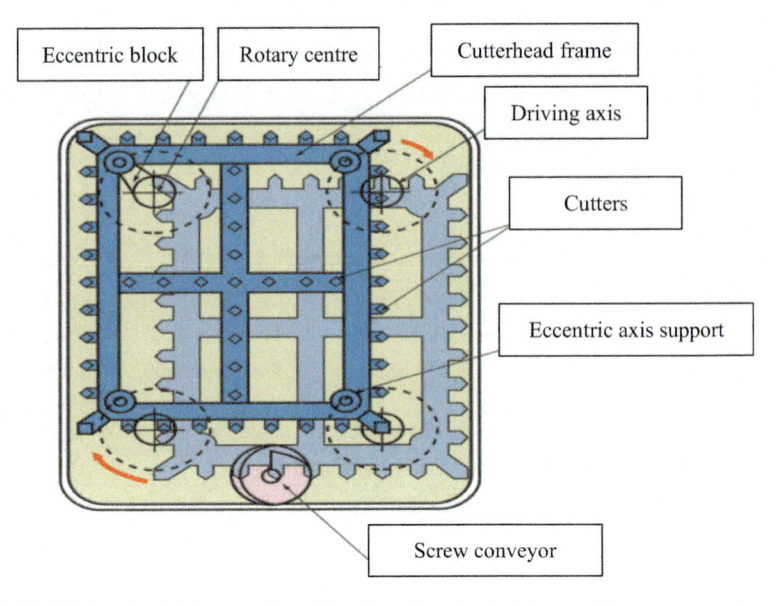

Fig. 5.16 Schematic of the eccentric multi-axis rectangular jacking machine

eccentric axis support. The eccentric axis support has a bearing inside. As soon as the driving axis rotates, the four edges of the entire cutterheads are then turned against the edges of the jacking shell. At the same time, the cutter is excavating, while the soil is mixed with a mixing rod located behind the cutterhead.

5.2.2.5 Composite Cutterheads

The eccentric multi-axis cutterhead can cut out any shape but has the weakness of insufficient mixing. It is, therefore, combined with circular cutterheads for designing pipe jacking machines with a variety of cross-sections. In this way, these cutterheads can operate well together by maximizing their respective strengths and minimizing their respective weaknesses.

Certain types of designs will be introduced in this sub-section. It should be noted that the following three types have already been granted national patents and have been authorized in China. The authors wish to draw some inspirations from the introduction of these designs.

(1) Combined cutterheads of a square pipe jacking machine

  A square pipe jacking machine (Patent No. 20102935.3) with combined cutterheads will be used as an example to describe the main structure and principle as shown in Fig. 5.17. The shell of the machine is divided into three parts by the compartment board, namely the excavation chamber in the front and the power chamber at the rear.

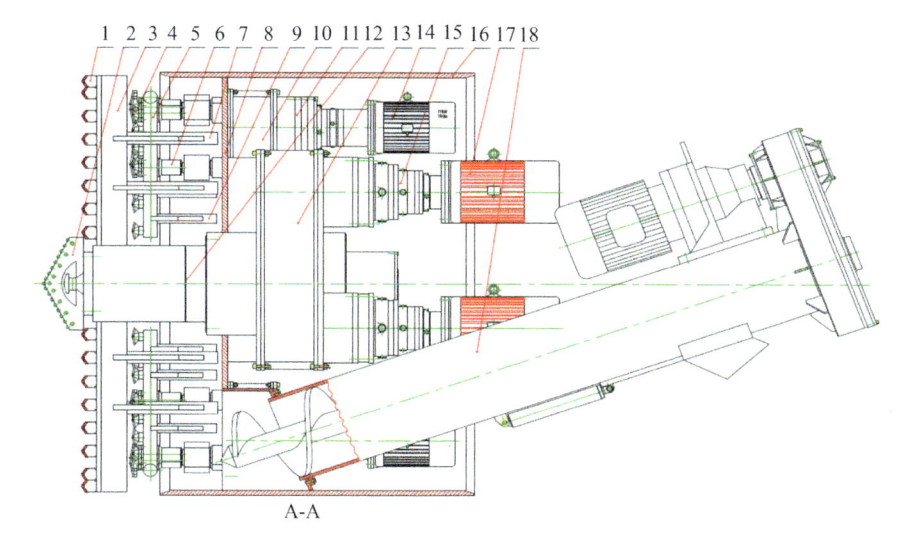

5.2 Types of Irregular-Shaped Pipe Jacking Machines

Fig. 5.17 Main structure of the square pipe jacking machine. 1—Larger cutter; 2—Center cutter; 3—Stiffener of large cutterhead; 4—Small cutter; 5—Panel of small cutterhead; 6—Eccentric axis; 7, 9—Mixing rod; 8—Compartment board; 10—Power device of small cutterhead; 11—Planetary deceleration of small cutterhead; 12—Main axis of large cutterhead; 13—Power device of large cutterhead; 14—Driving motor of small cutterhead; 15—Planetary deceleration of large cutterhead; 16—Machine shell; 17—Driving motor of large cutterhead; 18—Screw conveyer

Figure 5.18 shows that a large cutterhead is located in the center in the excavation chamber of the square pipe jacking machine. When the large cutterhead works, the four corners of the square cross-section are not covered. To make a full-section

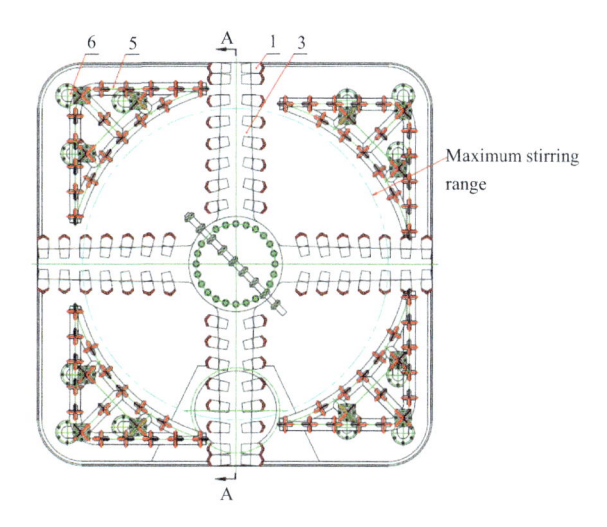

Fig. 5.18 Combined cutterheads in a square jacking machine. 1—Larger cutter; 3—Panel of large cutterhead; 5—Panel of small cutterhead; 6—Eccentric axis

excavation, four small cutterheads are placed at the four corners of the machine to form combined cutterheads together with the center cutterhead.

The general circular excavation by small cutterheads at the four corners still cannot reach 100% full-section coverage due to the near triangular areas (Fig. 5.18). Therefore, the small cutterheads are designed to be driven by an eccentric multi-axis drive and have a triangular shape with dimensions smaller than those of the near triangles at the four corners by one eccentricity. In this way, the excavating trajectory of the cutterhead is the same as the near triangular shapes at the four corners of the rectangular section, thus achieving 100% excavation of the square section.

Large and small cutterheads are equipped with mixing rods. Note that, the large and small cutterheads are overlapping in the axial direction, with the large cutterhead stretched to the front and the small cutterhead retracted to the back.

For the rectangular pipe jacking machine shown in Fig. 5.19, at each of the four corners, two of the three eccentric axes of the cutterhead can be driven by two electric motors, respectively. These two eccentric axes can be named as active eccentric axes. The remaining eccentric axis will not have a motor. It, therefore, has no power itself but is driven by the other two motors. As a result, it is termed as a passive eccentric axis.

Figure 5.20 shows the combined cutterheads with a size of 2.2 × 2.2 m of a square pipe jacking machine. The pipe jacking machine is displayed in Fig. 5.20a. Figure 5.20b shows the top view from the working shaft during the jacking operation, while Fig. 5.20c presents the inside view of the completed lining rings.

(2) Combined cutterheads of a rectangular jacking machine

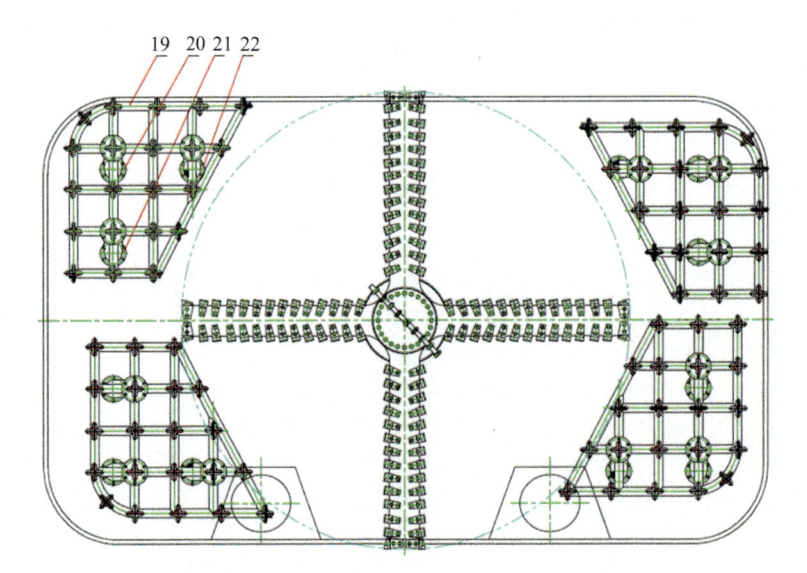

Fig. 5.19 Combined cutterheads for a rectangular jacking machine. 19—Panel of small cutterhead; 20–21—Eccentric axis; 22—Eccentric axis

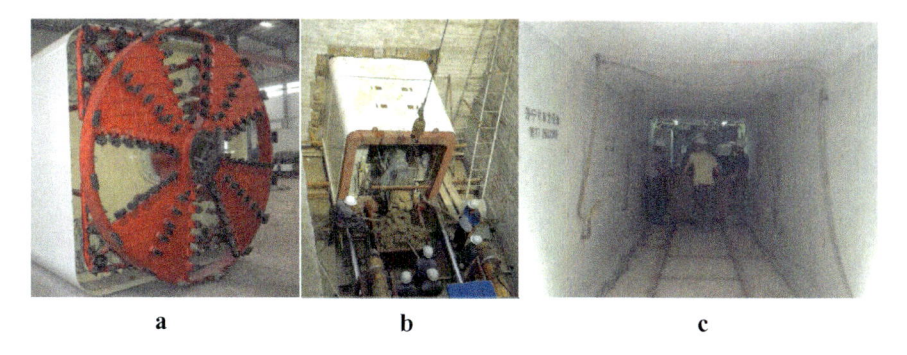

5.2 Types of Irregular-Shaped Pipe Jacking Machines

Fig. 5.20 Construction by a square jacking machine with square cutterheads (Yu et al. 2018)

Figure 5.21 shows a combination of two large cutterheads and six eccentric multi-axis small cutterheads (Patent No. 201120021923.7). This design aims to solve the problem of the full-section excavation for the large rectangular cross-section with a large ratio of length to width (2:1). The two large cutterheads can be driven by electric or petrol motors.

Figure 5.22 shows the combined cutterheads for a rectangular section. The center distance of the two large cutterhead is relatively small when the ratio of length to width of the rectangular section is 1.8:1.

Figure 5.23 shows the combined cutterheads for a rectangular section with a ratio of length to width is 1.7:1. The center distance of the two large cutterheads is even smaller, and the shape of the large cutterhead is of the 'X' shape rather than the '+' shape.

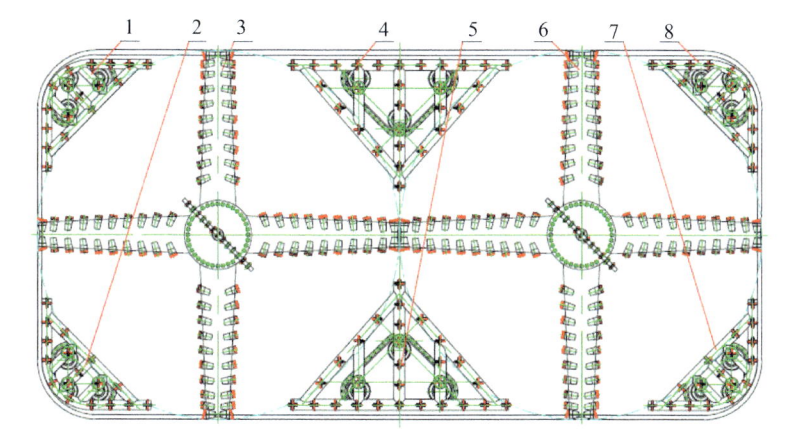

Fig. 5.21 Type I rectangular combined cutterheads. 1—Small cutterhead on the top left; 2—Small cutterhead on the bottom left; 3—Large cutterhead on the left; 4—Small cutterhead at top middle; 5—Small cutterhead at the bottom middle; 6—Large cutterhead on the right; 7—Small cutterhead on the bottom right; 8—Small cutterhead on the top right

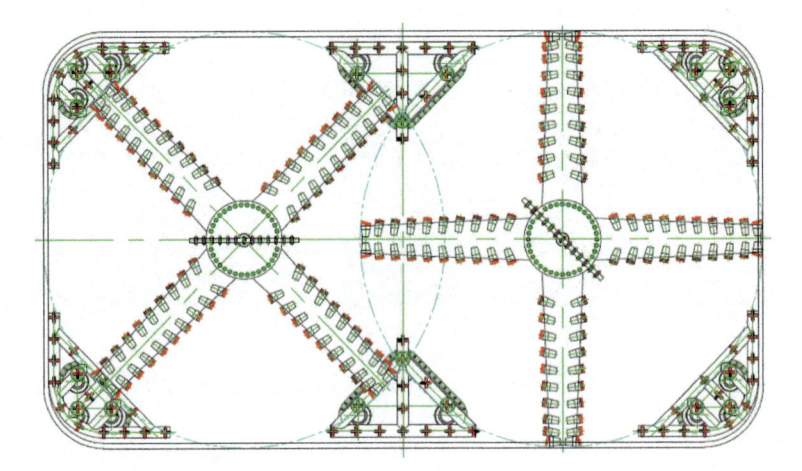

Fig. 5.22 Type II rectangular combined cutterheads

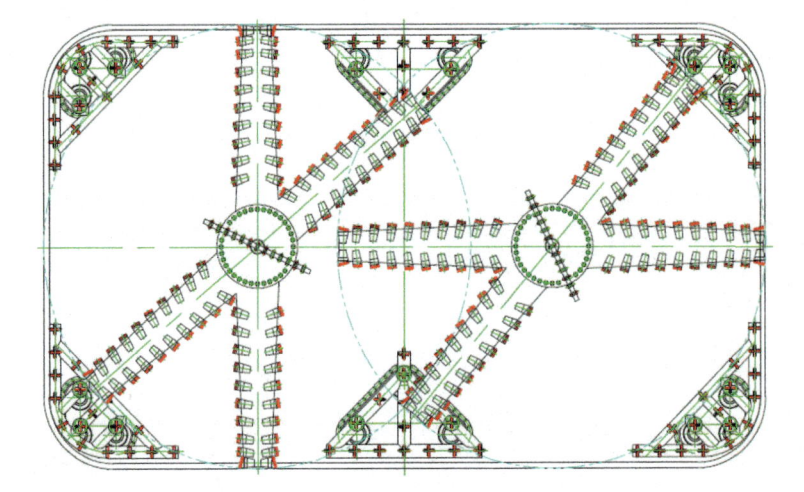

Fig. 5.23 Type III rectangular combined cutterheads

(3) Combined eccentric multi-axis and circular cutterheads for pipe jacking machine

The cutterheads shown in Fig. 5.24 consist of an eccentric multi-axis cutterhead and a circular cutterhead (Patent No. 201320305076.6) with an eccentricity e, which is the distance between the center of the eccentric multi-axis cutterhead and the center of the circular cutterhead. This eccentric multi-axis cutterhead is a single piece.

The structure of the eccentric multi-axis cutterhead is shown in Fig. 5.25. The frame of the eccentric multi-axis cutterheads is mainly composed of the outer frame,

5.2 Types of Irregular-Shaped Pipe Jacking Machines

Fig. 5.24 Eccentric multi-axis and circular combined cutterheads. 1—Eccentric multi-axis cutterhead; 2—Eccentricity *e*; 3—Circular cutterhead; 4—Trajectory of the eccentric multi-axis cutterhead

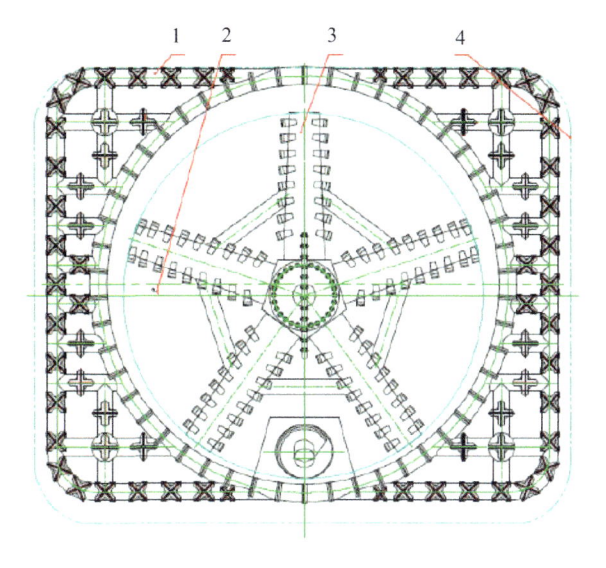

Fig. 5.25 Eccentric multi-axis cutterheads in Fig. 5.24. 5—Outer frame; 6—Eccentric axis sleeve; 7—Cross cutter; 8—Circular inner frame; 9—Straight cutter; 10—Rod

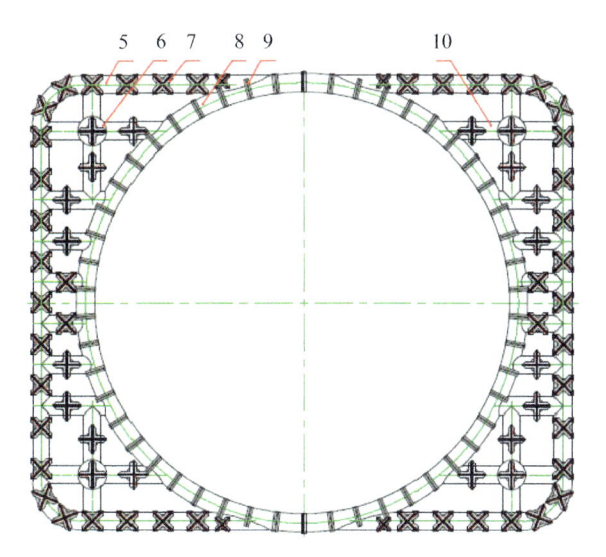

a circular inner frame, and a rod that binds them. An eccentric sleeve is provided at each of the four corners between the inner and outer frames. A cross-cutter and a straight cutter are provided at the front of the frame.

Figure 5.26 is the rear view of the jacking machine with combined eccentric multi-axis and circular cutterheads. The driving units of the eccentric multi-axis cutterheads are located at the four corners, while those of the circular cutterhead are at the center.

There are also some other examples of combined eccentric multi-axis and circular cutterheads, which are summarized in Table 5.3.

Fig. 5.26 Rear view of the jacking machine with combined eccentric multi-axis and circular cutterheads. 13—Driving unit for the eccentric multi-axis cutterhead; 14—Shell; 22—Power box for the circular cutterhead; 24—Driving motor for the circular cutterhead

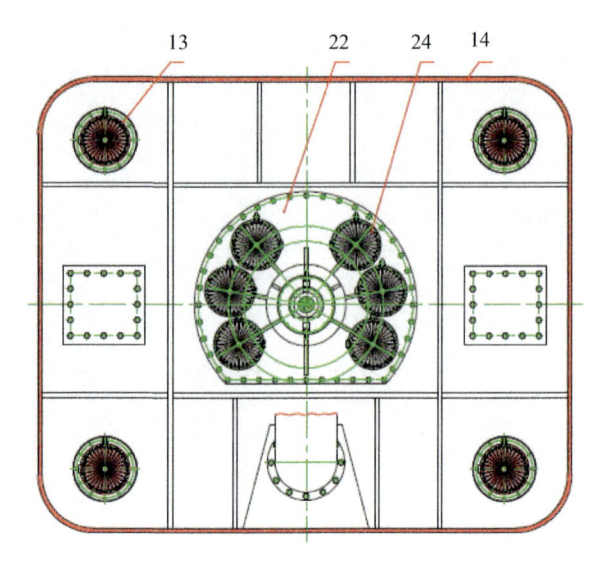

5.2.3 Square Planetary Gear-Type Jacking Machine

The square planetary gear jacking machine was developed and manufactured by Yangzhou Guangxin Machinery Co., Ltd and Shanghai Tunnel Engineering Co., Ltd. Compared with multi-cutterhead rectangular jacking machines, it is capable of conducting full cross-section excavation in a square shape without any blind areas. Compared with eccentric multi-axis jacking machines, it is capable of mixing muck in all directions. The square planetary gear jacking machine is a new rectangular jacking machine which has both the advantages and at the same time overcomes the weaknesses of multi-cutterhead rectangular jacking machines and eccentric multi-axis jacking machines. It has the simplest structure among all the planetary gear-type rectangular jacking machines. The square planetary gear jacking machine has already been granted national patents in China with the patent numbers of 201120338167.0 and 201220404306.X.

5.2.3.1 Working Principle of the Cutterheads of the Planetary Gear-Type Square Jacking Machine

The meshing transmission of any pair of gears can be considered as pure rolling of two circles tangent to each other, which are the pitch circles of the two gears. Figure 5.27 shows the meshing transmission of a pair of internal gears. When the planetary gear O_1 moves along a larger inner gear O, it not only revolves about the internal gear, but also rotates itself.

The radii of the pitch circles of the internal gear and the planetary gear are R and r, respectively. A is the eccentric point, which is fixed to the planetary gear. It can

5.2 Types of Irregular-Shaped Pipe Jacking Machines

Table 5.3 Some other examples of combined eccentric multi-axis and circular cutterheads

No	Example shapes	Types
1		Arch door type 1
2		Rectangular
3		Arch door type 2

also be extended to the outside of the pitch circle of the internal gear as the tip of the outer cutter of the cutterheads. The eccentricity $e = O_1 A$.

Based on the principle of kinematics, when the planetary gear is doing the above pure rolling, the coordinates of point A are expressed as follows:

$$\begin{cases} x_A = (R - r)\cos\alpha + e\cos\beta \\ y_A = (R - r)\sin\alpha - e\sin\beta \end{cases} \quad (5.3)$$

The arc length of the pitch circle O equals to that of the pitch circle O_1:

$$R\alpha = r(\alpha + \beta) \quad (5.4)$$

Fig. 5.27 Working principle of the cutterheads of the planetary gear-type square jacking machine

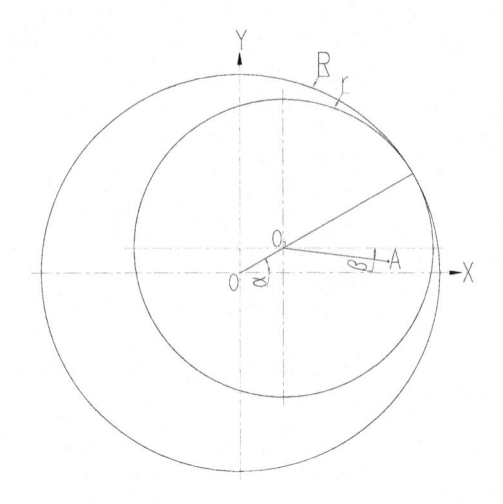

By re-arranging Eq. 5.4, we get.

$$\beta = \frac{R-r}{r}\alpha \tag{5.5}$$

Substituting Eq. 5.5 into Eq. 5.3, the coordinates of point A can be written as follows:

$$\begin{cases} x_A = (R-r)\cos\alpha + e\cos(\frac{R-r}{r}\alpha) \\ y_A = (R-r)\sin\alpha - e\sin(\frac{R-r}{r}\alpha) \end{cases} \tag{5.6}$$

Let $X = x_A/r$, $Y = y_A/r$, $\xi = R/r$, and $\eta = e/r$, then Eq. 5.6 becomes:

$$\begin{cases} X = (\xi-1)\cos\alpha + \eta\cos[(\xi-1)\alpha] \\ Y = (\xi-1)\sin\alpha - \eta\sin[(\xi-1)\alpha] \end{cases} \tag{5.7}$$

To make the trajectory of point A quadrilateral, ξ must take a value of 4 or 4/3. However, when ξ is taken as 4, the ratio of the internal gear to the planetary gear is too large, which is not allowed by the structure. For example, when we determined the size of the shell of the pipe jacking machine according to the geometry of the internal gear, it was found that the sizes of the planetary gear and the spindle of the cutterheads fixed to the planetary gear were too small to meet the requirements. If ξ is taken as 4/3, there is no problem. The η value depends on the following two factors:

- The structure. If the value of the η is too small, point A will move to the inside of the planetary gear O_1 or the internal gear O, which is not allowed by the structure
- The line type. If the η value is small, the four edges of the trajectory of point A will become concave. The smaller the η value is, the more concave the four

5.2 Types of Irregular-Shaped Pipe Jacking Machines

Fig. 5.28 The trajectory of point A when $\xi = 4/3$.
1—$\eta = 1.5$; 2—$\eta = 2.5$;
3—$\eta = 3.5$

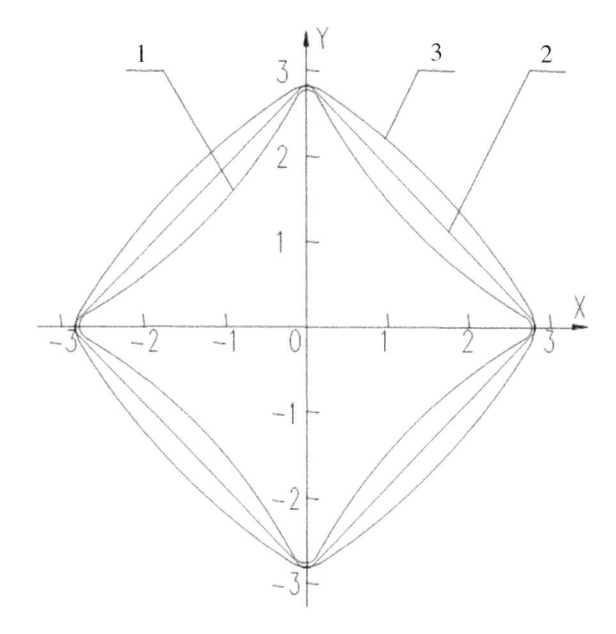

edges are (Fig. 5.28). Line 1 in Fig. 5.28 indicates the trajectory of point A when $\eta = 1.5$. If the value η is large, the four edges of the trajectory of point A will become convex. The larger the value is, the more convex the four edges are. Line 3 in Fig. 5.28 is the trajectory of point A when $\eta = 3.5$. Line 2 in Fig. 5.28 is the trajectory of point A when $\eta = 2.5$

There is no special requirement on the precision of the trajectory of point A. For normal soils, η is recommended to range from 2.4 to 2.5. For hard soils and rocks, considering the wear of the cutterheads caused by the soil, η can be considered to have relatively big values, ranging from 2.5 to 3.0. However, the actual values of η should be further verified in the projects. The diversity of the trajectories of the planetary gear is the difference between the cutterheads of a planetary gear-type square jacking machine and those of a Reuleaux triangular-shaped jacking machine, which will be discussed later. As the Reuleaux triangle has constant widths, the cutterheads of a square Reuleaux triangular-shaped jacking machine can only have one trajectory type. In other words, it is a type of the cutterheads for the planetary gear-type square jacking machine.

5.2.3.2 Design and Structure of the Planetary Gear-Type Square Pipe Jacking Machine

When designing the planetary gear-type square pipe jacking machine, a square with an edge length of B is selected (Fig. 5.29) and an internal gear is set with the same center as the square. Then, a planetary gear is set that meshes with the internal gear.

Fig. 5.29 Design of a planetary gear-type square pipe jacking machine

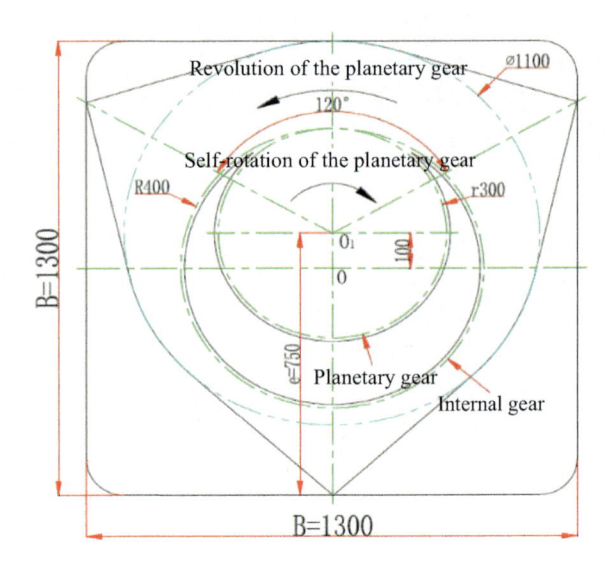

The ratio of the number of teeth in the internal gear to that of the planetary gear is 4:3. With the distance between the center of the planetary gear and the bottom of the square as the length of the stiffener, the three stiffeners are evenly distributed at an angle of 120°. Next, a circle is made with the center of the planetary gear as the center and the distance between the tangent point of the pitch circles of the internal gear and the planetary gear and the top of the square as the radius. Then, six lines are drawn from the vertices of the three stiffeners to be tangent to the circle. After the extra arcs are cut off, the outline of the cutterheads and the center lines of the three stiffeners are formed.

The main structure and components of the square planetary gear-type pipe jacking machine are shown in Fig. 5.30.

Figure 5.31 shows the front view of the planetary gear-type square pipe jacking machine. There are two screw conveyors beneath the cutterheads.

Figure 5.32 shows a square hole drilled by a 0.5 × 0.5 m planetary gear square pipe jacking machine in a Grade C30 concrete block. The cutterheads of a planetary gear-type jacking machine can also be composed of multiple square cutterheads with same or different sizes.

5.2.3.3 Combined Cutterheads for Planetary Gear-Type Square Jacking Machines

Different combinations of cutterheads will form different types and sizes of planetary gear-type square jacking machines. Table 5.4 summarizes the normal types of combined cutterheads for planetary gear-type square jacking machines.

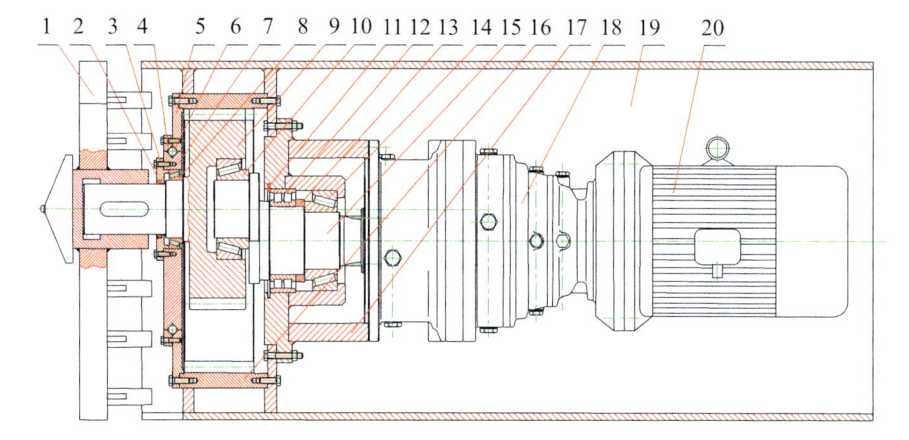

5.2 Types of Irregular-Shaped Pipe Jacking Machines

Fig. 5.30 Main structure of the planetary gear square pipe jacking machine. 1—Cutterhead; 2—Dust prevention seal; 3—End cap; 4—Press ring; 5—Cover of ring gear; 6—Steel ball; 7—V-shaped groove; 8–10 and 14—Tapered roller Bearing; 9—Planetary gear; 11—Shield ring; 12—Double row cylindrical roller bearing; 13—Washer; 15—Eccentric axis; 16—Gear ring; 17—Installation base; 18—Speed reducer; 19—Shell; 20—Electric motor

Fig. 5.31 Front view of the planetary gear-type square pipe jacking machine

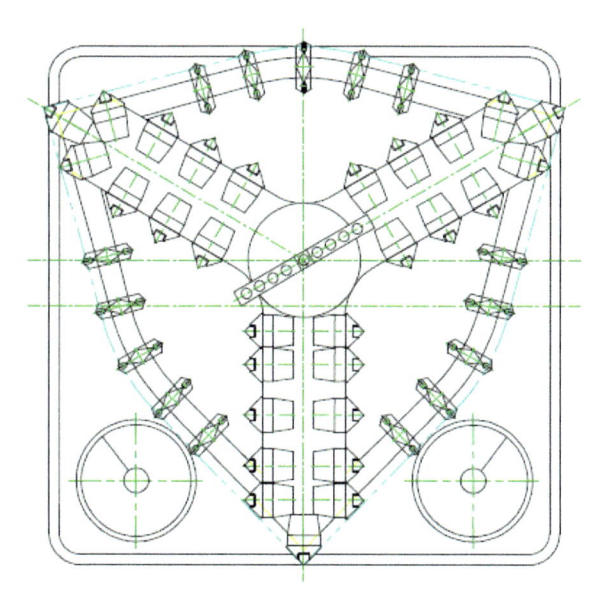

Fig. 5.32 Square hole drilled by a planetary gear-type square pipe jacking machine in a concrete block

5.3 Selection of Machine

5.3.1 Introduction

The principles of rectangular pipe jacking machines are basically the same as those of circular ones. However, the type of cutter, design of cutterheads, supporting systems, and the transportation system of the muck can be different, depending on the sizes of the rectangular pipes, the geological formations, and the existing site conditions.

The fundamental principle in the selection of a pipe jacking machine is to ensure that the project is completed in the most economical and fastest manner within the specified standard. The coordination between the construction method and the machine manufacturing is an important issue to be discussed.

In fact, the selection of the machine is a comprehensive decision with the considerations of project schedule, quality and standard, cost, and the maintenance of the machine. It should be noted that the construction by rectangular pipe jacking machine is a function of multiple disciplines, which requires knowledge from geotechnical, mechanical, hydraulic, electrical, materials engineering, civil engineering, management, and safety issue, etc.

5.3.2 Soil and Site Conditions

Soil and the existing site conditions or site constraints are parts of the basic information required for selecting the machine. The rectangular pipe jacking machines are designed based on the principle of earth pressure balance and are generally applied to soil conditions as per below:

5.3 Selection of Machine

- Clay and sandy soils with N values up to 50
- Gravel and pebble strata with a gravel content of less than 50% and a particle size of no greater than 1/3 of the inner diameter of the screw conveyor.

Based on experience obtained from practical projects, it is suggested that more boreholes are expected along the drive in order to cope with all the ground conditions accurately for the proper selection of the machine. During the jacking process, if the pipe jacking machine encounters special soil conditions, which are not shown in the soil report, the cost of construction will increase due to the additional work and delay of the project. For example, drawing the experience from a practical project in Singapore, when the pipe jacking machine advanced to a location, which is 10 m away from the receiving shaft, it could not move forward. After a time-consuming detailed study, it was discovered that a big rock, which was not shown in the soil

Table 5.4 Normal types of combined cutterheads for the planetary gear-type square jacking machines

Combination examples	Types
1	Type I (length-to-width ratio = 1:2)
2	Type II (length-to-width ratio = 2:1)
3	Type III (length-to-width ratio = 1:1)

(continued)

Table 5.4 (continued)

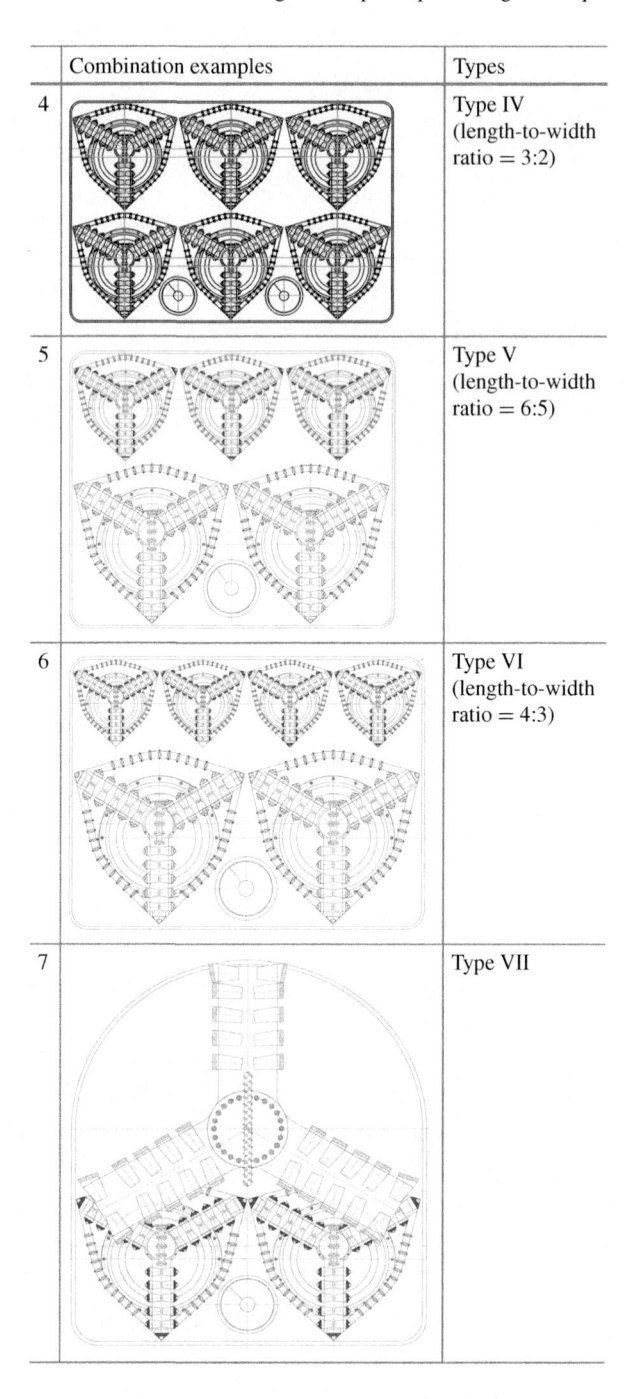

Combination examples	Types
4	Type IV (length-to-width ratio = 3:2)
5	Type V (length-to-width ratio = 6:5)
6	Type VI (length-to-width ratio = 4:3)
7	Type VII

report, was in front of the pipe jacking machine causing the machine to get stuck. To rectify this issue, the open cut method finally had to be adopted to complete the project, which incurred additional cost and caused delay to the project.

The existing site conditions are also an important factor to consider during the selection of the machine as it relates to both the construction method and the shafts. The site constraints include the existing road and traffic flow, the existing surrounding buildings, utilities, and other underground structures, etc.

5.3.3 Construction Method and Skill

As stated in Sect. 5.3.1, the design of a rectangular jacking machine has to suit the requirements of the construction method, while at the same time, the performance and efficiency of the machine will be reflected by the construction method. Therefore, it is imperative that the coordination between the construction method by the contractor and the design of the machine by the machine manufacturer has to be carried out diligently by understanding the full requirements to finalize the design of the machine and the construction method.

To understand the performance of the machine and the local construction method, it is highly recommended to check the track record of the machine manufacturer and pay site visits to actual projects completed using the machine from the same manufacturer.

Rectangular pipe jacking machines are earth pressure balance (EPB)-based machines, the strengths of which are summarized in Table 5.5. The machines should be designed for easy operation and low maintenance while at the same time have enough spare parts, which are particularly important if the machines are purchased from overseas.

The muck transportation system inside the tunnel in a rectangular pipe jacking project is related to the construction method. Soil and sand pumps have been applied

Table 5.5 Summary of the strengths of the EPB-based rectangular jacking machines

1	Greater stability during excavation for a wide range of soil types
2	Shallow over-burden as the height of the rectangular section is the shorter dimension of the cross-section to form the underground space of the rectangular tunnel
3	Increased safety and reliability as the screw conveyor system forms the fully enclosed excavation chambers
4	Easy discharge of pebbles with large sizes by the screw conveyor system
5	Easy delivery, stacking, and treatment of muck at a lower cost
6	Limited construction footprint, limited noise, and work carried out in the constrained areas and residential areas
7	Easy control of soil pressure, suitable for large and over-large dimensions of pipe jacking

Table 5.6 Strengths and weaknesses of different types of rectangular pipe jacking machines

Type Comparisons	Combined multi-cutterheads	Eccentric multi-axis	Square planetary gear type
Strength	• Excellent soil mixing • Ease of use for operation and control	• Full-section excavation • Suitable for a wide range of soil conditions • Simple structure and easy for maintenance	• Full-section excavation • Excellent soil mixing
Weakness	Limited blind areas	• Not easy to control alignment • Insufficient mixing compared with combined multi-cutterheads	• Complex manufacturing processes • Costly
Remarks	Has been widely used in the industry	Has been successfully applied to pipe jacking projects	Has not been applied to any projects

successfully as an efficient tool for muck discharge during the rectangular jacking process.

As the size and weight of a rectangular pipe are normally huge, the alignment and correction of deviation are essential during the jacking process. Soil and sand pumps have been performing well for correction of rolling deviations as well. For more details of soil and sand pumps, the interested readers can refer to the book published by Yu et al. (2018).

There are three types of rectangular jacking machines discussed in Sect. 5.2. Table 5.6 is a summary of the strengths and weaknesses of the three types of the machines during the selection of the machine.

5.3.4 Project Cost and Construction Schedule

The construction speed of the rectangular pipe jacking is much faster than that of the pipe roof method under the same project conditions and the jacking schedules for different projects can be found in Table 5.1 in Sect. 5.1.

Comparisons between the rectangular pipe jacking method and the pipe roof method are summarized in Table 5.7 based on a case study. It can be seen that the time saved by using the rectangular pipe jacking method is twice compared to the pipe roof method under similar project conditions. For more details of the comparisons between the two methods, interested readers can refer to Sun et al. (2016).

For the evaluation of the cost for the machine itself, Table 5.8 presents the factors linked to the selection of the machine.

5.3 Selection of Machine

Table 5.7 Comparisons between the pipe roof method and the rectangular pipe jacking method

No	Aspect	Rectangular pipe jacking project	Pipe roof project	Remarks
1	Site conditions and constraints	• Heavy traffic on the road above the jacking line • Proposed MRT (two circular lines) and a metro link tunnel below the jacking lines	• Heavy traffic on the road above the jacking line • Existing MRT (two circular lines) below the project • MRT via duct near by the project	Almost similar conditions for comparison
2	Major surrounding utilities	Existing drainage (4.2 × 2.5 m) above the jacking line	Existing joint bay above the jacking line	
3	Soil conditions	Sand and sandy soil	Clay, sand, and sandy soil	
4	Inner dimensions of the rectangular pipe section	4 × 6 (m × m)	4 × 6 (m × m)	
5	Jacking length (m)	242 (60.5 × 4)* Total jacking length equal to 242 m by 60.5 × 4	50	
6	Construction schedule (months)	11	22	Including shafts
7	Maximum settlement (mm)	25	30	
8	Environment	–	Underground space occupied by pipe roof permanently	

*It is a four parallel rectangular pipe jacking project as discussed in Sect. 5.4.1

Table 5.8 Factors for cost evaluation for the selection of the machine

No	Factor	Remarks
1	Maintenance cost	Important especially for machines purchased from overseas
2	Wear and tear	
3	Efficient muck delivery/transportation system and soil treatment if any	Soil and sand pump is an efficient option
4	Fast jacking speed	–
5	Electricity consumption	–
6	Quality and cost effectiveness	–
7	Suitability for complex soil conditions	–
8	Trained staff and skilled workers	–
9	Design standards	Different standards will lead to different designs of the machine with different costs

In conclusion, the consideration of the construction cost and schedule in a project must be in line with the selection of the machine, and an efficient jacking machine is the basic guarantee for good progress and reasonable cost of the project.

There are many factors affecting the cost and schedule of the project, which should be considered during the planning stage of the project. However, the points below focus on the factors related to the cost of the project during the execution stage.

- Machine cost and cost-effectiveness
- Construction schedule together with the construction method
- Time effect as it is related to all costs which will occur during the execution of the project, e.g., management cost, utilities and soil monitoring cost, labor cost, risk, safety, environmental friendliness, and maintenance cost during the project

A fast construction schedule will save project execution cost and reduce the risk.

5.3.5 Environmentally Friendly and Sustainable Development

The rectangular pipe jacking method is environmentally friendly for underground structures due to its construction process from shaft to shaft which has a limited site footprint. It is a construction method with safe working environment and minimal disruption and risk to the surrounding environment.

In comparison with the pipe roof technique for underground construction, the rectangular pipe jacking method promotes a cleaner environment for underground space occupation.

Meanwhile, re-using or re-cycling of the machine is one way of saving cost. Some manufacturers consider re-using their rectangular machines by converting them to circular machines with smaller sizes once the rectangular jacking project is completed. Re-cycling of the machines or machine parts is considered environmentally friendly and green underground construction.

In conclusion, the selection of the machine is a comprehensive consideration with knowledge required from different disciplines as discussed at the beginning of the section. The final decision should be made based on the final coordination among all specialists to facilitate the successful implementation of the project including sustainable development.

5.4 Case Studies

There are two case studies presented in this section. The first case study is a commercial underground space project in Foshan, Guangdong, China. The second case study

5.4 Case Studies

is an underground pedestrian project in Baotou, China. Both case studies adopted the same construction method of rectangular pipe jacking but under different conditions.

The commercial underground space project is composed of four parallel rectangular tunnels which has a cross-section of 6.9 m in width and 4.9 m in height. The jacking distance is 60.5 m, and the distance between each adjacent pair of the tunnels is only 500 mm. It is considered as the first rectangular pipe project practiced successfully under very challenging site conditions and constraints in Foshan, Guangdong, China (Sun et al. 2016).

The underground pedestrian project in Baotou, China, is the first rectangular pipe jacking project under the special soil condition of pebbles with sizes up to 300 mm. It has a cross-section of 6 m in width and 4 m in height and a jacking distance of 65 m.

The case studies provide an overview of the projects and discuss the key points related to the machine design and the construction method.

5.4.1 *Rectangular Pipe Jacking Project with Four Parallel Box Sections in Foshan, Guangdong, China*

5.4.1.1 Introduction

Rectangular pipe jacking has been widely used in the trenchless industry. In 2011, the project with four parallel box sections in Foshan was considered very challenging in view of the site conditions and constraints of the project. In today's terms upon reflection, it remains one of the most challenging projects.

Although this technology has already been further developed and the construction method is becoming more popular in the industry, the project is still considered as a successful example for study, discussion, and sharing.

The case study will discuss the rectangular pipe jacking method from design to construction of the project.

5.4.1.2 Project Description

(1) Background

This underground space project consists of four parallel rectangular tunnels with the cross-section dimensions of 6.9 m in width by 4.9 m in height and 60.5 m in length for each tunnel. It is located in Foshan, Guangdong, China.

Each rectangular tunnel was constructed by the rectangular pipe jacking method. The net distance between each adjacent pair of parallel sections was 500 mm as shown in Fig. 5.37.

After the completion of the four parallel rectangular tunnels, two side walls among three tunnels were opened in specific areas to form an underground retail mall, while the other tunnel was used as a motor vehicle passageway. Figure 5.33 shows a photograph taken from the retail mall.

(2) Site and soil conditions

The soil layers from top to bottom include clay, fine sandy silt, and sandy soil (Fig. 5.34). The ground is predominantly made up of fine sandy silt in the areas of the jacking zone. The depth of the underground water level ranges from 1.20 to 1.50 m. The elevation of the water level varies between 0.70 m and 1.14 m. Figure 5.35 shows the location of the road and the traffic flows, while further details of section A-A are shown in Fig. 5.36.

Four rectangular parallel tunnels act as an underground link between two parcels located on both sides of the street above the tunnels as shown in Fig. 5.36. The street is the main traffic road above the tunnels with heavy traffic flow in the city as shown in Fig. 5.36.

There is an existing in-service concrete drainage with a width of 4.2 m and a depth of 2.5 m located at a distance of 630 mm above the top of the four parallel

Fig. 5.33 Completed underground retail mall

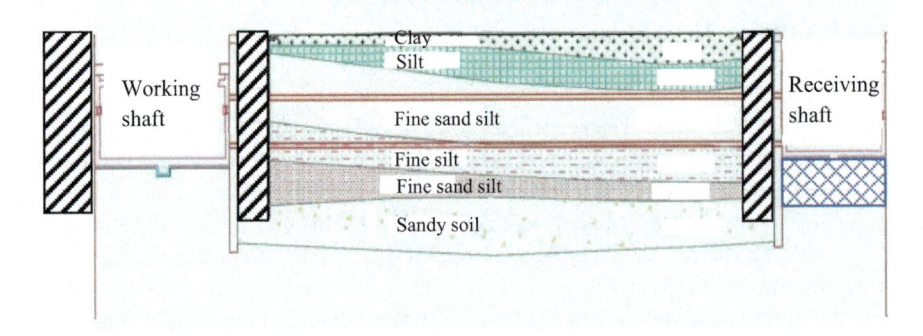

Fig. 5.34 Soil conditions

5.4 Case Studies

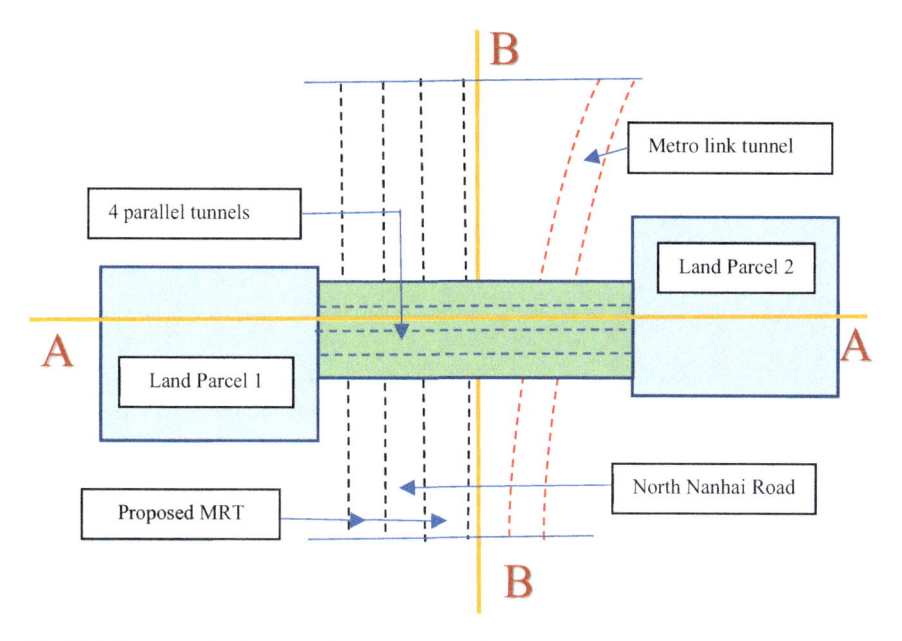

Fig. 5.35 Layout of the location of the four parallel underground tunnels

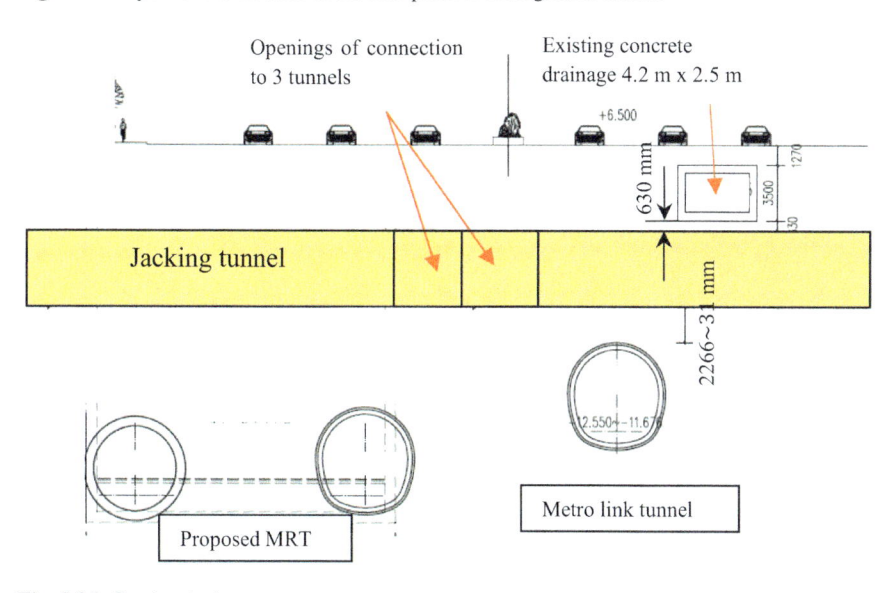

Fig. 5.36 Section A-A

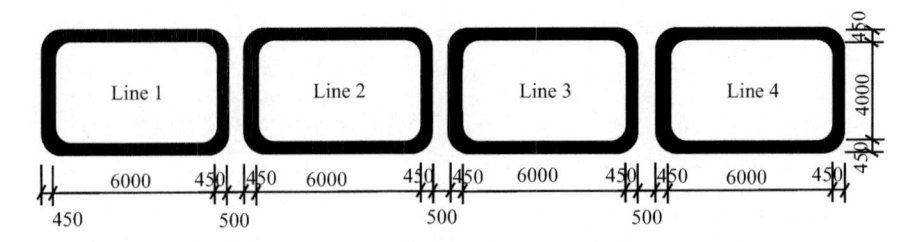

Fig. 5.37 Cross-section of the four parallel tunnels (units are in mm) (Section B-B)

rectangular tunnels as shown in Fig. 5.36. There are two proposed circular lines for the MRT underneath the rectangular tunnels and one metro link tunnel which is located at a distance of 2.5 m underneath the four parallel rectangular tunnels as shown in Fig. 5.36.

In summary, the construction of the four parallel underground rectangular tunnels is constrained by the existing concrete drainage from the top and MRT/metro link tunnel from the bottom. Moreover, the main street above the tunnels has heavy traffic flow as shown in Figs. 5.35 and 5.36. The existing in-service utilities in the surrounding are one more constraint to the project as there are many utilities nearby. The cross-section of the four parallel tunnels (Section B-B) is shown in Fig. 5.37.

5.4.1.3 Design Considerations

The following items are important to study and analyze for design works:

- Soil conditions
- Site conditions and constraints
- Construction method
- Shaft location, sizes, jacking force, and soil treatment if necessary
- Pipe design including materials, connection, and water proofing
- Environmental protection
- Design standard, schedule, and cost

As per the site conditions introduced in Sect. 5.4.1.2, the open cut method and the pipe roof method are not suitable for this project due to the site constraints.

To minimize the diversion of existing underground network and the disruption of heavy traffic on the road during construction, it is important to control the soil movement and the ground settlement to avoid any leakage from the existing drainage located at the top of the rectangular tunnels and reduce the risk during construction. Through a comprehensive study and analysis of the site constraints together with the project schedule, quality and standard, project cost, the rectangular pipe jacking method was considered a better solution for this project.

5.4 Case Studies

Table 5.9 Parameters of the machine

No	Description	Value
1	Outer dimensions of the machine	6920 × 4920 mm
2	Length of the machine	5500 + 1500 mm
3	Torque	420 kNm
4	Disposal volume by screw conveyor	0.93 m^3/min
5	Maximum deviation correction (left and right)	1.3°
6	Maximum deviation correction (top and bottom)	2°
7	Total installed power	476 kW
8	Total weight	About 210 tons

(1) Rectangular pipe jacking machine

The principle of the rectangular pipe jacking is the same as the normal circular pipe jacking except for the shape of the rectangular pipe sections. The earth pressure balance (EPB) type is designed for the rectangular pipe jacking machine. It has the advantages of easy operation and short construction time. It is also an efficient and safe construction method for underground structures especially in congested urban areas.

The design of the rectangular pipe jacking machine is described below:

- Multi-cutterheads are specially designed for this project. There are six cutterheads having the same size and layout, similar to Fig. 5.5 to minimize the blind areas during excavation. The distribution diagram of the blind areas is similar to Fig. 5.12, with the blind areas highlighted in red
- The shell of the jacking machine is assembled consisting the front upper, front lower, rear upper, and rear lower parts (Fig. 5.6) for easy installation and transportation
- Parameters of the machine are summarized in Table 5.9

(2) Rectangular pipe section

As shown in Fig. 5.37, the rectangular reinforced concrete pipes have a thickness of 450 mm and a length of 1.5 m.

As the net distance between each adjacent pair of pipes was only 500 mm, the uneven soil pressure from both sides of the tunnels during jacking needed to be considered in the design. The holes provided in the pipe sections for grouting reduce the frictional resistance and protect the completed tunnel during the jacking of tunnel 2. Meanwhile, the stranded holes were provided in the pipes to improve the longitudinal stiffness of the rectangular pipes by post tensioning. The purpose is to minimize the influence on the Mass Rapid Transit (MRT) below the tunnel.

(3) Soil and sand pump

During the construction using the rectangular pipe jacking machine, it is very important to control the alignment deviation due to the large size and mass

Sand and soil pump   Hydraulic power source

Fig. 5.38 Soil and sand pump

of the rectangular machine. To control and correct the deviation, a specially designed sand and soil pump was used for deviation correction especially for correction of clockwise and anti-clockwise rolling deviation during jacking. Figure 5.38 shows the pictures of a sand and soil pump.

The soil and sand pump is designed based on the concept of sleeve extraction. It is used to improve soil conditions during jacking, to fill the voids between the soil and the tunnel to minimize ground settlement, or to counteract the tendency of the shield rotation. It can also be used for soil discharge and delivery. For details of the soil and sand pump, interested readers are invited to refer to the book by Yu et al. (2018).

5.4.1.4 Construction Considerations

(1) Jacking sequence

As the net distance between each pair of adjacent rectangular tunnels was 500 mm only, the influence of the jacking procedure needed to be considered to control the alignment in the right direction and to minimize the influence of soil disturbance by uneven soil pressure on both sides. Therefore, the jacking sequence was specially arranged to start from Line 1, followed by Line 3, Line 2, and lastly Line 4. Figure 5.39 shows the completed four parallel rectangular tunnels with the jacking sequence.

(2) Grouting

Grouting is an important technique in the process of rectangular jacking. Grouting holes were considered during the design and manufacturing of the rectangular jacking machine and rectangular pipe cross-section to ensure smooth jacking and deviation correction when necessary.

As the net distance between the adjacent tunnels was 500 mm, to minimize the influence of the soil disturbance, pre-grouting for the completed tunnel

5.4 Case Studies

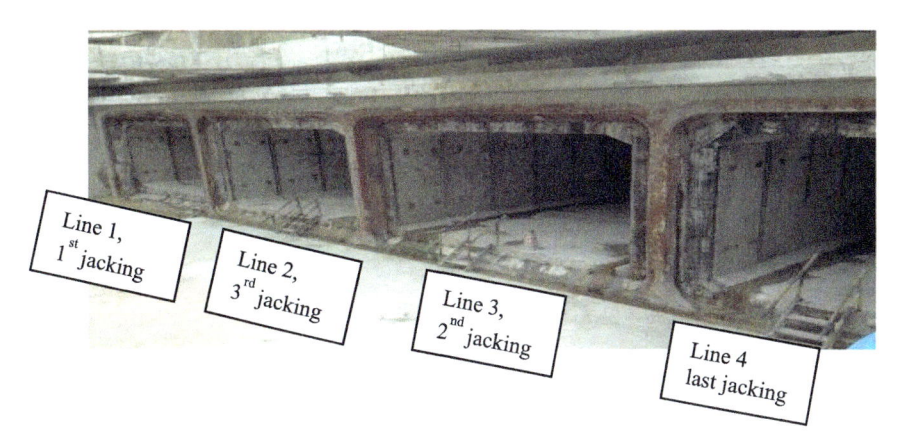

Fig. 5.39 Construction sequence

must be carried out. For example, grouting was conducted to protect Tunnel 1 and Tunnel 3 before the jacking of Tunnel 2 (Wang 2012).

Grouting technology and experience of the operation should be based on the soil conditions. The details of the grouting experience will not be discussed here. However, interested readers can refer to Sect. 5.4.2.5 for details of the grouting technology in a different case study.

(3) Settlement

Settlement is a sensitive controlling factor due to site constraints and the surrounding in-service utilities in the project which are introduced in Sect. 5.4.1.2. The maximum settlement was 25 mm, and the maximum displacement (horizontal settlement) was 32 mm for Tunnel 1, which satisfied the design allowance of ±50 mm. For Tunnel 2, the maximum settlements were +4.15 mm and −26.58 mm, respectively, which also satisfied the design requirements.

The structural deformations in the horizontal direction were monitored at the rectangular cross-sections of Tunnel 1 and Tunnel 3 during the jacking of Tunnel 2. They were recorded as +1.66 mm and −1.06 mm for Tunnel 1 and +0.97 mm and −0.94 mm for Tunnel 3. Note that ' + ' means that the deformation is away from Tunnel 2, while '−' means that the deformation is toward Tunnel 2 (Wang 2012).

(4) Jacking speed

From the construction record, the jacking speed is summarized in Table 5.10. It was found that the jacking speed was satisfactory compared with the common practice using circular pipes. Meanwhile, the jacking speed needed to be controlled to minimize the influence of soil disturbance by uneven soil pressure on both sides. For Tunnel 2, the jacking speed was controlled to be within 20–30 mm/min (Wang 2012).

Table 5.10 Jacking schedule

Items	Line 1	Line 2	Line 3	Line 4
Length (m)	60.5	60.5	60.5	60.5
Start date	Dec-9-11	Mar-22-12	Feb-13-12	Apr-28-12
Completion date	Jan-13-12	Apr-12-12	Mar-6-12	May-15-12
Average jacking speed (m/day)	1.78	2.75	2.75	No record
Maximum jacking speed (m/day)	4.5	6	7	7.5

5.4.1.5 Conclusions

This case study entails an innovative practical project, in which four parallel rectangular box sections were constructed, with 500 mm net distance between the rectangular tunnels and surrounding site constraints.

It is believed that there is still great potential for development if further studies are carried out. Interested readers can also refer to the case study by Sun et al. (2019), in which two parallel tunnels are constructed by monitoring the soil movements on site during the full jacking process.

5.4.2 Rectangular Pipe Jacking Project in Complex Soil Conditions in Baotou, Inner Mongolia, China

5.4.2.1 Background

This case study will discuss the design of a rectangular jacking machine, its corresponding construction method via a practical project, which is the construction of Baotou Alding Street (Inner Mongolia University of Science and Technology) underground passage. Meanwhile, the valuable experience and lessons obtained from the practice of the project will be discussed (Yu et al. 2018).

Baotou City Alding Street underground pedestrian passage was a rectangular pipe jacking construction project with a height of 4.3 m, a width of 6.5 m, and a jacking distance of 64 m. It was the first project to adopt the rectangular pipe jacking machine in complex soil conditions consisting of sand pebbles in China.

5.4.2.2 Parameters of the Rectangular Jacking Machine

The technical parameters of the rectangular jacking machine for the Baotou project are summarized in Table 5.11.

5.4 Case Studies

Table 5.11 Technical parameters of the rectangular pipe jacking machine

No	Aspect	Details	Remarks
1	Inner dimensions of the pipe section	5500 mm in width, 3300 mm in height, 300 mm in radius of the arc at each of the four inner corners	
2	Outer dimensions of the pipe section	6500 mm in width, 4300 mm in height, 700 mm in radius of the arc at each of the four outer corners	
3	Soil conditions in jacking zone	Calcium formations and sand pebbles, individual pebbles with diameters up to 300 mm	Refer to Figs. 5.40 and 5.44
4	Design of the cutterheads/arrangement of cutterheads	• Three large cutters, diameter of 2600 mm • Three small cutters, diameter of 2160 mm • The large cutter is driven by 330 kW motors at the speed of 0–5 rpm, torque is 580 kNm, and alpha coefficient is 3.3 • The small cutter is driven by 230 kW motors at the speed of 0–5 rpm, torque is 380 kNm, and alpha coefficient is 3.8 • All six cutterheads are variable-voltage/variable-frequency (VVVF) speed adjustment	Refer to Fig. 5.41
5	Deviation cylinders and the arrangement	• Twelve deviation cylinders with 200 tons each • Four at top and bottom each • Two at each side of left and right • Maximum correction angles are as below: • 2.6° for up and down • 1.8° for left and right	
6	Cylinder arrangement to loosen soil	A cylinder is provided behind two blind spots at the top of the machine each to loosen the soil	
7	Characteristics of wear-resistance and impact-resistance of cutter design	Rollers to be provided on each cutter to crush possible large pebbles	
8	Major cutters arrangement	By front and rear double-layer design	
9	Connection of cutters	The cutter should be welded rather than assembled due to limited jacking length and complex soil conditions	

(continued)

Table 5.11 (continued)

No	Aspect	Details	Remarks
10	Opening of cutterhead	The opening of the cutterheads must be protected by preventing pebbles with particle size larger than 300 mm from entering the excavation chamber	
11	Grouting holes	Grouting holes are designed for the spindles and the cutterheads for grouting to improve the soil conditions	
12	Diameter of screw conveyor blades and pitch of screw	670 mm diameter of the screw conveyor blades and 454 mm pitch of screw in the design to ensure that pebbles with particle size smaller than 250 mm can be discharged directly	
13	Wear-resistant welding	The screw and the barrel of the screw conveyor must be stacked with wear-resistant welding to reduce the abrasion under the soil conditions	
14	Control panel	• Accurate data collection and display • Easy to read and operate • Latest electronic products and technologies adopted	Refer to Fig. 5.42
15	Assembly of shell of machine	The front and rear shells of the pipe jacking machine are designed as top and bottom portion for easy delivery	
16	Total installed power	About 530 kW	
17	Excavation areas	More than 88% of the excavation area is achieved compared with the total area of the cross-section	

Fig. 5.40 Sample hole near the working shaft

5.4 Case Studies

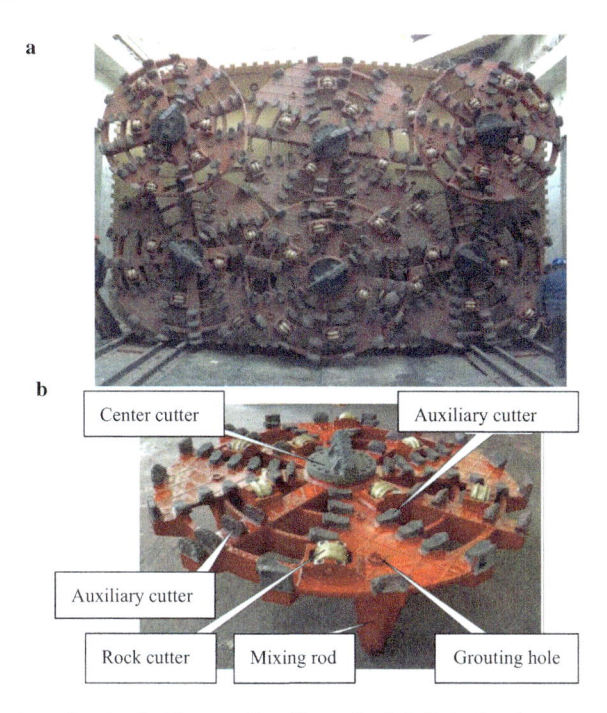

Fig. 5.41 **a** Rectangular pipe jacking machine (front view); **b** Cutterhead

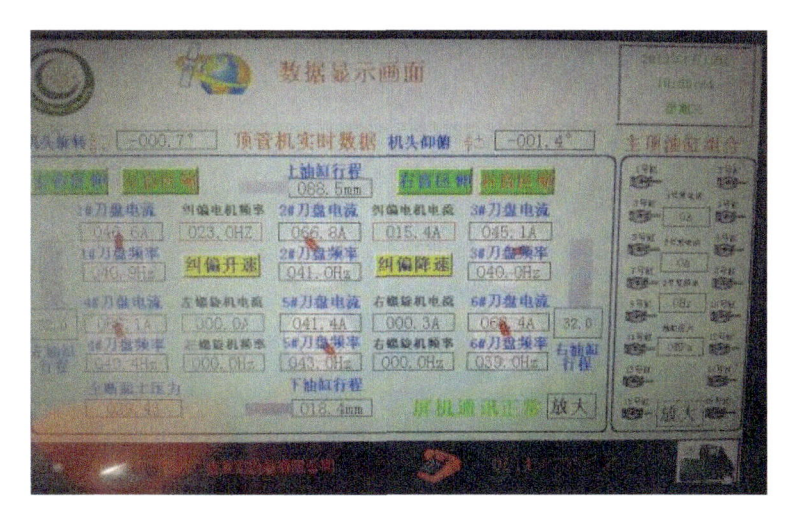

Fig. 5.42 Display of the control panel

5.4.2.3 Design of the Rectangular Jacking Machine

The main features of the 4.3 × 6.5 m rectangular pipe jacking machine specially designed for the complex soil conditions of the project are described in four major aspects below.

(1) Adoption of the advanced earth pressure balance principles

The design for the 4.3 × 6.5 m rectangular pipe jacking machine adopted the most popular earth pressure balance principle. During the jacking process, the soil pressure P in its excavation chamber is always controlled to be within the value which is designed and tested including the underground water pressure. The tolerance is controlled within the range of ∓ 20 kPa. In addition, the amount of soil discharge is always controlled to be between 95 and 100% of the jacking space volume.

(2) Unique full-section soil pressure acquisition and display technology

To truly achieve earth pressure balance according to accurate soil pressure data collection and display technology, the 4.3 × 6.5 m rectangular pipe jacking machine had a special pressure gage at the tail of the deviation cylinder as shown in Fig. 5.43.

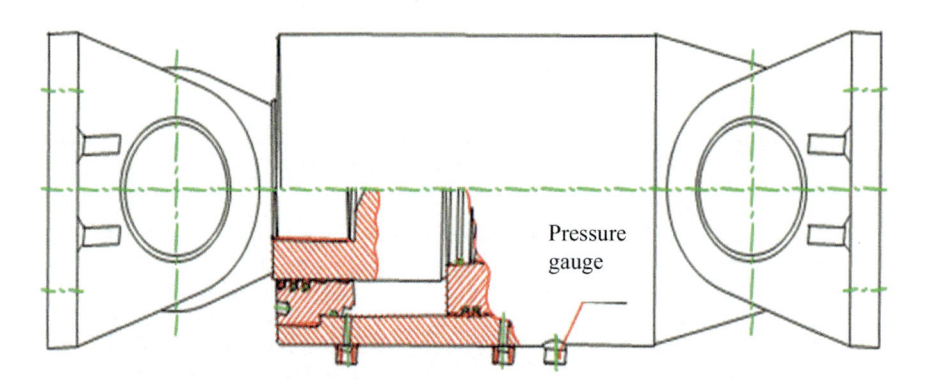

Fig. 5.43 Correction cylinders and pressure gages

Fig. 5.44 Soil conditions and a big, discharged pebble

5.4 Case Studies

The soil pressures of the full section of the pipe jacking machine measured by the pressure gages of each correction cylinder are grouped, then transmitted using the pressure sensors, and finally displayed on the control panel. As long as this pressure is controlled within the set range, true earth pressure balance can be achieved.

The ordinary diaphragm-type soil pressure gages, installed in the compartment board, could only be used as reference. They not only provided inaccurate measurements, but also could not be utilized to determine the soil pressures along the full section of the pipe jacking machine. As a result, real earth pressure balance cannot be established.

(3) Multi-grouting channels

There were many grouting holes considered during the design stage of the 4.3 m × 6.5 m rectangular pipe jacking machine to improve the soil condition. They were located not only at the center cutter, but also on the panel of the cutterheads. This design allows the grouting to be transferred to the cutterheads when the cutterheads are in contact with the excavation surface. When the cutterhead was excavating the sand pebbles, the grouting was mixed well with the sand pebbles. This naturally improved the conditions of the sand pebbles to suit the jacking process.

(4) High ratio of excavation area

The arrangement of the six cutterheads for the pipe jacking machine is shown in Fig. 5.12. The area of the full cross-section of the pipe jacking machine is 28.15 m^2. The excavation area of the cutterheads is 24.88 m^2, and it accounts for 88.38% of the total section area.

The areas which cannot be excavated by the cutterheads are the blind areas as indicated in red in Fig. 5.12. From Fig. 5.12, the blind areas are primarily located in the inner ring of the pipe jacking shell with only two thin blind areas in the middle part of the pipe jacking machine. It should be noted that if a large range of blind spots appear in the middle of the pipe jacking machine, they will block the flow of the soil, ultimately leading to the jacking failure.

This design minimizes the proportion of blind areas. Under complex soil conditions, a larger blind area not only increases the on-coming resistance, but also makes it difficult to control the direction of the pipe jacking machine and can cause a series of other problems such as difficulty in soil discharge.

5.4.2.4 Construction Experience

The early stage of the construction consists of the following three stages:

(1) In the first stage, a total jacking distance of 14 m was achieved, with an average daily jacking of 2 m. This progress was considered slow. In a reported incident, a steel bar used as a reference point for monitoring surface settlement, dropped through the ground 'mysteriously.' This incident was later understood to be due to the presence of a large void that was formed under

the at-grade nonvehicular and pedestrian lanes. Fortunately, it was the winter season in December, and there was a thick layer of frozen soil under the road surface. Otherwise, it might have caused an accident. This issue was caused by over-excavation.

(2) Within a month of the second stage, a total of 12 m was completed. During this period, apart from the slow jacking speed, there were other issues such as the machine heating up, which was caused by high electricity current, huge jacking force, holes/voids formed in certain areas. In addition, the excavation chamber showed high pressure, and there were also some operational problems. During this time period, the project was still considered to be in the trial and exploration stage. Meanwhile, re-thinking of the construction method in the project was carried out.

Special meetings were held during the time, but no proper conclusions were drawn from the meeting as no rectangular pipe jacking had ever been carried out, especially in complex soil conditions in Baotou, China. It was not feasible to carry out the rectangular pipe jacking for the project only based on construction experience in typical soil conditions. Therefore, extensive work had to be carried out under the guidance of a specialist from the industry. The construction method also had to be re-studied and adjusted accordingly.

After studying and analyzing the issues including slow jacking speed, hot cutterhead due to high electricity current, huge jacking force, two main concerns were identified. Firstly, in the soil excavation chamber of the pipe jacking machine, a large amount of the sand was consolidated and could not enable good plasticity and mobility for the soil in the excavation chamber. Secondly, the shell of the machine, especially the part that adhered to a thick layer of soil, not only caused an increase in the jacking force, but also resulted in voids (or hollow areas) in the soil. Three main solutions were proposed, and actions were taken to address these issues:

(2.1) A large amount of thin mud was injected into the excavation chamber of the pipe jacking machine, so that the consolidated soil was softened by soaking the soil in the thin mud to make it flow. As the soil conditions were primarily calcium formations and sand pebble formations without ground water, a small amount of improved mud dehydration would produce re-consolidation. That was why the electric current in the cutterheads increased, causing an increase in heat. The heat again aggravated the solidification in a vicious circle.

(2.2) Two sets of soil and sand pumps as shown in Fig. 5.38 were applied in the construction. By using the pump for grouting the space and possible holes around the rectangular pipes, ground settlements were minimized.

(2.3) A large amount of clay was prepared to make a thick filling grouting and a thin soil improvement mud.

(3) Finally, a variety of monitoring methods were employed to ensure that the above preparations from 2.1 to 2.3 were done in place before the jacking re-commenced . Since the pipe jacking machine was already under the main

5.4 Case Studies

road for re-starting the jacking at this stage, any small mistake may lead to a major accident. After three days' trial on site, the pipe jacking process was finally under control.

5.4.2.5 Grouting Technology and Method

During the jacking process, grouting was essential for the successful implementation of rectangular pipe jacking. There are three types of grouting adopted by the project, namely grouting for soil improvement, grouting for friction reduction, and grouting for synchronous filling. In other words, it was not possible to successfully complete the project by rectangular pipe jacking without the proper grouting technology and method. In this case study, we focus more on the soil improvement mud with testing results.

(1) Grouting for soil improvement

As there was no ground water in most of the sand pebble layer as shown in Fig. 5.44, in order to achieve good plasticity, fluidity, and impermeability of soil in the excavation chamber, the soil had to be improved. A series of testing was carried out to improve the soil conditions with the testing results summarized in Table 5.12.

It was proven that the plastic flow characteristics of the soil were improved, and the cohesion increased significantly when the water content in the soil that modulated the improved grouting reached 10% and then mixed into 7% to 15% of loess mud,

Table 5.12 Summary of testing of grouting

Testing groups	Water: bentonite	Rubble: bentonite	Density (ton/m^3)	Slump (cm)
Group 1	**1000:1500**		1.065	
		6000:400		0
		6000:500		4
		6000:600		11–13
Group 2	**1000:2000**		1.085	
		6000:400		2.5
		6000:500		13–14
Group 3	**10,000:3000**		1.126	
		6000:500		3
		6000:600		15
		6000:700		12–15
Group 4	**10,000:4000**		1.161	
		7000:10,000		18
		6000:500		6–12
		6000:600		12

respectively. However, for the slump determination, when the soil self-mobility was inadequate, the soil showed limited slump. Analysis revealed the reason behind this, which was that a higher solid content of loess mud caused an increase in the cohesiveness and density of the soil and led to poor self-mobility of soil.

It was recommended that more water and bentonite were to be mixed; e.g., each 1 m^3 of the improved mud had around 300 kg of loess, 50 kg of bentonite, 1 kg of carboxy methyl cellulose (CMC), and 900 kg of water in this project. The soil modified mud modulated in this project is shown in Fig. 5.45.

The grouting of the improved mud was conducted through the injection holes, located at the center cutter and each cutterhead panel, to the excavation surface directly. Then, the improved mud and excavated sandy soil were constantly mixed using the mixing rod in cutterheads. Only in this way, the original sand pebble could be improved to achieve good plasticity, mobility, and impermeability. The injection of the improved mud was done through a single-cylinder hydraulic plunger slurp pump shown in Fig. 5.46a, while the effect of the improvement can be seen in Fig. 5.46b.

Fig. 5.45 Improved mud

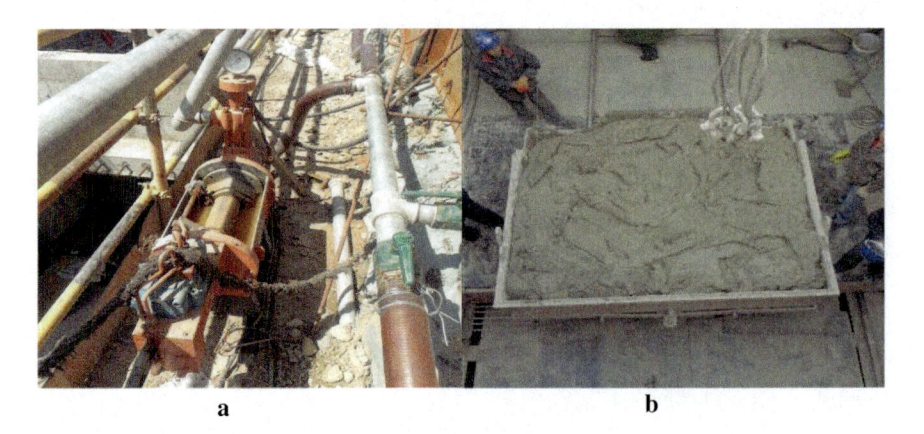

Fig. 5.46 **a** Single-cylinder hydraulic plunger grouting pump; **b** Improvement effect

Figure 5.46b shows the improved mud with good plasticity, mobility, and impermeability on a screw conveyor. For EPB-based pipe jacking under the sand pebble soil condition, soil improvement is the key to the success of pipe jacking.

It should again be highlighted that good plasticity, mobility, and impermeability are the three basic factors to ensure the quality of the soil improvement.

(2) Grouting for reducing frictional resistance (friction reduction mud)

The pipe surface friction reduction mud was to reduce the frictional resistance between the outer surface of the pipes and the sandy soil during pipe jacking. The main component of pipe surface friction reduction mud was bentonite. To improve its performance, several different additives must also be added according to different soil conditions which are the common practice in the project according to the soil conditions.

(3) Filling mud

The so-called filling mud was very thick loess, which was injected from the special injection hole at the back of the jacking machine to fill the gaps due to various reasons during the jacking process to prevent ground movement, settlement, and bulging.

The ground bulge is formed due to the use of the filling mud being different from that caused by excessive soil pressure in the excavation chamber. The soil in the former bulge area was dense and did not settle after the bulge was formed, while that in the later shows the opposite.

As the filling mud is very thick, the injection pressure was relatively high, generally ranging from 3 to 4 MPa. When it spread to the outer circumference of the tunnel, it will not only fill the gaps, but also penetrate the sand pebbles, forming an impermeable membrane on the surface of the sand pebbles. This membrane made it not easy for friction reduction mud to penetrate the sand pebbles, thus greatly enhances the effect of friction reduction. This is both a new discovery and the reason why the jacking force of the rectangular pipe jacking machine in the sand pebbles only has a slow increase. This method has great potential to be applied to all pipe jacking projects in sand pebbles in the future.

However, the injection pressure for the filling mud should not be too high, especially in the case of deep over-burden. The higher the pressure, the larger the jacking force will be. Therefore, special attention should be paid to this issue.

Another function of the filling mud is to correct the rolling deviation of the rectangular pipe jacking machine. There were many grouting holes in the cutterheads as introduced in Sect. 5.2, located at the top, bottom, left, and right of the cutterheads of the pipe jacking machine. If the pipe jacking machine had the tendency to turn clockwise, the filling mud should be injected to the grouting holes located at the bottom right of the front shell of the pipe jacking machine. The jacking machine would then slowly turn counterclockwise. This method was used to correct the 70 mm side bias successfully, indicated as 'A' in Fig. 5.48 in this project.

Figure 5.47 is a soil and sand pump as introduced in Sect. 5.4.2.4. It was filled with thick mud. It used the unique concept of sleeve extraction, which made it possible to pump thick mud.

Fig. 5.47 Soil and sand pump adopted in the project

Fig. 5.48 Rolling deviation of the jacking machine

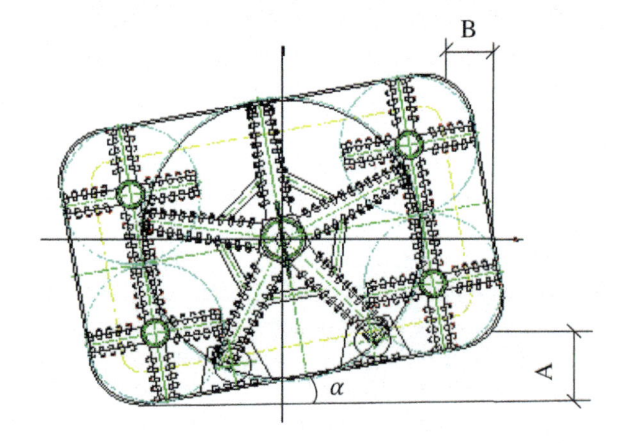

In this project, jacking for the first 26 m was not smooth as it was considered a period of trial. There were many difficulties and problems encountered. The lessons and experiences we learned from the project were that the key point for the success of the project was to get a proper grouting technology for projects with such soil conditions, and filling mud was the most effective method.

After 26 m trial distance with proper grouting technology, the rectangular pipe jacking project in sand pebbles was successfully implemented, from the original uncontrollable situation to a stable jacking process. Without proper grouting practice as discussed, there would not be impressive jacking performance with maximum ground settlement of 30 mm and deviation of 10 mm recorded in the project.

Over-turning in the horizontal direction is not a common occurrence in rectangular pipe jacking. However, it is not easy to make corrections once it happens. In the past, normally the ingot steel and regulus lead would be provided as a balance loading to overcome a turnover, but it would not be an effective way in practice. No methods were considered possible for correction until the soil and sand pump and grouting technology were practiced in the project.

Fig. 5.49 Mud from the soil and sand pump

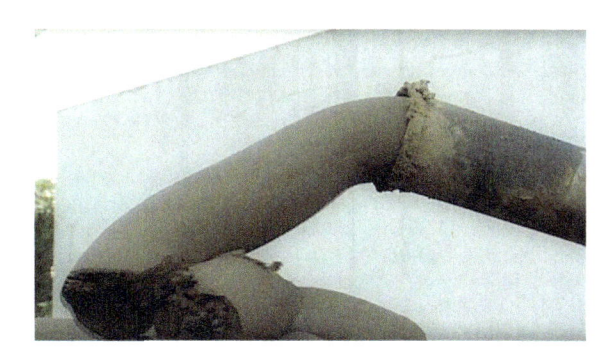

Once a counterclockwise over-turning occurred as shown in Fig. 5.48, it would be efficiently corrected by grouting the holes on the left side of the rectangular jacking machine.

Two key points must be noted. One is to correct the deviation immediately once it occurred, and the other was that the water content in the mud needed to be controlled. If water content is too high in the mud, it is easy to flow away and does not have efficient supporting effect. Figure 5.49 shows the idealized discharged mud from the soil and sand pump, which is good for deviation rectification.

Beyond this, the same principle of grouting can be adopted to maintain the right location of the rectangular pipe jacking machine during the construction except that the deviation occurs vertically toward the ground.

5.4.2.6 Conclusions

Rectangular pipe jacking can be applied for projects in complex soil conditions consisting of sand pebbles with individual particle size up to 300 mm.

Grouting is an efficient construction technology not only for reduction of frictional resistance and soil improvement but also for deviation correction during the jacking process.

References

ASTM D1586/D1586M (2018) Standard test method for standard penetration test (SPT) and split-barrel sampling of soils. ASTM International
CREG (2021) CREG delivers the world's first super-large cross-section rectangular hard rock pipe jacking machine. http://www.crecg.com/english/2691/2743/10163653/index.html. Accessed 14 Nov 2021
Li JB (2017) Key technologies and applications of the design and manufacturing of non-circular TBMs. Engineering 3:905–914

Sun T, Zhang DG, Yu BQ (2016) Cases study on application of rectangular shield jacking and TBM machine. Paper presented at the international No-Dig 2016, Beijing, China, 10–12 Oct 2016

Sun Y, Wu F, Sun WJ, Li HM, Shao GJ (2019) Two underground pedestrian passages using pipe jacking: case study. J Geotech Geoenviron Eng 145(2):05018004. https://doi.org/10.1061/(ASCE)GT.1943-5606.0002006

Tunnel (2017) Rectangular pipe jacking. In: International/TBM. https://www.tunnel-online.info/en/artikel/tunnel_Rectangular_Pipe_Jacking_2822434.html. Accessed 14 Nov 2021

Wang JX (2012) Key construction techniques for rectangular pipe jacking with large cross section and shallow overburden. In: Symposium on rectangular pipe jacking with large cross section and shallow overburden. China Railway No. 2 Group Co., Ltd, Foshan, May 2012 (in Chinese)

Yu BQ, Jia LH, Fan L (2018) Pipe jacking technology. China Communications Press. ISBN 978-7-114-15014-2 (in Chinese)

Chapter 6
Irregular-Shaped Shield

6.1 Introduction

Systems of excavation, thrust, muck removal, and erection of lining segments are key processes of conventional boring using shield tunneling machines. For an irregular-shaped (noncircular) shield, the process is principally the same except that the systems of excavation and erection of lining segments have to be changed to cater for the irregular shape of the eventual tunnel. The rectangular shield (also known as the box shield) is more often adopted in industry practice and is considered the most common shape for an irregular-shaped shield.

In this chapter, rectangular shield method is introduced with a focus on the newly developed erection system for the rectangular shield. The construction process using the specifically designed rectangular shield will be illustrated in a step-by-step manner from the stage of preparation to completion. Innovations are discussed in relation to development and applications, such as:

- Maximization of the space for erecting lining segments
- Grouting system in the shield tail
- Square column for erecting segments

The advantages of the rectangular shield method including the innovative segment erector system with its simple operational procedure will considerably reduce the construction time and minimize leakage and other potential risks during the execution of the project due to the limited connections of the two-piece only segments compared with conventional multi-segments.

The introduced rectangular shield method in this chapter applies to utility tunnels, pedestrian crossings, subway entrances/exits and connection tunnels between existing buildings beneath the ground. The lining rings formed by two simple U-shaped (upper portion and lower portion) segments can be applied to any rectangular tunnels with a maximum size of up to 9 m in width and up to 5 m in height. However, as an example, a rectangular shield with a width of 6.6 m and a depth of 3.3 m will be discussed in this chapter.

In the second part of this chapter, a project involving a rectangular utility tunnel will be discussed. The tunnel, which had a slightly arched top, was constructed using the rectangular shield method in Shandong, China in 2015. It was considered as the first rectangular shield project using such special construction method (Gong 2015).

Over time, various systems have been further developed for the rectangular shield. For example, a new erector system for the rectangular shield will be introduced in Sect. 6.2. The case study in this chapter adopted the rectangular shield method and the project was executed six years ago. At that time, the method was considered to be innovative. Even now, it is still considered as a good experience in the industry for discussion and sharing.

Meanwhile, the hybrid method using both rectangular jacking and shield will be introduced through an example of a completed project in Sect. 6.3.4.

6.2   6.0 m × 3.3 m Rectangular Shield

The 6.0 m × 3.3 m rectangular shield including the erector system for lining segments represents one of the latest design and research outcomes described in this book. It can be applied to the construction of tunnels with any rectangular cross-sections with maximum external dimensions of 9.0 m in width and 5.0 m in depth.

6.2.1   Rectangular Lining Rings and Seals

Figure 6.1 shows the overview of the rectangular lining ring with U-shaped segments. The internal dimensions of the rectangular lining rings are as follows: 6.0 m in width, 3.3 m in depth, 0.5 m in thickness, 1.2 m in length, 200 mm × 45° for the inner chamfer and 500 mm in radius of the outer chamfer.

Fig. 6.1 Overview of the rectangular lining rings with U-shaped segments

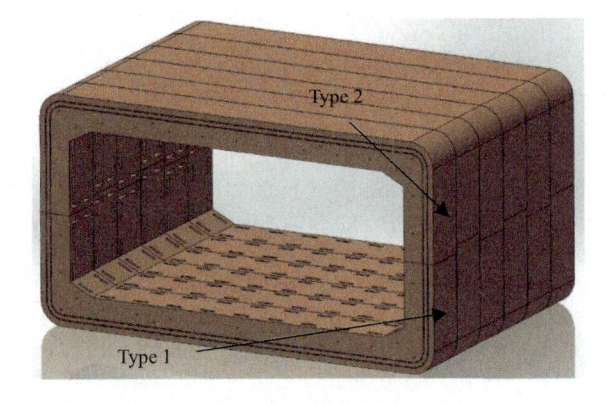

6.2 6.0 m × 3.3 m Rectangular Shield

The lining rings of the tunnel are divided into two U-shaped parts in the design, namely, the upper segment and lower segment. Both upper and lower segments consists of two types, type 1 and type 2 (Fig. 6.1). The interfaces of upper and lower segments in adjacent two lining rings are staggered in 200 mm. The U-shaped lining segments are designed as symmetrical. Therefore, they can be used as either the upper or lower lining segment in the tunnel lining rings from the installation point of view.

Figure 6.2a shows the front view of a type 1 lining segment with an internal depth of 1550 mm. There are two sealing grooves on the front surface of the segment. This surface usually faces the working shaft. Figure 6.2b shows the rear view of the type 1 lining segment without sealing grooves. The rear surface usually faces the shield machine, and it bears the thrust of the shield hydraulic cylinders.

Figure 6.3 shows more details of the type 1 lining segment. There are 22 holes for connecting bolts and operational spaces for connections on inner sides, respectively, four positioning holes for lifting and two layers of seals on the surface at a distance of 100 mm along the joint surface.

Figure 6.4a, b are the front and rear views of a type 2 lining segment. Except for an internal height of 1750 mm and two seals on the two joints along surfaces, the rest are designed the same as a type 1 lining segment.

Fig. 6.2 a Front view of a type 1 lining segment, b back view of a type 1 lining segment

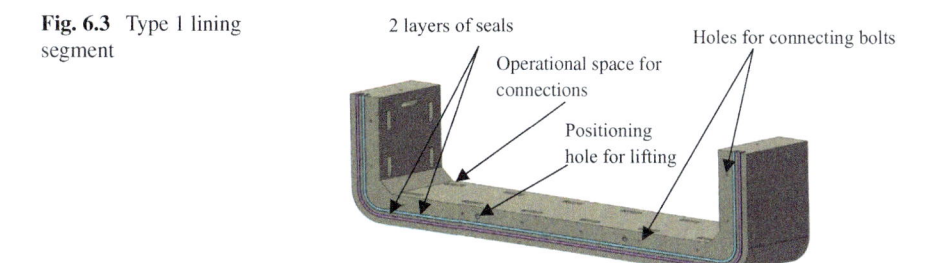

Fig. 6.3 Type 1 lining segment

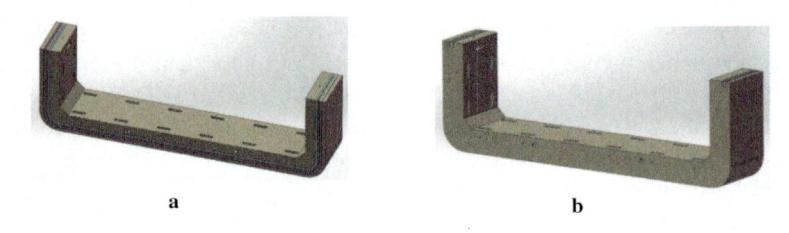

Fig. 6.4 a Front view of a type 2 lining segment, **b** rear view of a type 2 lining segment

6.2.2 Configuration of the Segment Erection Shield

The configuration of the segment erection shield of the 6.0 m × 3.3 m rectangular shield is shown in Fig. 6.5. It consists of a segment shield, a square column, shield tail for grouting, and articulated edge to the shield machine.

The segment shield is designed for easy transportation as shown in Fig. 6.6. It consists of upper, lower, and two side parts and can be easily assembled on site.

The bottom part of the shield is shown in Fig. 6.7. The aforementioned grouting lines are located at two sides of the bottom corners. There are ten hydraulic thrust cylinders, and those located on the two sides are mounted at relatively high positions. The square column is mounted in its seat. The movable track can be moved within the mounting slot. In addition to strengthening the bottom rib plate, it also functions

Fig. 6.5 Configuration of the segment erection shield of the rectangular shield

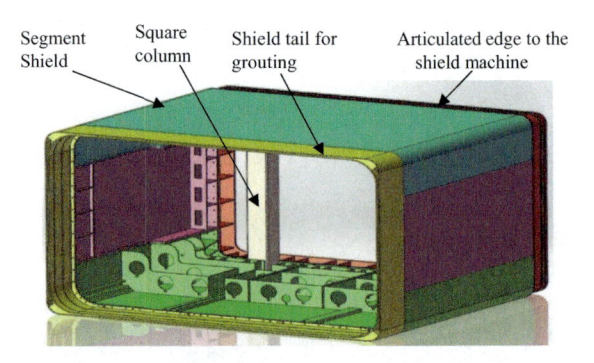

Fig. 6.6 Upper, lower, and two side parts for the segment shield

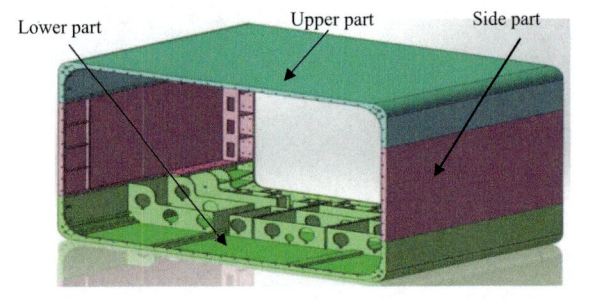

6.2 6.0 m × 3.3 m Rectangular Shield

Fig. 6.7 Lower part of the segment shield

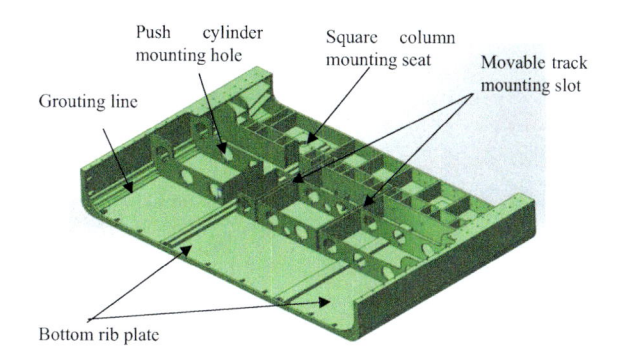

as the track of the lower lining segment. The grooved rib plate is also the rail track of the roller of the frame for lifting lining segments.

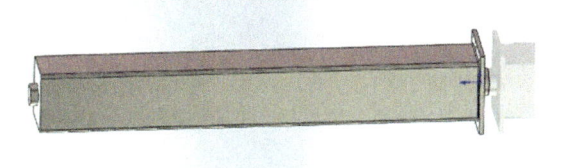

The upper part of the shield is shown in Fig. 6.8. 5 slabs with a thickness of 20 mm are welded on the top plate for improved strength. There are ten hydraulic thrust cylinders for the shield and the cylinders on the left and right sides are located slightly toward the bottom. The top cylindrical end of the square column is inserted in the mounting seat.

Figure 6.9 is a horizontal elevation of the square column. The right side is the bottom of the column with a square flange facing down to connect with the lower portion of the column. The left is the top portion of the column. The design of this square column is one of the innovations in the design of the rectangular shield. Because of the use of the column with a rectangular cross-section, the big arm does

Fig. 6.8 Upper part of the segment shield

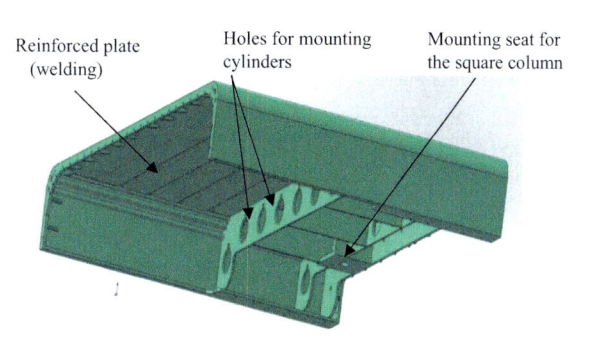

Fig. 6.9 Square column

Fig. 6.10 Side of the shield

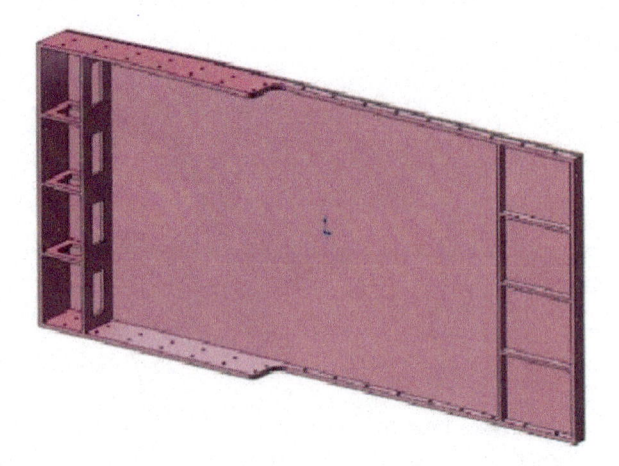

not rotate even if a single column is used. However, if a circular cross-section of the column is adopted, two columns must be used instead.

Figure 6.10 shows a side portion of the shield which has the same specifications when installed on both sides of the shield. The space between the two side parts of the shield is the maximum for erecting segments. Therefore, rotation will occur within this range when erecting the upper and lower lining segments.

This is considered as one of the innovations in the design. As a result, all hydraulic thrust cylinders can be held in place, and the space for erecting lining segments is maximized. Consequently, the rotation of the lining segments becomes simple and easy, and the speed of erection is greatly accelerated.

The shield tail for grouting is shown in Fig. 6.11. Since the segment shield is divided into four parts, the shield tail for grouting and the articulated edge to the shield machine are designed as a full-ring piece for structural stability.

Figure 6.12 is an enlarged view of one of the four corners of the connection between the shield tail for grouting and the segment shield. It is equipped with holes for connecting bolts, seal grooves, and pin holes for positioning. The grease filling

Fig. 6.11 Shield tail for grouting

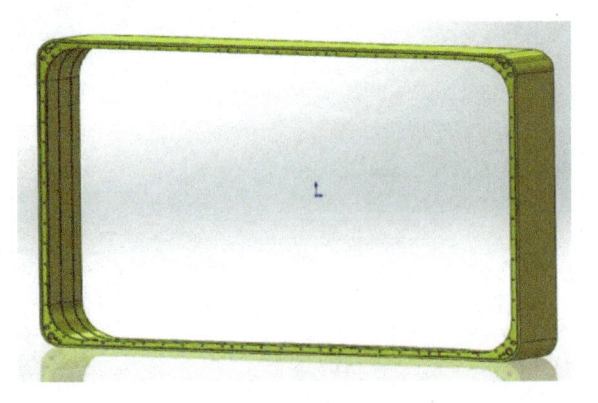

6.2 6.0 m × 3.3 m Rectangular Shield

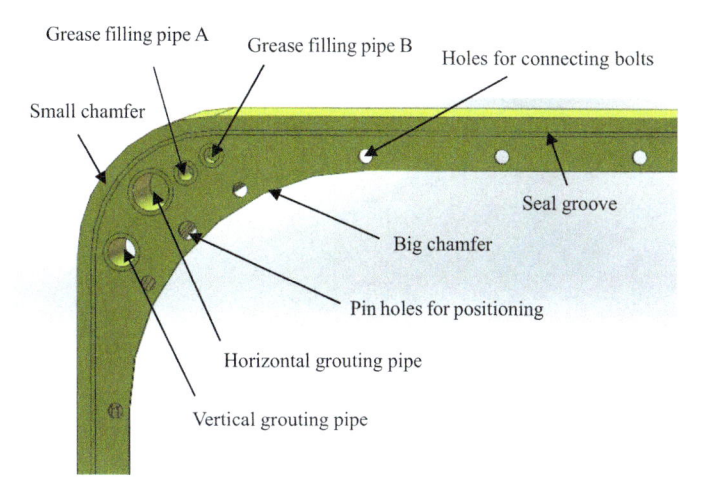

Fig. 6.12 Enlarge view of a corner of the connection between the shield tail for grouting and the segment shield

points A and B are steel pipes with a diameter of 40 mm, which are used to fill grease into the gaps between the shield tail brushes. The horizontal grouting pipe is made of steel with a diameter of 90 mm, which is utilized for grouting purpose along the shield machine horizontally. The vertical grouting pipe is also made of steel with a diameter of 70 mm, which is used to grout along the shield machine in the four corners and the vertical direction.

At the inside corner of the shield is a big chamfer, which remains concentric with the arc on the outside of the lining segment. The outside of the shield is a small chamfer. As this design creates a bigger space between the big chamfer and the small chamfer, it is possible that the grouting pipes with diameters of 70–90 mm be accommodated. This specific concept is exclusive for the rectangular shield only, and it cannot be achieved for circular shield.

Figure 6.13a shows that the exit for horizontal grouting is blocked by a sealing plate, while Fig. 6.13b reveals the grouting pipes after removing the seal plate. Apart

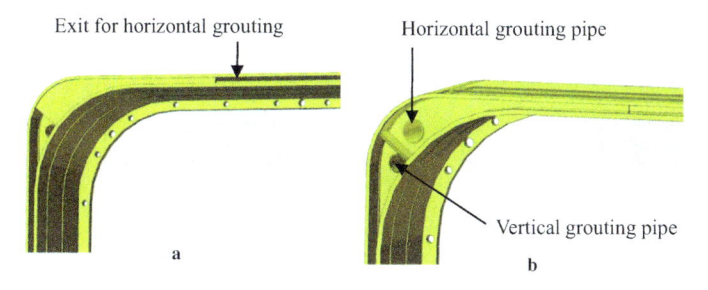

Fig. 6.13 Exit for horizontal grouting: **a** before, and **b** after removing the seal plate

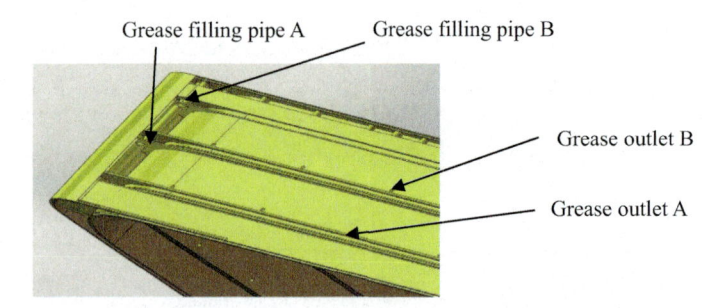

Fig. 6.14 Grease filling pipes A and B and their outlets at a corner of the shield tail for grouting

from the sealing plate, there is a partition between the horizontal and vertical grouting outlets. Therefore, grouting in the horizontal direction can only be done through the horizontal direction.

This grouting system is another innovation in this design which makes the grouting more uniform and effective. Two grease filling pipes A and B are shown at each corner of the shield tail for grouting in Fig. 6.14. Pressure is produced constantly from the four channels of the grease filling pipes A and B around the shield tail through their respective channels A and B for injection of grease, and the grease is then injected through the grease outlets A and B into the gaps between the brushes at the shield tail. The above design is also considered an innovation.

6.2.3 Transportation of Lining Segments Inside the Tunnel

Figure 6.15a is a common transport wheel for both upper and lower lining segments, while Fig. 6.15b is a transport wheel only for upper lining segments. There is an additional connection part in Fig. 6.15b, which is used to connect the battery-operated cart, compared with the wheel in Fig. 6.15a.

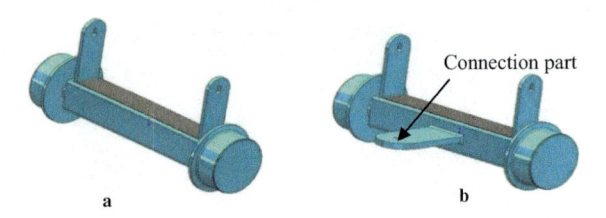

Figure 6.16a, b are the assembled delivery wheels for lower and upper lining segments, respectively. In both Fig. 6.16a, b, the front and rear wheels are bolted through the joint holes of the lining segment themselves.

Fig. 6.15 **a** Delivery wheel for both upper and lower lining segments, **b** back wheel for upper lining segments

6.2 6.0 m × 3.3 m Rectangular Shield

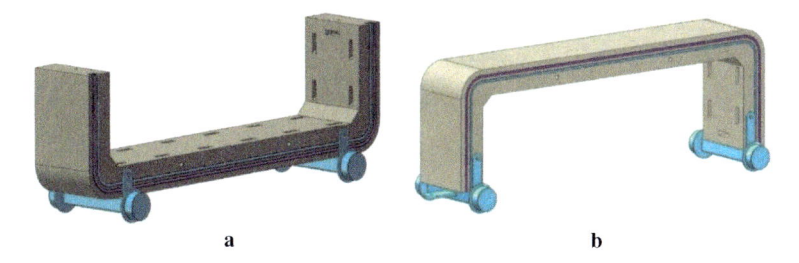

Fig. 6.16 **a** Assembled delivery wheels for lower lining segments, **b** assembled delivery wheels for upper lining segments

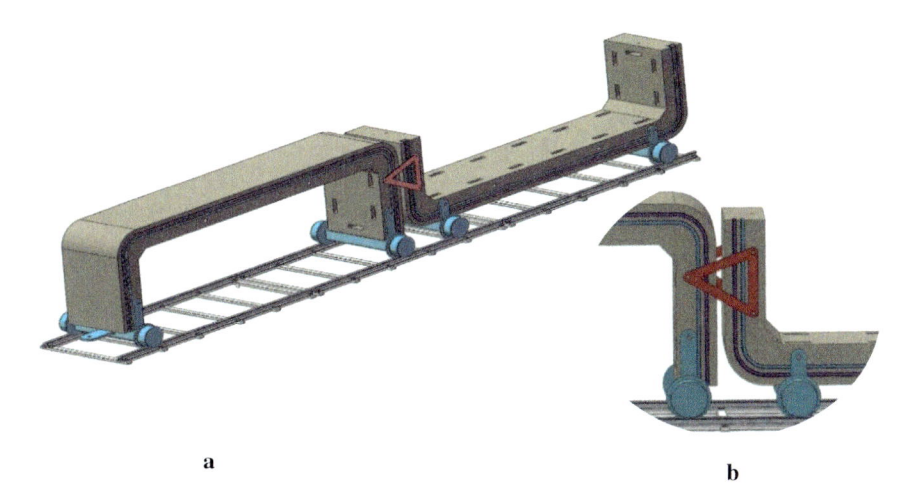

Fig. 6.17 **a** Linked upper and lower lining segment for transportation, **b** close-up view of the connection of the upper and lower lining segments

Figure 6.17a shows the transport mode where the upper and lower lining segments are linked together. As the lower lining segment is installed first, it must be in the front while the upper lining segment is at the back.

As the center of gravity of the upper lining segment is relatively high compared with the lower one, a triangular frame is specially designed to connect the upper and the lower segments as shown in Fig. 6.17b. It is installed on both sides using the joint holes of the lining segments.

6.2.4 Erector System for Lining Segments

The erector system for lining segments is a key feature of a rectangular shield. Figure 6.18 shows a schematic of the designed erector system for lining segments. It

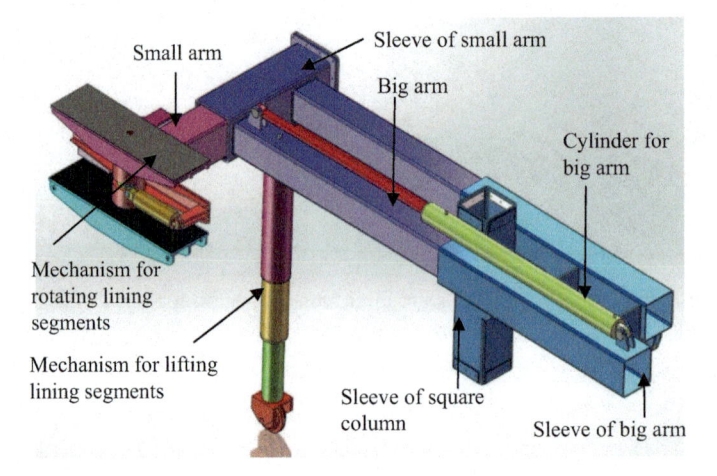

Fig. 6.18 Erector system for lining segments

consists of a mechanism for rotating lining segments, a small arm, the sleeve of the small arm, a big arm, the sleeve of the big arm, a cylinder for the big arm, the sleeve of the square column, and a mechanism for lifting lining segments.

6.2.4.1 Mechanism for Rotating Lining Segments

The mechanism for rotating the lining segments is shown in Fig. 6.19. The shell of the mechanism is connected to the small arm through four bolt holes at the rear, and the extension and retraction of the small arm drives the rotating mechanism to move to left and right. The top beam is used to erect the upper lining segment which

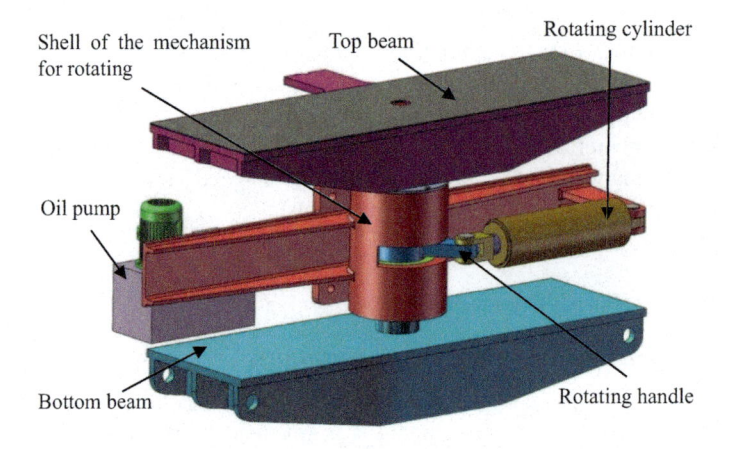

Fig. 6.19 Mechanism for rotating lining segments

should hold the lining segment in place during the erection. The protruding plate at the rear is used for easy positioning, and the surface is anti-slip with a rubber plate. When the rotary cylinder extends or retracts, the top and bottom beams can be rotated simultaneously by the rotating handle which is connected to the internal spindle. The maximum rotation angle is 90°. The four holes at the bottom beam are used to lift the lower lining segment after connecting to the steel wire rope separately. The oil pump is set up separately due to low oil consumption required by the rotation of the cylinder. As the oil pump is installed on the lower shell, it is difficult to move the oil pipe. Therefore, the oil pump is installed behind the shell of the rotating mechanism. At this point, only a very thin power cable is connected to the oil pump as it is much easier to move the cables than to move the oil pipe.

6.2.4.2 Small Arm

Figure 6.20a shows an extended small arm assembly, consisting of a small arm, a small arm sleeve, and a cylinder for the small arm. The big arm is welded together with the sleeve of the small arm to simplify the design. Figure 6.20b is a retracted arm assembly.

Figure 6.21 is the cylinder for the small arm. Figure 6.21a shows the full extension, while Fig. 6.21b shows the full retraction of the cylinder. The stroke of the cylinder is 1100 mm. A special feature of the cylinder is that its piston rod is fixed, while the cylinder barrel is movable, driving the small arm.

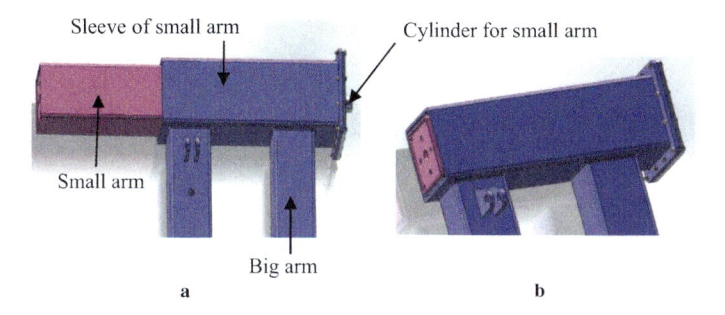

Fig. 6.20 **a** Extended small arm, **b** retracted small arm

Fig. 6.21 **a** Cylinder for the small arm in full extension, **b** cylinder for the small arm in full retraction

6.2.4.3 Big Arm

Figure 6.22 is the sleeve of a big arm that is welded into the sleeve of the square column. The big arm has been inserted into the sleeve of the big arm to prepare for installation of the big arm pad as shown in Fig. 6.23.

Figure 6.24 shows the fully installed big arm pad located at each side of the big arm within the sleeve. From the above explanation, it can be seen that the big arm, the small arm, and the sleeves of the big arm and the small arm are in contact through the pads. The square column and the column sleeve are also in contact through the pad. The pad is made of poly tetra fluoroethylene which not only has high strength and small resistance but also make processing easier.

As shown in Fig. 6.25, the symmetrical arrangement of the big arm cylinders improves the function and performance of the structure.

Fig. 6.22 Sleeve of the big arm

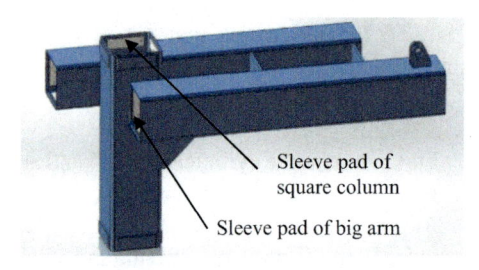

Fig. 6.23 Big arm inserted to its sleeve

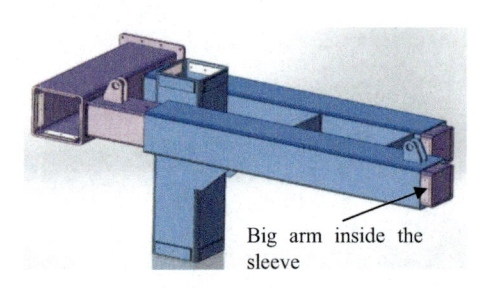

Fig. 6.24 Completion of the installation of the big arm pad

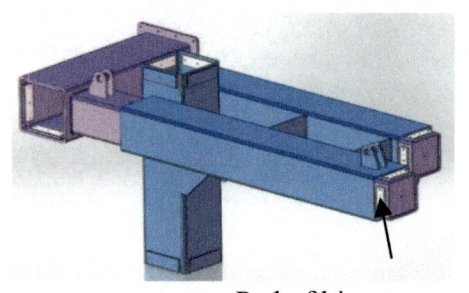

6.2 6.0 m × 3.3 m Rectangular Shield

Fig. 6.25 Symmetrical arrangement of the cylinder for the big arm

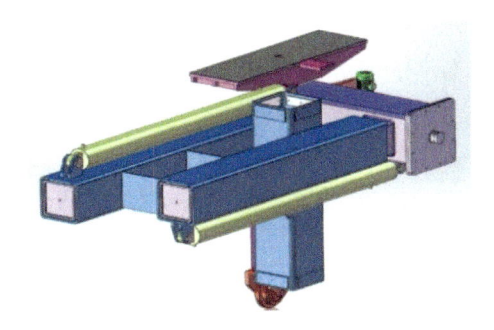

6.2.4.4 Mechanism for Lifting Lining Segments

Figure 6.26a shows the tubes when it is fully extended. The extension and retraction of the tubes, i.e., the lifting and lowering of the lining segments, are controlled by one two-stage hydraulic cylinder.

Figure 6.26b shows the outer tube when it is fully retracted. There is a single acting cylinder used for lifting of the lining segments. The retraction of the cylinder depends on the erector system for the lining segments and/or the self-weight of the lifted lining segment.

Figure 6.27 shows the rotating process of the pipe section to 90°. The mechanism for lifting the lining segments can always maintain the fixed value of the cantilever length of the big arm. Therefore, it can erect heavy lining segments with large cross-sections.

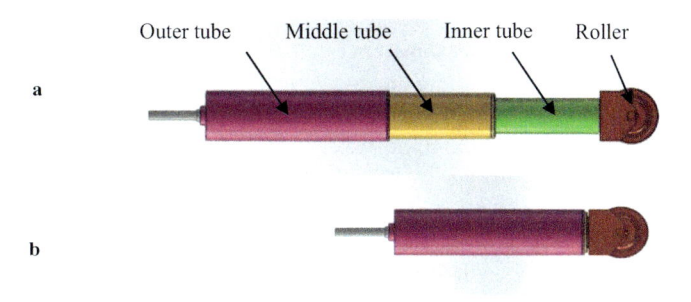

Fig. 6.26 **a** Tubes in full extension, **b** outer tube in full retraction

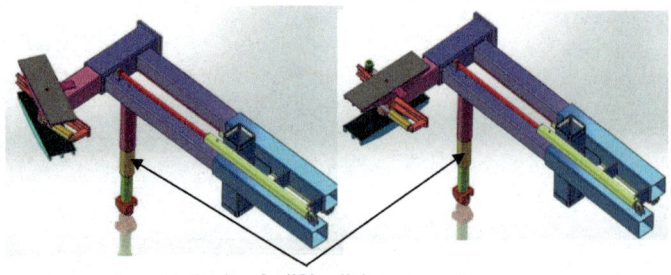

Mechanism for lifting lining segments

Fig. 6.27 Rotating process of the mechanism

Fig. 6.28 Top hydraulic thrust cylinders

6.2.5 Other Accessories

6.2.5.1 Top Hydraulic Thrust Cylinders

A top hydraulic thrust cylinder is shown in Fig. 6.28, which is a two-stage isothrust cylinder. Two-stage means that there are one large and one small piston rods during extension. The stroke of the large piston rod is 1300 mm while that of the small piston rod is 1200 mm. Thus, the total stroke is 2500 mm. There are a total of ten cylinders.

Generally, the thrust from a two-stage cylinder is not equal. The large piston rod produces a big thrust, while the smaller piston rod induces a small thrust. However, the design of the two-stage cylinder presented in the chapter has almost the same thrust, which is 200 t. It is also called an equal thrust cylinder. For more details, readers are referred to Yu et al. (2018).

6.2.5.2 Bottom Hydraulic Thrust Cylinders

Figure 6.29 shows a bottom hydraulic thrust cylinder. It is a conventional cylinder with a stroke of 1400 mm, a thrust of 200 t, and a total of ten cylinders.

Fig. 6.29 Bottom hydraulic thrust cylinder

Fig. 6.30 Moveable track

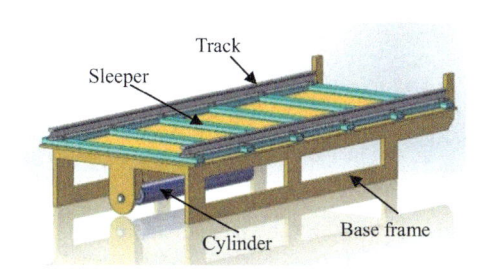

6.2.5.3 Moveable Track

Figure 6.30 shows the moveable track. Rail sleepers and tracks are mounted on the top of the base frame, while the cylinders are mounted under the base frame. The vertical plates on both sides of the base frame are placed in the mounting groove of the moveable track in the lower part of the shield. The stroke of the cylinder is 1400 mm.

6.2.6 Erection Process of Lining Segments

Preparatory work has to be carried out before erecting the lining segment. The first step is to retract all cylinders of the erector system (Fig. 6.31) back to their original positions.

The bottom ten hydraulic thrust cylinders (Fig. 6.32) are retracted, while the moveable track is maintained.

However, the ten hydraulic thrust cylinders on the top cannot be retracted as shown in Fig. 6.33. Otherwise, the shield will be backed off which will cause the ground to collapse.

First, install the lower lining segment to push it into the area to be installed, as shown in Fig. 6.34.

The process of pushing the lower lining segment into the area to be installed can be viewed from a different direction as shown in Fig. 6.35.

Fig. 6.31 Initial stage of the erection system for lining segments

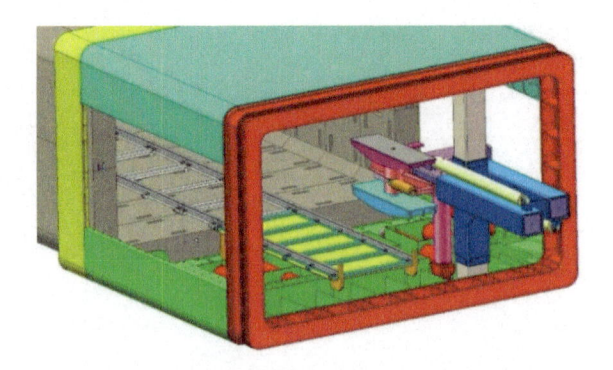

Fig. 6.32 Bottom ten hydraulic thrust cylinders

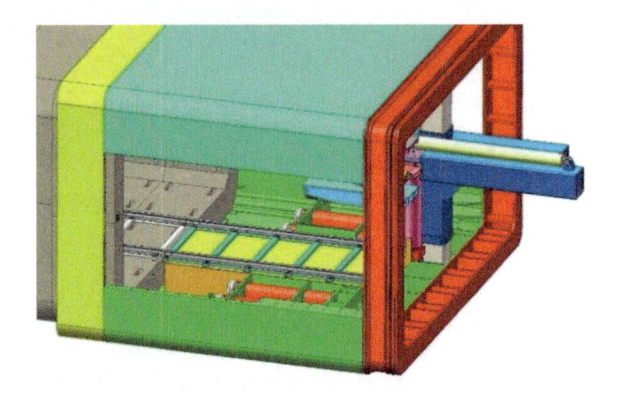

Fig. 6.33 Top ten hydraulic thrust cylinders

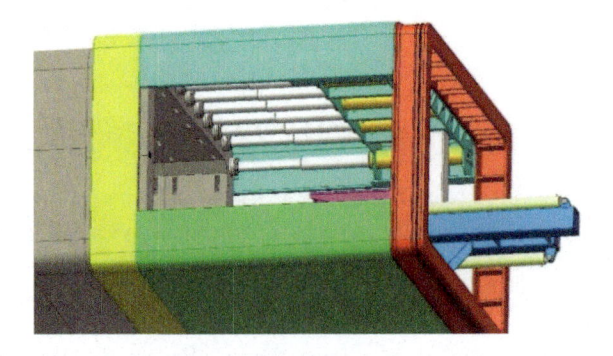

The erector is hung on a steel mesh (not shown in the figure), while the center of the mechanism for rotating lining segments and center of the lower lining segment are ensured to be on the same vertical line. Then, the lower lining segment is lifted up a little bit as shown in Fig. 6.36.

After the lower lining segment is lifted, the transport wheels of the lining segment will be lifted off the track as well (Fig. 6.37). After that, they are manually removed as shown in Fig. 6.38.

6.2 6.0 m × 3.3 m Rectangular Shield

Fig. 6.34 Erecting the lower lining segment

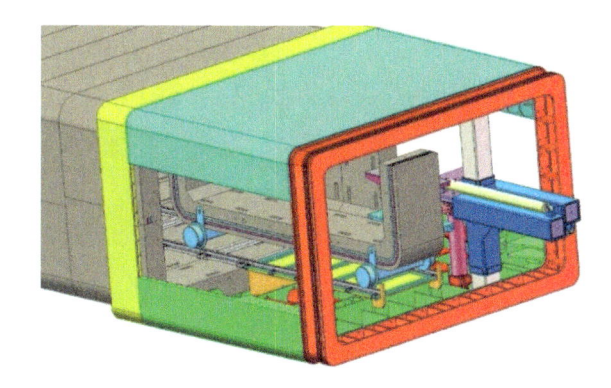

Fig. 6.35 View of pushing the lower lining segment

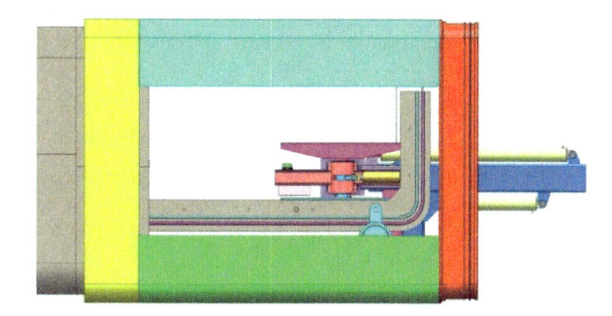

Fig. 6.36 Lifting the lower lining segment

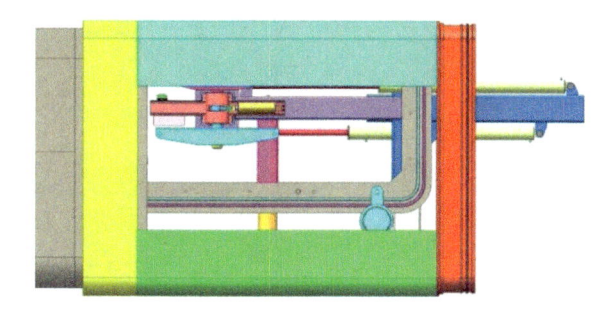

Fig. 6.37 Delivery of the lower lining segment

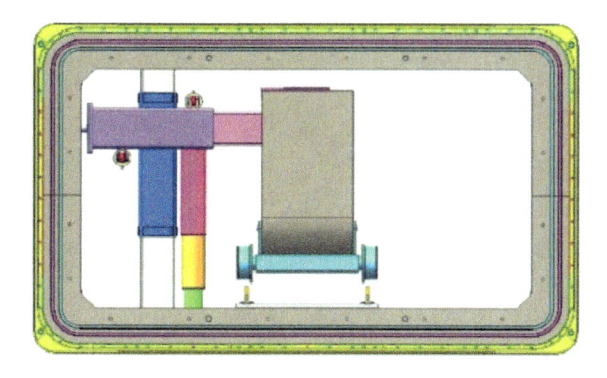

Fig. 6.38 Wheel removal

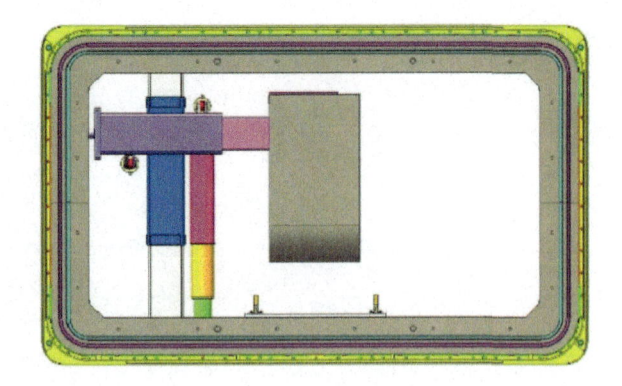

The lining segment is lifted within the range of the two side parts of the segment erection shield, as shown in Fig. 6.39. While doing these operations, the movable track must be retracted to allow space for the lining segment.

The lining segment is rotated by the rotating mechanism as shown in Fig. 6.40.

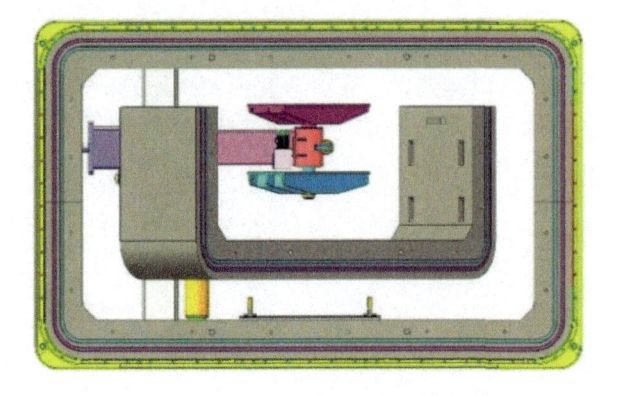

The lining segment is further rotated to the right position as shown in Fig. 6.41.

The lining segment is then lowered down as shown in Fig. 6.42a until the lining segment is placed on the rib of the bottom plate (Fig. 6.42b).

Fig. 6.39 Lifting the lower lining segment into the segment erection shield

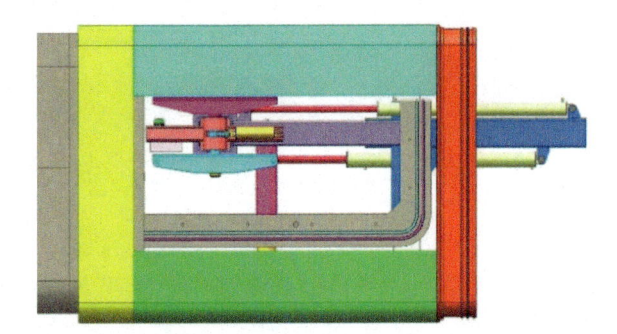

Fig. 6.40 Rotation of the lower lining segment

6.2 6.0 m × 3.3 m Rectangular Shield

Fig. 6.41 Lower lining segment rotated to the right position

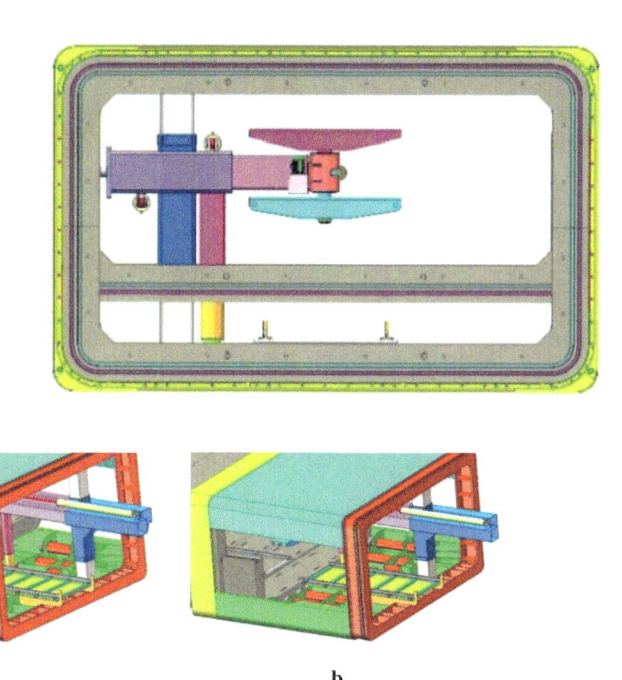

Fig. 6.42 **a** Lowering down the lower lining segment; **b** lower lining segment in place

Before the bottom hydraulic thrust cylinders are started, the two pins are aligned and inserted partially into the fitting hole (the original lifting hole) of the erected lower lining segment, as shown in Fig. 6.43.

The new lower lining segment is attached to the previous erected lining segments and fastened with the connecting bolts, as shown in Fig. 6.44.

Fig. 6.43 Alignment insertion of the two pins

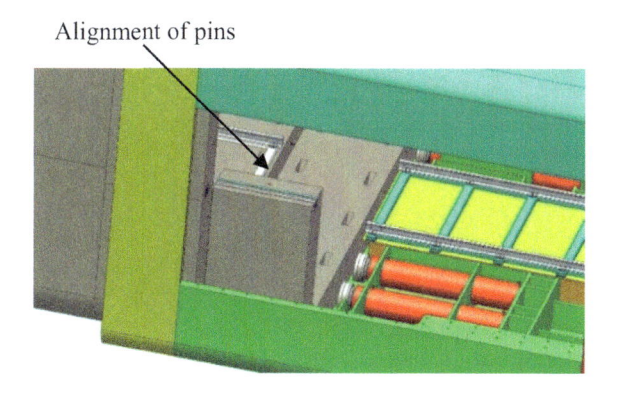

Fig. 6.44 Erection of lower lining segment

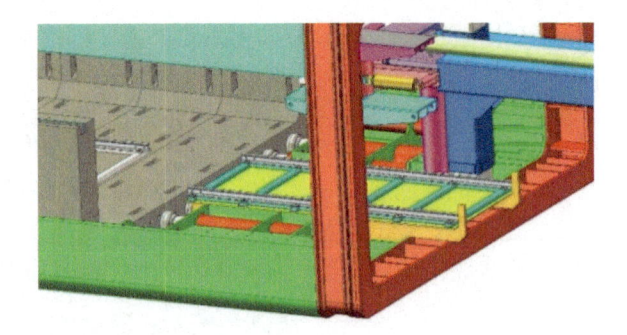

After the new lower lining segment is erected, the original 1.2 m track is removed as shown in Fig. 6.45. Then, a new track with a length of 2.4 m is installed as shown in Fig. 6.46.

After another lining segment is installed, the 2.4 m track is replaced with a 3.6 m track. The track is no longer replaced until a 4.8 m track is installed, i.e., after the four lining segments are erected and the installed track is 4 0.8 m.

Fig. 6.45 Removal of the 1.2 m track

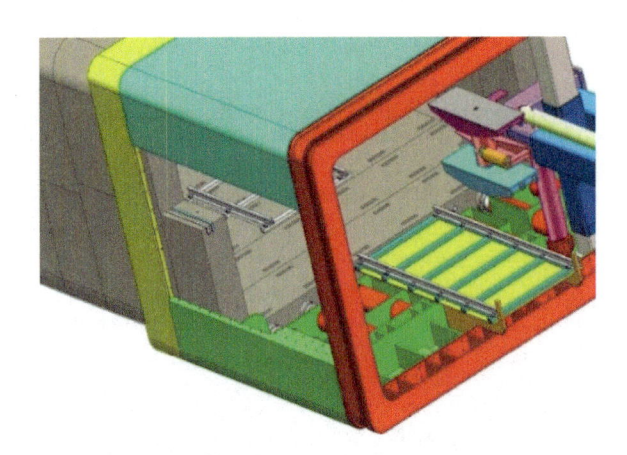

Fig. 6.46 Replacement of a 2.4 m track

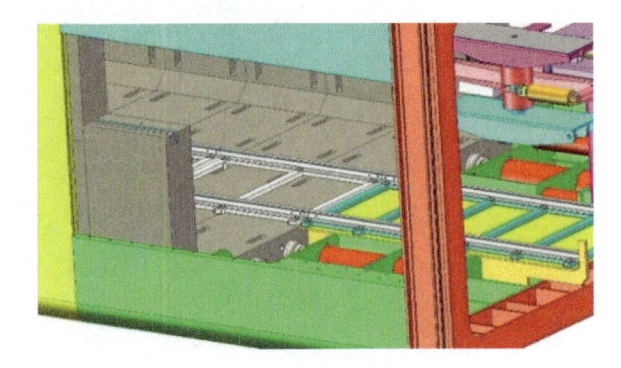

6.2 6.0 m × 3.3 m Rectangular Shield

After the erection of the lower lining segment is finished, the erection of the upper lining segment starts as shown in Fig. 6.47. Before erection, the erector system must be retracted to its original position, the bottom hydraulic thrust cylinders are extended, and the top hydraulic thrust cylinders are retracted.

Then the upper lining segment is moved into the erection area as shown in Fig. 6.48.

Figure 6.49 shows a side view of the upper lining segment in the areas to be installed.

Fig. 6.47 Erection of the upper lining segment

Fig. 6.48 Upper lining segment to be installed

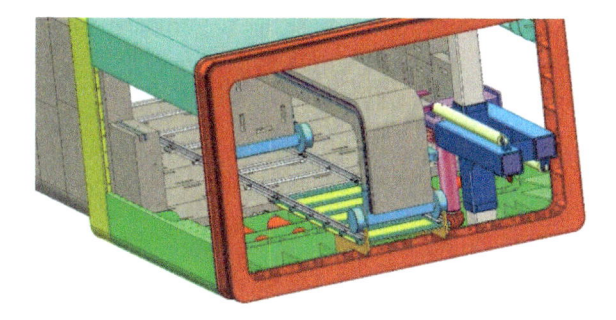

Fig. 6.49 Side view of the lining segment to be installed

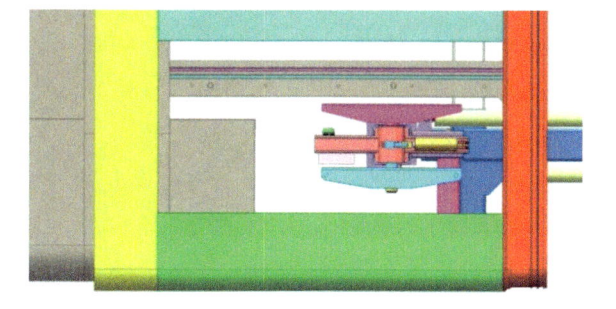

Fig. 6.50 View of the upper lining segment from the shield machine

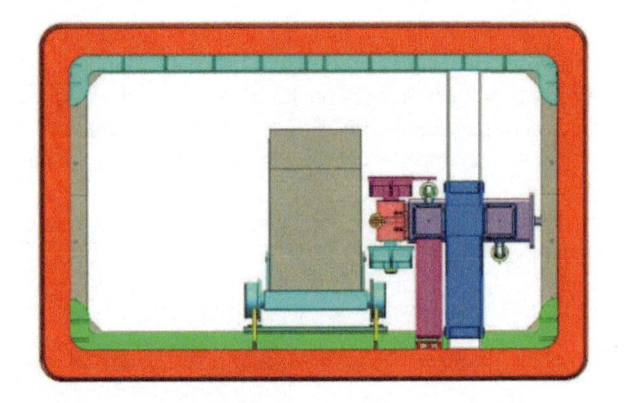

Figure 6.50 shows the view of the upper lining segment from the direction of the shield machine. After the cylinders are fully retracted, there will be no interference with the upper lining segment.

It is necessary to align the upper plate beam of the mechanism when rotating the lining segment with the top inside of the lining segment during the erection of the upper lining segment as shown in Fig. 6.51.

The erector system is lifted up first while the upper lining segment is slightly moved up. Then, the transport wheel for the lining segment is removed as shown in Fig. 6.52. Note that there is no pin on the upper lining segment.

Then, the upper lining segment is rotated within the side panel as shown in Fig. 6.53. Meanwhile, observation and monitoring have to be carefully carried out to avoid collision between the upper and the lower lining segments.

As can be seen from Fig. 6.54, the upper lining segment is moved toward the erected lower lining segment after it is turned to the proper position and stopped when the horizontal gap between the upper and lower lining segments is about 50 mm.

The upper lining segment is then slowly moved up, and the relative position of the upper and lower lining segments is monitored during the lifting process. Necessary

Fig. 6.51 Alignment of the rotating mechanism with upper lining segment

6.2 6.0 m × 3.3 m Rectangular Shield

Fig. 6.52 Removal of transport wheels for upper lining segment

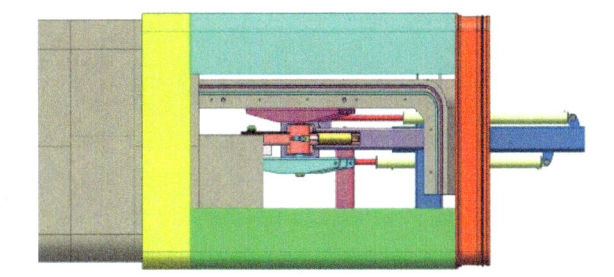

Fig. 6.53 Rotation of the upper lining segment

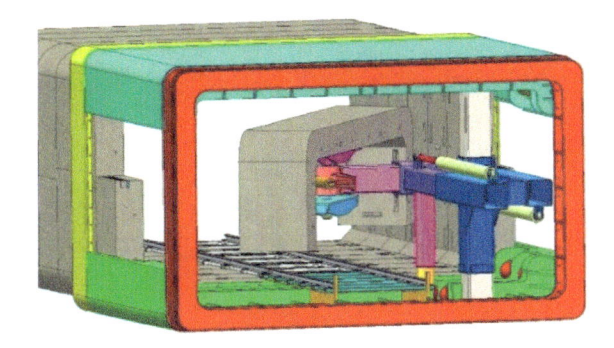

Fig. 6.54 Moving the upper lining segment to the lower one

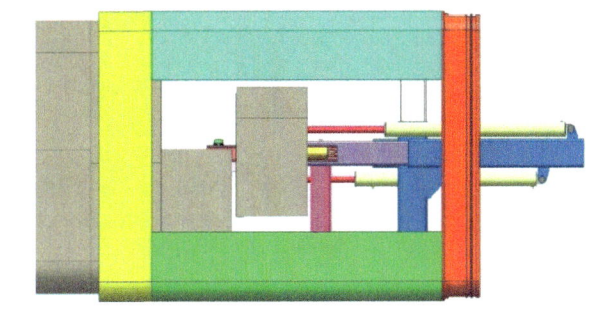

adjustments have to be followed up in a timely manner for deviation rectification. The upper lining segment will be lifted only after the deviation has been rectified.

The process is shown in Figs. 6.55, 6.56, and 6.57.

When the vertical gap between the upper and lower lining segment is about 20 mm, and the top of the upper lining segment does not touch the upper part of the segment erection shield, the big arm will be slowly extended (Fig. 6.58) until the upper lining segment is close to its counterpart in the previous lining ring. When the upper lining segment fall close to the lower lining segment, all the connecting bolts are fastened. The final stage is to extend the top hydraulic thrust cylinder to hold the upper lining segment in place. The process is shown in Figs. 6.59, 6.60, and 6.61.

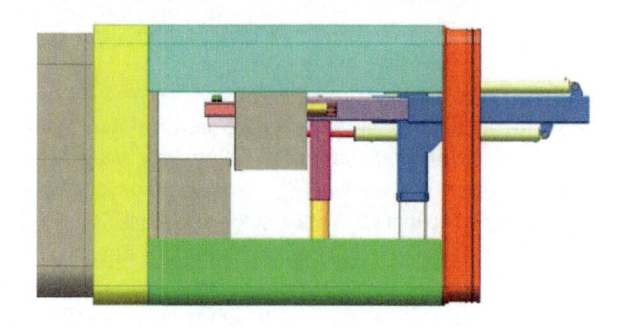

Fig. 6.55 Moving the upper lining segment up slowly

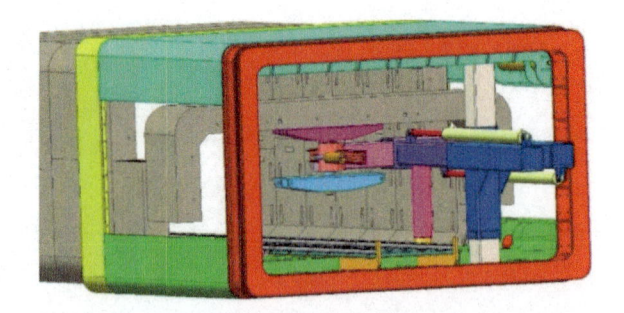

Fig. 6.56 Observation of deviation for the upper lining segment during the lifting process

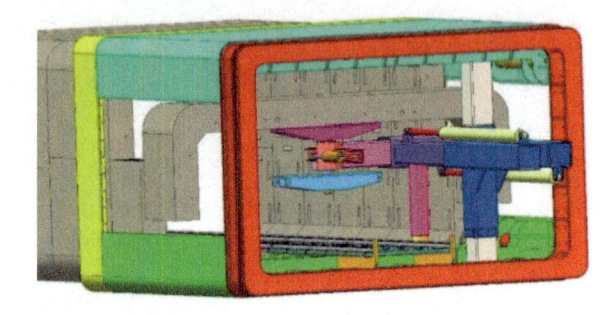

Fig. 6.57 Deviation rectification

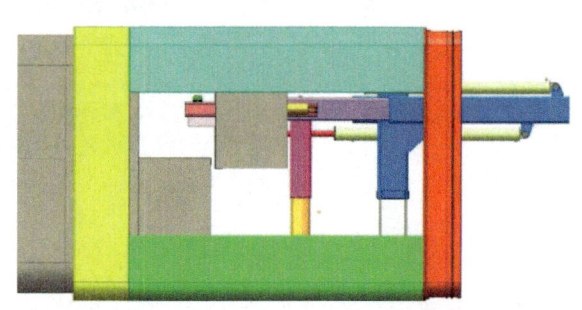

Fig. 6.58 Lifting up the upper lining segment after deviation rectification

6.2 6.0 m × 3.3 m Rectangular Shield

Fig. 6.59 Moving the upper lining segment close to its counterpart in the erected lining ring

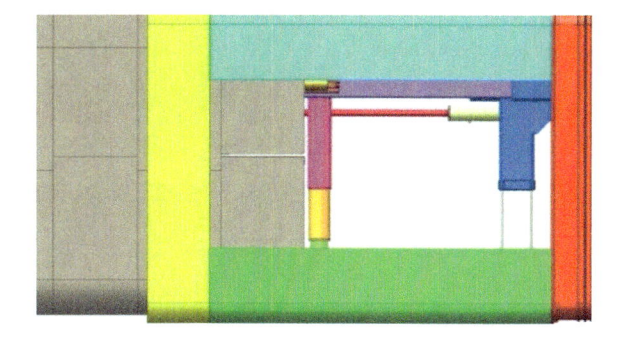

Fig. 6.60 Upper lining segment in place

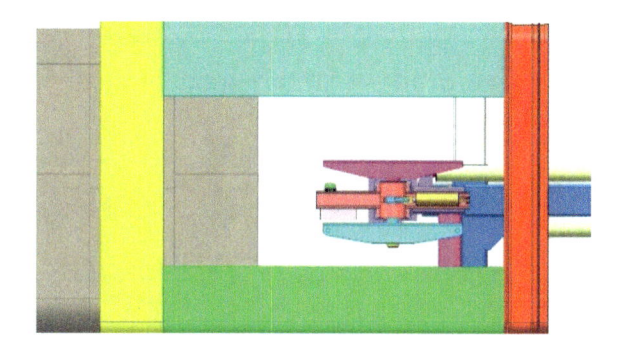

Fig. 6.61 Extension of the top hydraulic thrust cylinders to hold the upper lining segment

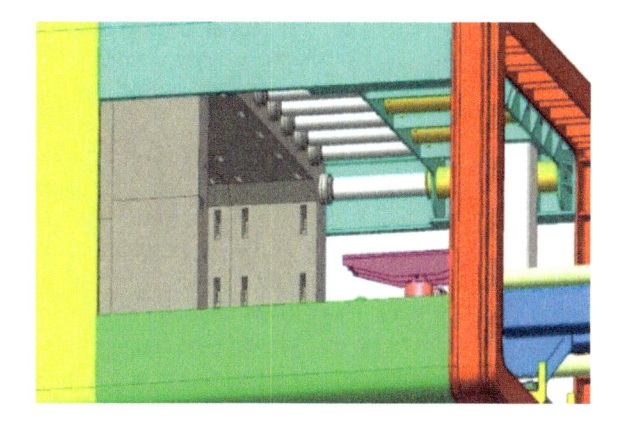

At this point, the erection of the new ring of lining segments is completed, ready to move forward as shown in Fig. 6.62.

Using detailed specifications and illustrations, the entire procedure of the erection of lining segments in the shield was introduced. Construction methods will be further improved and developed through engineering best practices in time to come. At the time of writing, the patent application of this method is currently being processed in China.

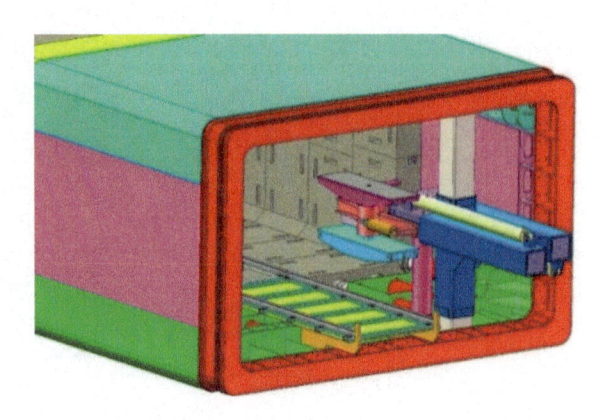

Fig. 6.62 Overview of the completion of one ring of lining segments

6.3 Case Study—An Irregular-Shaped Shield Project in Shandong, China

6.3.1 Project Background

The underground utility tunnel project, located at Jibei New District, Renxing Road, Jining City, Shandong province, is considered as the first rectangular shield construction project in the world (Gong 2015). The construction method was designed by Shandong Wan Guang Construction Engineering Co., Ltd China, and Guang Xin Heavy Industry Co., Ltd, Yangzhou, China. The construction was carried out by Shandong Wang Guang Construction Engineering Co. Ltd in 2015.

The cross-section of the 366 m long tunnel has a width of 2.7 m and a height of 3.0 m. The tunnel mainly goes through the loess soil with an overburden of 4.5–6 m. Figure 6.63a, b shows the part view of the completed tunnel and the working shaft of the irregular shield respectively. From Fig. 6.63b, several lower U-shaped segments were already installed in the working shaft. The number of the lower segments were determined according to the requirements of the muck delivery system and segment lifting. The location and elevation of the lower segments must meet the design requirements. The first lower segment was connected to the last ring of the tunnel, while the last lower segment was supported by several hydraulic cylinders. The cylinders rested against the reaction frame, which contacted closely to the reaction wall. When the tunnel construction was completed, the hydraulic cylinders were simply retracted, and these lower lining segments were removed.

6.3.2 U-shaped Lining Segments and Connections

The lining segments and connections of the utility tunnel are shown in Fig. 6.64, with Fig. 6.64d showing the three-dimensional schematic of the tunnel. The specifications

6.3 Case Study—An Irregular-Shaped Shield Project in Shandong, China 225

Fig. 6.63 **a** Part view of the completed utility tunnel; **b** working shaft for shield construction

of the lining rings, which are formed by upper and lower U-shaped segments, are summarized in Table 6.1.

The design and application of the two-piece upper and lower U-shaped lining segments in this project is an innovation based on the concept of the conventional segment erection knowledge. It reduces the number of connections among all segments compared with the conventional multiple segments in shield or tunnel boring machine (TBM) construction. Consequently, it reduces leakage and risk of construction besides improving the overall stability of the tunnel.

As the upper and lower U-shaped segments can be rotated in the tunnel during erection, it is much more advantageous in terms of construction speed and soil settlement control. In general, it overcomes the construction limitations of pipe jacking. The major advantages of using U-shaped segments are summarized in Table 6.2.

From Fig. 6.64d, there are two types of both the upper and lower segments, i.e., type 1 and type 2. Note that Fig. 6.64d shows 2 lining rings. Take the upper lining segment for example, the bottom surfaces of the upper and lower segments are 150 mm below and above the center line, respectively. Similarly, the top surfaces of the upper and lower segments are 150 mm above and below (staggered along) the center line, respectively. There are sealing grooves on the interfaces between the adjacent lining rings and between the upper and lower segments. In addition, both sides of a lining ring are chamfered with dimensions of 230 mm × 50 mm (Fig. 6.64c). This is to minimize the space requirement for rotating the lining segments, thus reducing the gap at the shield tail. The U-shaped lining segments used in the tunnel construction are shown in Fig. 6.65.

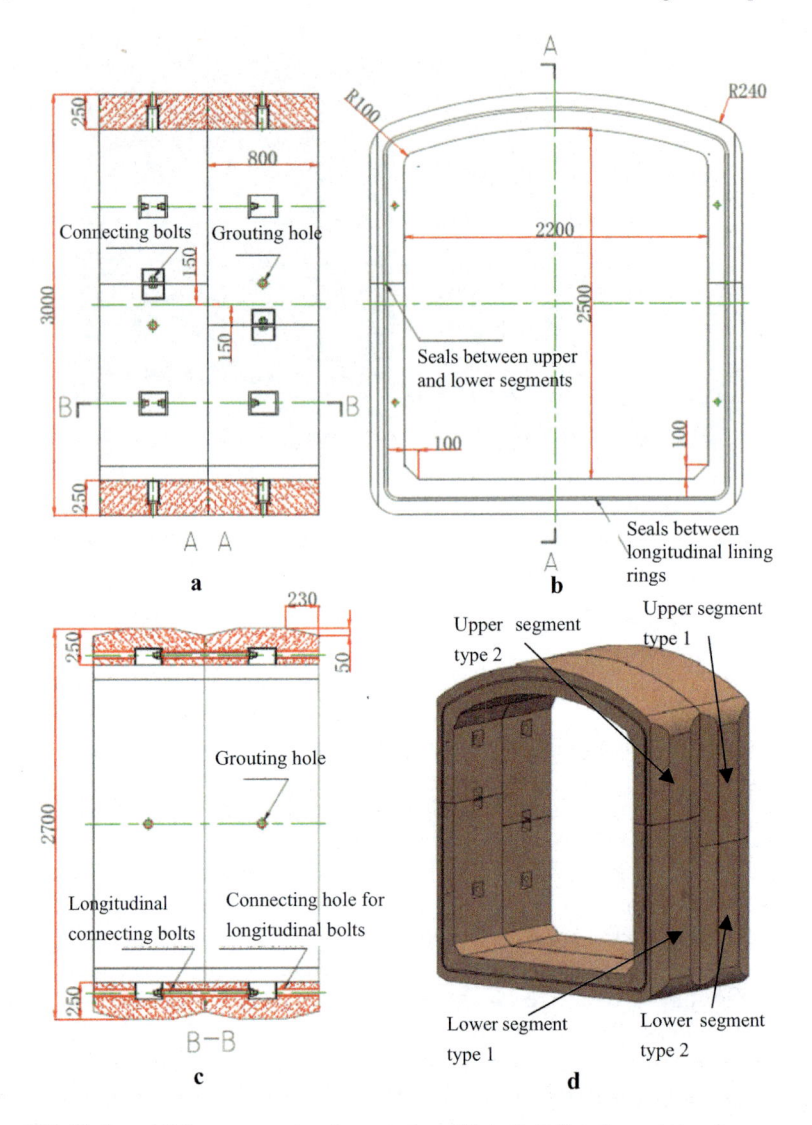

Fig. 6.64 U-shaped lining segments and connections (Note that all units are in mm)

Table 6.1 Specifications of lining rings

Parameters	Values
Outer dimensions (mm)	3000×2700
Inner dimensions (mm)	2500×2200
Thickness (mm)	250
Width (mm)	800
Radius of the top arc (mm)	3437.5

6.3 Case Study—An Irregular-Shaped Shield Project in Shandong, China

Table 6.2 Advantages of using U-shaped segments

Shield	Advantages
Leakage	Reduced in comparison with conventional segment connections
Construction safety	Workers physically shielded from soil face and construction risk reduced due to fast schedule of construction
Soil stability	Better control of soil movement and ground settlement
Quality control	Easy to control as U-shaped segments are precast
Productivity	High compared with conventional segment connections
Environmental control	More environmentally friendly with lower noise level and less dust

Fig. 6.65 U-shaped lining segments on site

6.3.3 Procedure of the Shield Construction

The procedure of the shield construction is described in detail with the illustrations from Figs. 6.66, 6.67, 6.68, 6.69, 6.70, 6.71, 6.72, 6.73, 6.74, 6.75, 6.76, 6.77, 6.78, 6.79, 6.80 and 6.81.

It is to be noted that the pipe jacking process is also introduced in this section for better understanding of the hybrid method using both the jacking and shield methods, which will be discussed in Sect. 6.3.4. In the hybrid method, the pipe jacking method should be carried out in the first construction phase, followed by the shield method in the second phase.

A pipe jacking machine used during the construction stage is shown in Fig. 6.66. The muck produced in the jacking process by the earth pressure balance shield is transferred to the soil and sand pump, installed at the shield front via the screw conveyor and then to the muck truck via the soil transporting pipe. The pipe and the bend are joined using clamp connectors. The soil transporting pipe is supported by frames with rollers at certain distance intervals in the tunnel rectangular pipes. The control console and the electrical cabinet are located in the segment erection shield. The shield thrust cylinder is placed at the shield tail, and it is kept fully extended

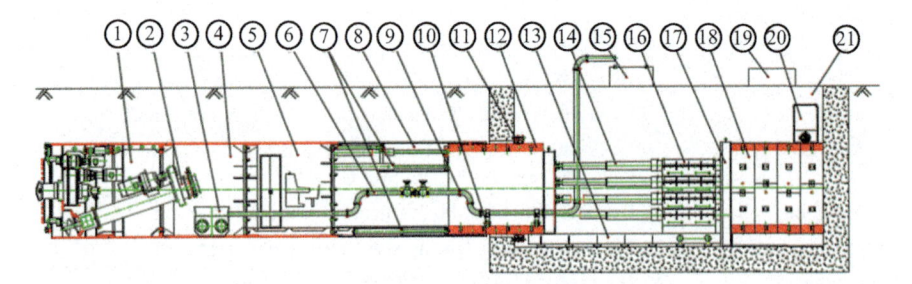

Fig. 6.66 Configuration of the pipe jacking assembly during construction stage. 1—Shield machine; 2—Screw conveyor; 3—Soil and sand pump; 4—Shield front; 5—Segment erection shield; 6—Soil transporting pipe; 7—Shield thrust cylinder; 8—Shield tail; 9—Clamp connector; 10—Frames with rollers; 11—Entrance seal; 12—Rectangular pipe jacked in place; 13—Rail track in the working shaft; 14—Main thrust cylinder; 15—Muck truck; 16—Frame of the main thrust cylinder; 17—Reaction frame; 18—Multiple rectangular pipes; 19—Grouting device; 20—Main hydraulic pump station; 21—Working shaft

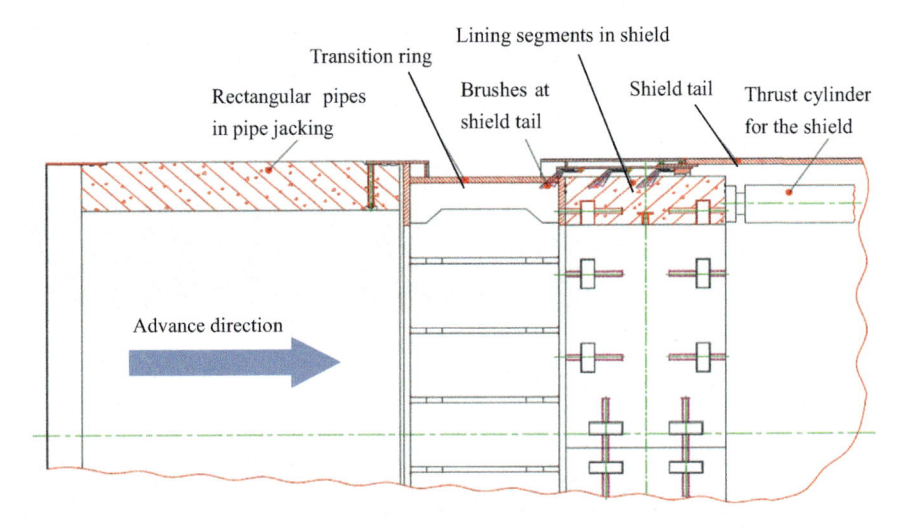

Fig. 6.67 Connection between the pipe jack and the shield stages

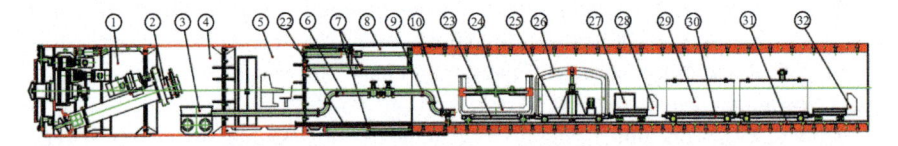

Fig. 6.68 General concept of the shield construction stag. 1—Shield machine; 2—Screw conveyor; 3—Soil and sand pump; 4—Shield front; 5—Segment erection shield; 6—Soil transporting pipe; 7—Shield thrust cylinder; 8—Shield tail; 9—Clamp connector; 10—Frames with rollers; 22—Working platform; 23—Cart for transporting lower segments; 24—Lower lining segment to be transported; 25—Cart for transporting upper segments; 26—Upper lining segment to be transported; 27—Assembly unit for synchronized grouting; 28—Battery-operated truck; 29—Soil container; 30—Muck truck; 31—Rail track in the tunnel; 32—Battery-operated truck

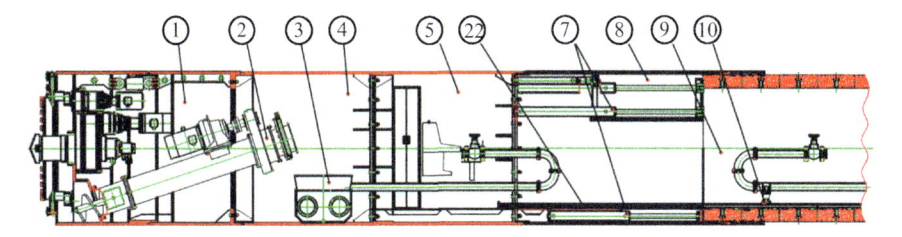

6.3 Case Study—An Irregular-Shaped Shield Project in Shandong, China 229

Fig. 6.69 Preparations for lining installation

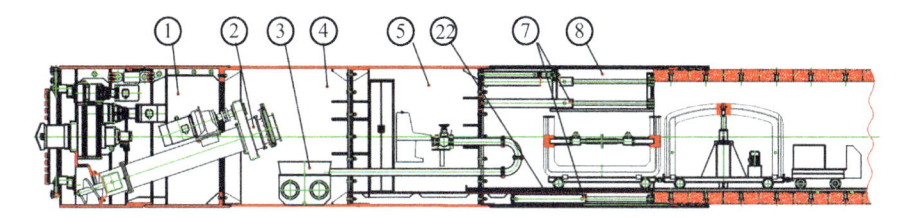

Fig. 6.70 Lower lining segment on the working platform

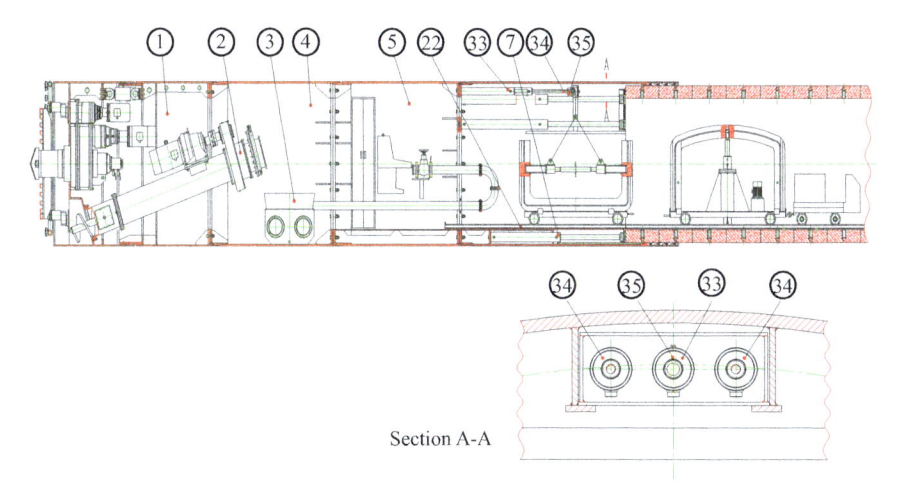

Section A-A

Fig. 6.71 Preparations for lifting the upper lining segment. 33—Cylinder for the steel wires; 34—Cylinders on both sides of the pulley; 35—Pulley with steel wires

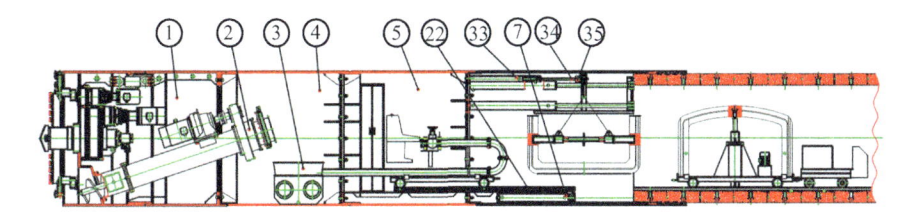

Fig. 6.72 Space for installing the lower lining segment

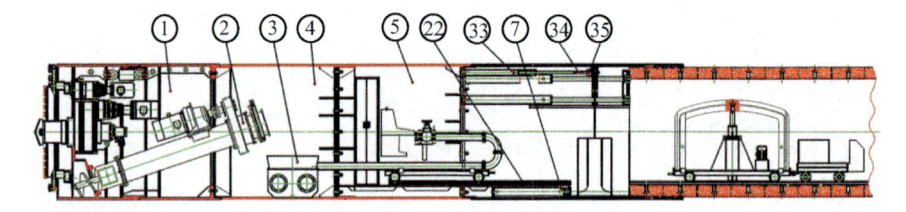

Fig. 6.73 Lowering down the lower lining segment for installation

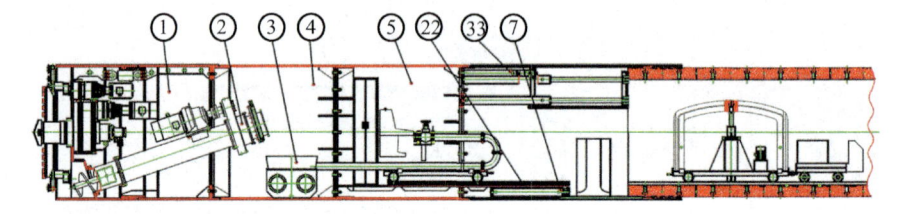

Fig. 6.74 Pushing back the lower lining segment

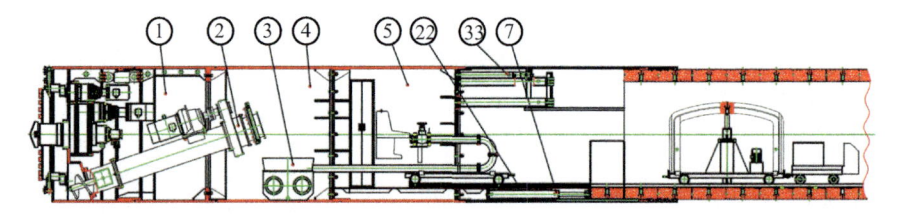

Fig. 6.75 Connecting the lower lining segment to its counterpart in the previous ring

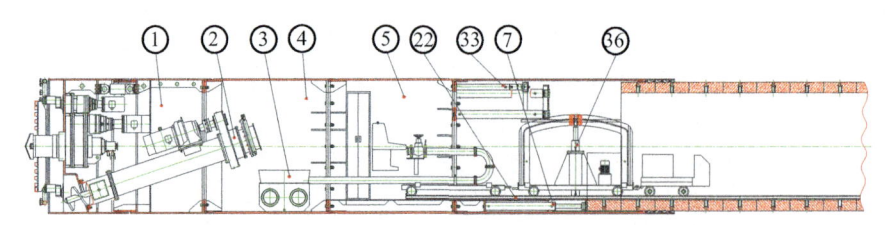

Fig. 6.76 Transportation of the upper lining segment by the battery-operated truck. 36—Cylinder for upper lining segments

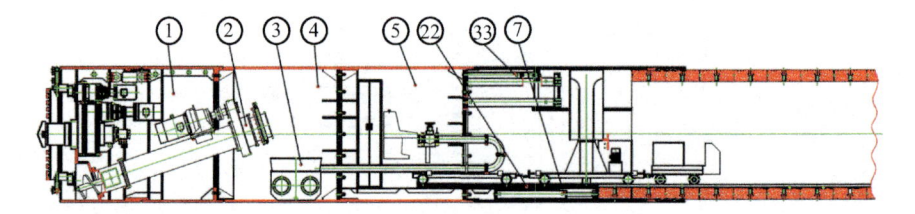

Fig. 6.77 Lifting the upper lining segment for installation

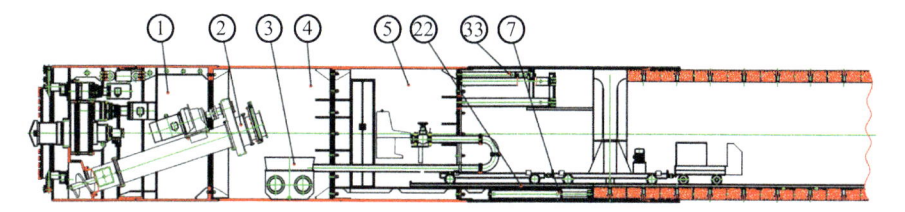

6.3 Case Study—An Irregular-Shaped Shield Project in Shandong, China 231

Fig. 6.78 Joining the carts for transporting the upper and lower lining segments

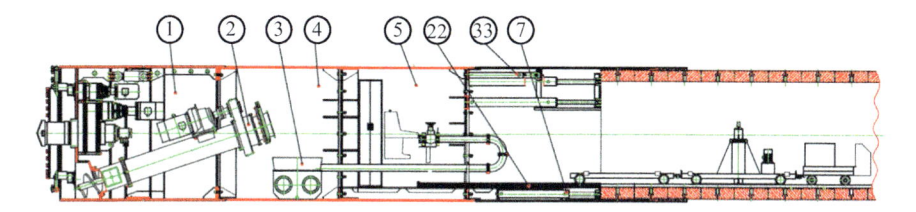

Fig. 6.79 Completing the installation of one ring of lining segment

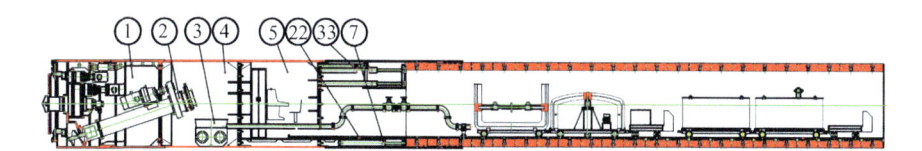

Fig. 6.80 Preparations for erection of the next lining ring using the rectangular shield

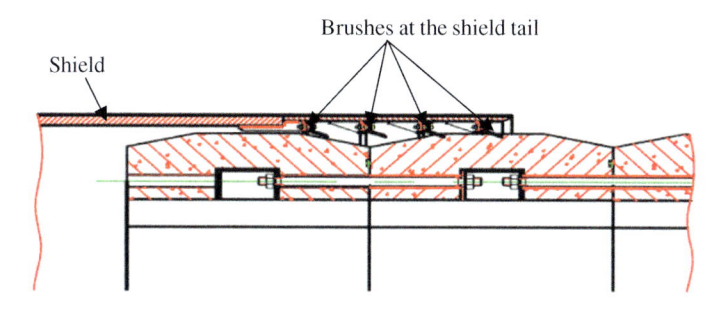

Fig. 6.81 Seals at the shield tail

during pipe jacking. Rectangular pipes are erected on site and hoisted into the working (launching) shaft. If the stroke of the main thrust cylinder is long enough, several rectangular pipes can be hoisted into the working shaft and then jacked forward once. To make the schematic simple, the connections of the rectangular pipes (lining rings in the shield construction) are not shown.

The working shaft is used for both jacking and shield construction. It is a bit too long when used for jacking. Therefore, multiple rectangular pipes must be put at the rear of the main thrust cylinder. Note that at this stage, the rectangular pipes just serve as a support structure and must be removed when switching to shield construction. Therefore, no seals are required to be installed.

There is not much difference between the pipe jacking during the construction stage and the conventional rectangular pipe jacking. Compared with the shield construction stage, the construction speed is much faster, but its jacking length is constrained by the relationship between the huge frictions and the jacking force. When the pipe jacking force reaches the maximum bearing capacity of the rectangular sections and/or the reaction wall in the working shaft, the pipe jacking construction must be stopped and switched to the shield construction stage.

Figure 6.67 shows the connection between the pipe jacking and the shield stages.

The concept of the shield construction stage is shown in Fig. 6.68. The process for transporting the muck out of the tunnel is the same as that in the pipe jacking stage. When one ring of the rectangular lining segments is erected in place, the muck truck is moved by the battery-operated truck to the working shaft to be hoisted to the above ground.

The two carts for transporting the upper and lower lining segments are dragged by the battery-operated truck. The assembly unit for synchronized grouting is mounted on the battery-operated truck.

At the beginning of the shield construction stage, the two gate valves of the soil transporting pipe at the shield tail are first closed, followed by loosening of the clamp connector between the two gate valves. Then the two segments of the soil transporting pipe are rotated around 180° in both directions, as shown in Fig. 6.69, to leave enough space for rotation and installation of the lining segments.

Figure 6.70 shows a lower lining segment which is transported to the working platform.

As shown in Fig. 6.71, the connection between the carts for transporting the upper and lower lining segments is removed and the upper lining segment is slightly moved back. The two hydraulic cylinders used for the pulley with steel wires are extended to the appropriate position, and the one for the steel wires is also extended. The steel wires, hoisting sling, and sling for lifting the lower lining segments are put together as shown in Fig. 6.71.

Then, after the hydraulic cylinder for steel wires is retracted to lift the lower lining segment to a certain height, the bottom hydraulic cylinder for the shield is retracted, while the cart for transporting lower lining segments is manually pushed forward to the end. As the working platform, the rail track and the bottom hydraulic cylinder for the shield are connected; they move with the bottom hydraulic cylinders, therefore leaving room for installation of the lower lining segments (Fig. 6.72). At this point, the top hydraulic cylinder for the shield should not be retracted. Otherwise, the shield machine and the shield will move back.

6.3 Case Study—An Irregular-Shaped Shield Project in Shandong, China

The lower lining segment is manually rotated 90° in the specified direction, and the notches for installing the seal ring should be exposed. Then, the hydraulic cylinder for the steel wires is extended to drop the lower lining segment to the bottom of the shield tail (Fig. 6.73).

The bottom hydraulic cylinder for the shield is slowly extended to push the lower lining segment back, as shown in Fig. 6.74.

When the lower lining segment is pushed into position, the upper lining segment is connected to its counterpart in the previous lining ring using the double-end connecting bolts, washers, and nuts. At the same time, the sealing strips of the upper and lower lining segments are glued in the sealing grooves (Fig. 6.75). So far, the installation of the lower lining segment is completed, and the next step is to install the upper lining segment.

The top hydraulic cylinders for the shield are fully retracted first, and the battery-operated truck is used to push the upper lining segment to an appropriate position as shown in Fig. 6.76. Then the oil pump above the upper lining segment is started. After the hydraulic cylinder for the upper lining segment is moved to a certain height, the upper lining segment is manually rotated at a slow pace in the specified direction, while the battery-operated truck is moved back and forth as needed.

Until rotated 90°, the upper lining segment is lifted again. When the upper and lower lining segments have a gap (Fig. 6.77), the upper lining segment is moved closer to its counterpart in the previous lining ring by the battery-operated truck, followed by placement of the upper segment on the lower segment.

When placed in position, the upper lining segment is connected to its counterpart in the previous lining ring using the double-end connecting bolts, washers, and nuts. Lastly, the carts for transporting the upper and lower lining segments are joined together as shown in Fig. 6.78.

The empty carts are pulled to the working shaft to transport the upper and lower segments for the next lining ring. At the shield tail, the sealing strips of the upper lining segment are glued in the sealing grooves. Then the top hydraulic cylinder for the shield is extended until it reaches the upper segment, followed by retraction of the bottom hydraulic cylinder for the shield. When there is a gap between the bottom hydraulic cylinders and the lower segment, the sealing strips of the lower segment are glued in the sealing grooves. After that, the bottom hydraulic cylinders for the shield are extended until they reach the lower segment (Fig. 6.79).

As indicated by Fig. 6.80, the soil transporting pipe is reconnected, ready for erecting the next lining ring using the rectangular shield. During the excavation, the muck is pumped out with the soil and sand pump and then transferred to the muck truck through the soil transporting pipe and lastly transported out by the battery-operated truck.

As there are chamfers on both sides of the lining sections, leading to wavy surfaces, a seal ring at the shield tail must be specifically designed to cater for this feature. Figure 6.81 shows the specifically designed seal rings without compression. They are V-shaped seal ring. As shown in Fig. 6.81, there are four seals in total, which can ensure that there are two seals at any location.

Table 6.3 Summary of the details of Phases 2 and 3 of the project

Project Items	Phase 2	Phase 3
Jacking length (m)	102	310
Shield length (m)	208	36
Average jacking speed (m/day)	10–15	
Average shield speed (m/day)	8–12	

6.3.4 Introduction of Hybrid Method Using Irregular-Shaped Pipe Jacking and Irregular-Shaped Shield Methods

Further examples of using a combination of the jacking and shield construction method will be introduced in this subsection. This method is useful to demonstrate the diversity, in terms of the construction speed, advantages, and applications to the projects.

The irregular pipe jacking method and the shield method can be applied in practical projects separately or together as described in Sect. 6.3.3. In this case study, only the shield method was applied to a trial project.

The subsequent projects, which belong to the same group of the project in this case study, have been undertaken. We refer to these projects as Phase 2 and Phase 3 here. The hybrid method using the irregular pipe jacking method and the shield method has been applied successfully to Phase 2 and Phase 3. This hybrid method aims to make full use of the advantages of both the irregular pipe jacking and the shield methods. For example, the construction speed of the pipe jacking method is faster than that of the shield method but is constrained by the jacking distance. The shield method is more efficient not only for long distance tunneling but also for curved tunnels if compared with the pipe jacking method.

The experience from Phase 2 and Phase 3 of the project as mentioned above demonstrates that the jacking speed by the pipe jacking method and the shield method is about 10–15 m and about 8–12 m daily on average, respectively. The jacking lengths by the pipe jacking method and the shield method are 102 m and 208 m, respectively, for Phase 2; 310 m and 36 m, respectively, for Phase 3 as shown in Table 6.3. More details may need to be further documented in future when more information and records become available from Phase 2 and Phase 3 of the project.

6.3.5 Site Photos

Some site photos (Figs. 6.82, 6.83 and 6.84) from Phases 2 and 3 of the project, using the hybrid method, i.e., irregular pipe jacking and shield technique, are presented below.

Fig. 6.82 Working shaft for the hybrid method using the pipe jacking method and the shield method

6.3.6 Conclusions

The project is considered as the first irregular shield technique using simple U-shaped lining segments to form the lining rings. This was carried out via simple operations based on the innovative erection system.

The underground utility tunnel project and the subsequent Phases 2 and 3 of the project in this case study are useful in demonstrating the diversity in terms of the irregular pipe jacking method, the irregular shield method, and a combination of both as a hybrid method. The irregular shield method has great advantages in terms of construction cost, schedule, settlement control, and space utilization.

It should be noted that more and more projects using the irregular pipe jacking method have been completed in the industry, in particular, in China recently. This shows increasing market demands driving the irregular pipe jacking and shield industry. Interested reader can also refer to http://en.crectbm.com/ for the latest information.

Fig. 6.83 Conversion portion of the connection between the pipe jack and the shield stages in Phases 2 and 3 of the projects

Fig. 6.84 a View from the tunnel section by the pipe jacking method to that by the shield method (toward the receiving shaft); **b** tunnel section constructed in the shield stage (toward the working shaft)

6.4 Summary

The simple erector system for irregular-shaped tunnels with lining rings formed by two-piece U-shaped lining segments introduced in this chapter is an innovation in the trenchless industry. It has already been applied successfully in many projects including the one presented in the case study. Meanwhile, the hybrid construction technique using the irregular pipe jacking and shield methods was a new trial during Phases 2 and 3 of a larger project associated with this case study. This application was considered efficient in making full use of the advantages of the two different methods.

The latest design of the erector system used for lining segments demonstrates the importance of the valuable experience accumulated through successful completions of practical projects. The benefits derived from the erector system are: maximized space for erection of lining segments, increased efficiency of grouting, and the use of square column in the segment erection shield, which significantly improve the efficiency of the irregular shield construction.

References

Gong SY (2015) Two-piece rectangular shield method opens a new era of Jining underground utility tunnels - construct underground utility tunnels, no more open cut of road surfaces. Qilu Evening News, C05, 18 Nov 2015

Yu BQ, Jia LH, Fan L (2018) Pipe jacking technology. China Communications Press. ISBN: 978-7-114-15014-2 (in Chinese)

Chapter 7
Pipe Jacking Performance: Mechanistic Behavior and Maintenance Challenges—An Artificial Intelligence-Based Approach

7.1 Introduction

The mechanistic behavior of a tunnel boring machine (TBM) starts drawing great attention from scientists and engineers in recent years because it appears to affect the efficiency of tunnel excavations and project costs. Factors affecting the mechanistic behavior of tunnel boring machine primarily include geology and jamming phenomena. The following section details how the two factors influence the mechanistic behavior of TBM.

Sudden changes in geological conditions (e.g., karst cavern and fault zone) not only threaten the safety of tunnel excavations but also cause geohazards such as water ingress and surface subsidence. Identifying the type of geomaterials encountered during tunnel excavation can reduce the potential of geohazard and prevent unplanned downtimes and operation costs. However, it is, in fact, a challenging task. Tunneling operational parameters, including thrust force, face pressure, cutter-wheel torque, advance rate, etc., are highly variable because of their dependence on a few influencing factors, like surrounding geology and lubrication. Despite that, tunneling parameters that relate closely to the surrounding geology have proliferated by modern TBMs and present a substantial opportunity for scientists and engineers to link tunneling parameters to a change in surrounding geology, leading toward the ability to predict the mechanistic behavior of the machine.

'Clogging' is a common challenge encountered during tunneling in clayey soils, which can affect tunnel construction, cause unplanned downtimes, and result in significant extra project costs. Clogging refers to the adherence of fine-grained soils to cutters at the cutterhead, opening on the cutterwheel, screw conveyor, and/or conveyor belt. A notable reduction in performance would be expected due to reduced advance rates and the time required for cleaning if clogging occurs and cannot be thoroughly migrated. There are four potential mechanisms that affect adhesion of clay to a cutter, namely adhesion of clay particles, inherent cohesion, bridging of clay particles over a cutterwheel opening, and an inability for the clay to dissolve in water (see Fig. 7.1).

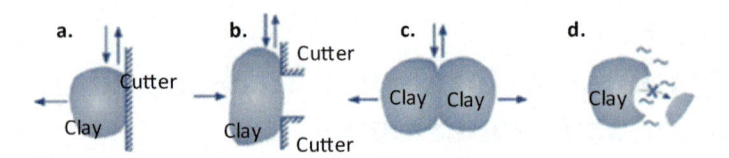

Fig. 7.1 Four potential mechanisms affecting adhesion of clay to a cutter: **a** adhering, **b** bridging, **c** cohering, and **d** no dissolving (after Thewes and Burger 2005)

7.2 Mechanistic Behavior of a Tunnel Boring Machine

7.2.1 Operational Parameters

While many geophysical imaging techniques are currently available for pre-tunneling geological characterization (e.g., water body and karst cavern), there remains great motivation for the industry to develop a real-time identification technology that links complex geological conditions to tunneling parameters for machine operators to adopt proper and timely countermeasures toward preventing geohazard from happening.

7.2.1.1 Face Pressure Response

The cutterwheel, while tunneling within clayey soils, is pressed into the cutting face, leading to local variation of total jacking loads. The increase in jacking loads is attributed to the increasing face resistance. When traversing into the fine-grained soil governed gravel or sand deposit, the roller disks easily sink into the cutting face, resulting in full contact between the cutterwheel and the ground and subsequently developing additional face resistance due to the increasing contact area (see Fig. 7.2a). However, the local variation of total jacking loads would go into a decline when driving back into coarse-grained soils (see Fig. 7.2b). Provide that the roller disks are designed for cutting in coarse materials, driving into the fine-grained soil governed gravel or sand deposit not only has a notable impact on total jacking loads but also could result in a slower advance rate. The shield machine has to reduce its advance rate when traversing through fine-grained governed soil in order to provide sufficient time for the slurry circulating system to transport tunneling spoil. This also eliminates further increase in the face resistance and prevents the risk of stalling of the cutterwheel. The above-said phenomenon also indicates that different cutter configurations can affect the advance rate.

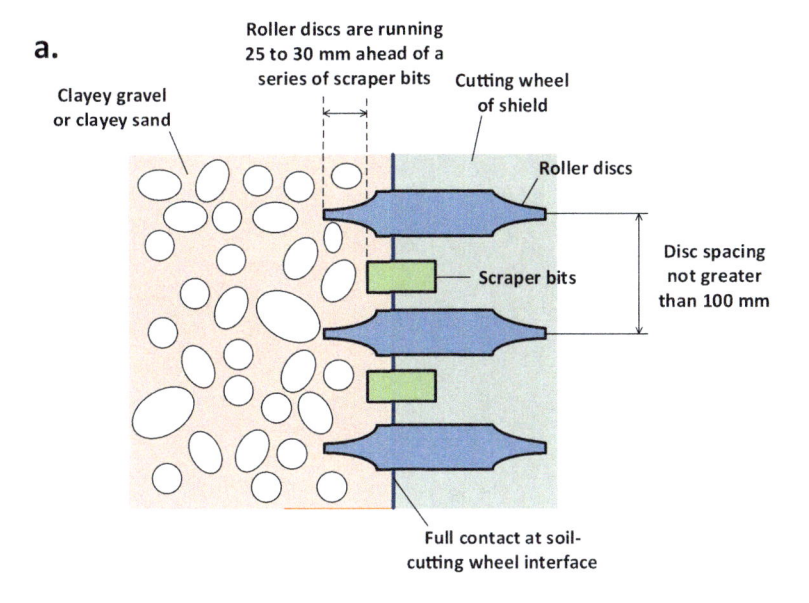

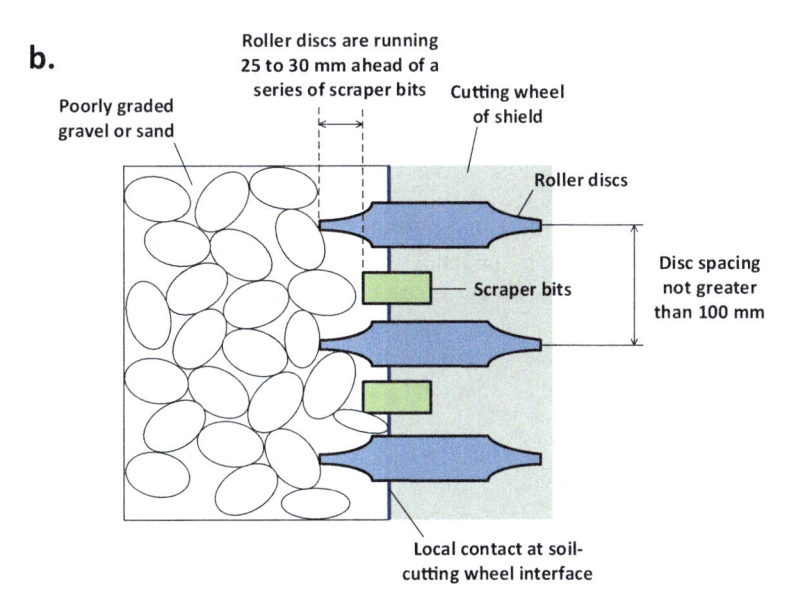

7.2 Mechanistic Behavior of a Tunnel Boring Machine

Fig. 7.2 Sketch of soil–cutterwheel interface: tunneling in: **a** coarse-grained soils with more than 12% fines and **b** coarse-grained soils (after Cheng et al. 2017)

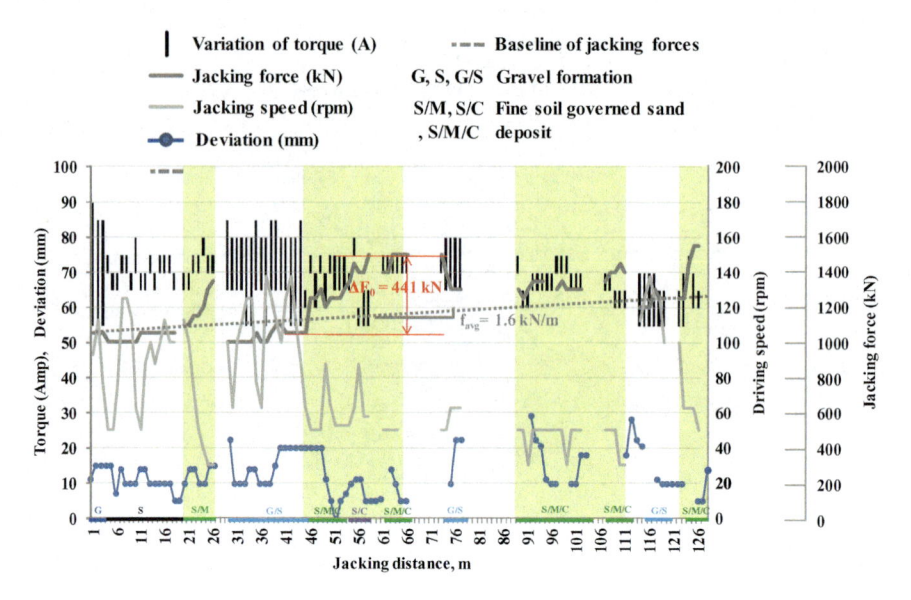

Fig. 7.3 Soft ground tunneling activities where 'black bar' represents the variation of cutterwheel torque and 'white area' and 'yellow area' indicate gravel and clayey gravel respectively (Cheng et al. 2017)

7.2.1.2 Cutterwheel Torque Response

In spite that it is unlikely to develop local variation of total jacking loads while driving in coarse-grained soils (see 'white area' in Fig. 7.3), the cutterwheel torque may vary very significantly compared to tunneling in fine-grained soils (see 'yellow area' in Fig. 7.3). The presence of gravels or cobbles can aggravate the variation of cutterwheel torque, and it can reduce when traversing back to fine-grained soils. Given that the variation of cutterwheel torque is more distinct for tunneling in gravel than in other soils, it, while introducing data-driven intelligent techniques for the prevention and mitigation of geohazard, can be utilized to identify 'gravel'. Furthermore, according to the authors' observations, tunneling in 'sand' can make the cutterwheel torque constantly high. There appears a discrepancy in the response of cutterwheel torque between 'gravel' case and 'sand' case. Such constantly high cutterwheel torque may be useful in identifying 'sand' during tunnel excavations.

7.2.1.3 Advance Rate Response

In the past, cutterheads are fitted with roller disks to grind down stones and boulders. Nowadays, cutterheads are fitted with a combination of roller disks and scraper bits. Generally, the roller disks are spaced up to 100 mm apart and run 25–30 mm ahead of the scraper bits. The roller disks are responsible for grinding down hard stones and boulders, while the scraper bits should shave off weak materials, including gravel,

7.2 Mechanistic Behavior of a Tunnel Boring Machine   243

sand, silt, or clay. If the advance rate is higher than 10 mm per rotation, stones will not be crushed by the roller disks when they are ripped off the cutting face, leading to the local collapse of the cutting face and endangering face stability. This also indicates that the use of slower advance rates, while tunneling in such a material, prevents the face instability and the advance rate may also be an assessor for the change in surrounding geology.

7.2.2 Soil Characteristics

Clayey soils tend to trigger clogging, which has a significant impact on tunneling activities. In contrast to rocks, soils do not contain any binding minerals. In spite that the majority of noncohesive soils are not subjected to any binding forces, the grains in cohesive soils are bound by cohesion forces. The presence of cohesion, in turn, determines soil excavation during tunneling process. While noncohesive soils are excavated grain by grain, there are two potential mechanisms for transforming uncritical soils into sticky material. Firstly, the cutter rings of the disks during tunneling are pressed into the cutting face, pushing the plastically deformable soil to both sides where it is cut by the drag picks in the form of 'lumps'. Water can transform the consistency of cut lumps and the soil at the cutting face into a sticky consistency. In this circumstance, uncritical soils can turn into sticky material provided that water is available. Secondly, parts of the fines contained in the lumps or the soil at the cutting face may disintegrate and accumulate in the supporting fluid or inflowing ground water. To this end, lumps with sticky outer layers may take place in addition to dry soil lumps. In this context, clogging poses a higher risk to the tunnel excavation than disintegrated fines in suspensions. Despite negative impacts, induced by disintegrated fines, at the final stage of the spoil disposal process, clogging takes place throughout the process from the excavation at the cutting face up to separation and transportation for spoil disposal. This is to say, one and the same sticky material may trigger clogging of the tunneling system at different stages. In light of this, primary clogging may be present at the cutterwheel and will then cause secondary clogging at sieves of the separation or at spoil conveyor belts.

Although previous studies performed over the past 10 years have largely enhanced our understanding of the clogging process and the role of key influencing factors, the impact of clogging on the response of tunneling parameters remains unclear and still poses a serious technical and economic risk. Therefore, it is rather important to forecast the probability of occurrence of the following cases:

- dry or slightly sticky lumps
- lumps with sticky outer surface
- sticky (critical) material
- disintegrated fines in supporting fluids

Furthermore, the prevention and mitigation of clogging throughout tunneling process is still highly dependent on the operator's experience.

7.2.2.1 Particle Size Distribution

It is widely recognized that the shear strength of coarse-grained soils with less than 5% fines is derived on the basis of interparticle friction and geometrical interference, including dilation, particle crushing, and rearrangement, rather than cohesion. This, in turn, mitigates the clogging potential throughout tunnel construction. However, the clogging potential could be increased provided that coarse-grained soils contain more than 5% fines. In light of this, soil particles larger than 0.075 mm become part of clogging material causing the blockage, whereas particles smaller than 0.075 mm become abrasive material causing wear of cutting tools. On the other hand, fine-grained soils always aggravate the clogging potential. A comprehensive assessment of the particle size distribution is, therefore, a prerequisite for a rigorous evaluation of clogging behavior.

7.2.2.2 Soil Plasticity and Consistency

Three water-related parameters typically applied to define the state of fine-grained soils are the natural water content ω_n, liquid limit ω_L, and plastic limit ω_p. An increasingly popular approach of assessing clogging potential is through a combination of the plasticity index I_p and the consistency index I_c:

$$I_c = \omega_L - \omega_n / \omega_L - \omega_p \tag{7.1}$$

According to Thewes (1999), clayey soils with I_p higher than 20% and I_c ranging from 0.75 to 1.25 have the highest potential to develop clogging. Notwithstanding that, more recent investigations have shown that extensive clogging can even present in clayey soils with I_c ranging from 1.25 to 1.50 (Thewes 1999). Hollmann and Thewes (2013) examined 150 samples of sticky material obtained from open and slurry-supported shield tunneling projects. The 150 samples varied between very soft (corresponding to 23% of samples; $I_c = 0.4$–0.5), soft-medium (corresponding to 58% of samples; $I_c = 0.5$–0.75), and stiff (corresponding to 19% of samples; $I_c = 0.75$–1.0) consistency, which are in good agreement with those documented by Feinendegen et al. (2011). Tunneling in clayey soils with 'soft' consistency is thus considered high risk for the development of clogging (see Fig. 7.4). In contrast to clayey soils with soft consistency, those with stiff consistency is likely to show relatively low clogging potential due to their lower ω_n. In the case of clayey soils with very soft consistency, their shear strength is not high enough to resist the shear stresses exerted in the excavation chamber toward reducing the potential to develop clogging.

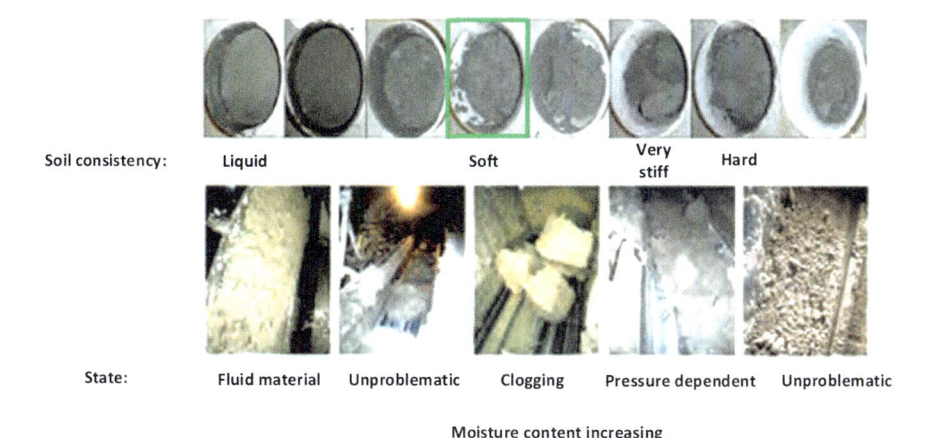

7.2 Mechanistic Behavior of a Tunnel Boring Machine

Soil consistency:	Liquid	Soft	Very stiff	Hard
State:	Fluid material	Unproblematic	Clogging	Pressure dependent	Unproblematic

Moisture content increasing

Fig. 7.4 Illustration of clayey soils' consistency evolution over water content and aggregate formation of the clayey soil with 'soft' consistency (after Weh et al. 2009)

7.2.2.3 Free Water

Free water, including ground water ingress and water contained in bentonite slurry, and the soil conditioning agent, also plays a key role in developing clogging during tunnel excavation. It is likely for cohesive soil with clay fraction higher than 10% to be transformed into clogging material. Nevertheless, the time spent to achieve such transformation is related closely to the natural soil consistency and the availability of free water, which is, in turn, dependent on the type of tunnel excavation method. The great amount of free water for slurry-supported shield tunneling within clayey soils causes a higher clogging potential. The authors' experiences are that even very stiff consistency clays can ultimately be transformed to a soft consistency. In contrast, provided that open shield tunneling is used, the amount of ground water depends heavily on both the inflow rate and construction downtime. Previously sticky clays in the excavation chamber could transform into nonsticky material as their adhesion reduces over time. These results lead us to summarize that the availability of free water must also be concerned when assessing the potential of clogging throughout shield tunneling.

7.2.2.4 Approaches for Assessing Clogging Potential

Various approaches have been proposed to assess the potential for clogging to develop such as the use of plasticity index measurements, semi-empirical diagrams, and laboratory-based drilling tests. Figure 7.5 shows the variations of adhesion and undrained shear strength with the increasing water content. Hu and Rostami (2020) highlighted the importance of soil conditioning during tunneling and the role of soil rheology in tuning the desired characteristics of the conditioned soil. Those authors

Fig. 7.5 Variations of adhesion and undrained shear strength with increasing water content (or liquid limit) (after Jancsecz 1991)

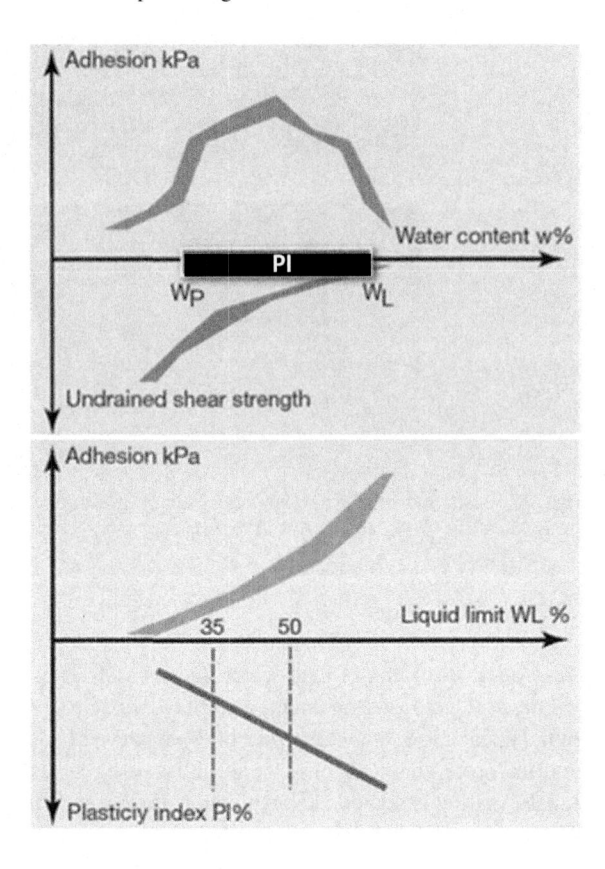

established a relationship between soil rheological parameters, soil type, and conditioning parameters for soft ground tunneling using a novel device. Using a new framework and new devices, Pelia et al. (2015) noted that the effectiveness of a polymer in clay conditioning is highly dependent on the plasticity index of the clay. For low-plasticity clay, the use of polymers can cause an increase in the volume of foam required due to the water absorption effect of the polymer itself. Notwithstanding that, this can also lead to a more homogeneous conditioned soil with long-lasting mechanical properties. Differentiation and characterization of clays and silts have also been explored extensively by Wong et al. (2017).

Alberto-Hernandez et al. (2017) utilized the relationship between cohesion (soil-soil strength) and adhesion (soil-structure strength) to assess clogging potential, though this method is restricted to situations where soil cohesion is in excess of adhesion. Hollmann and Thewes (2013) documented influencing factors for the development of clogging and presented a new classification diagram (see Fig. 7.6) which uses changes in water content to estimate changes in the consistency of fine-grained soils. Thewes and Hollmann (2016) explored the risk of clogging in various ground conditions and for different shield types and presented a summary of methods to

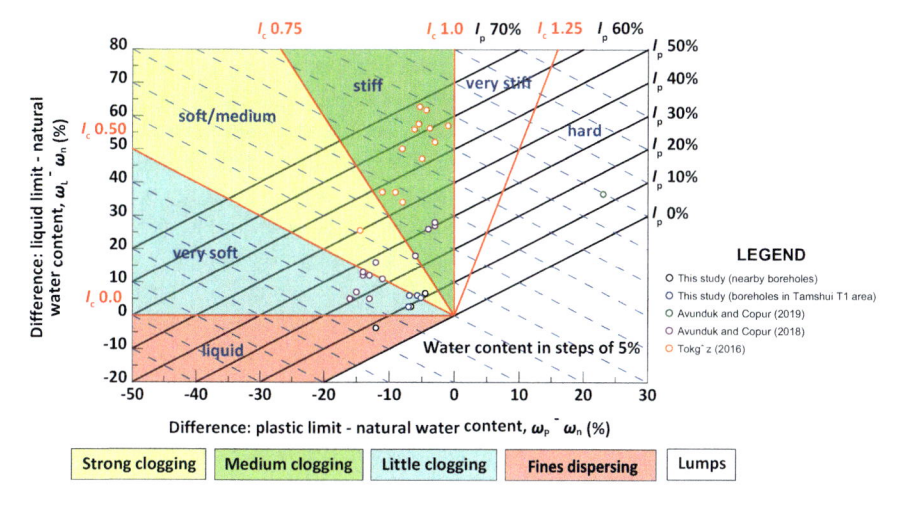

7.2 Mechanistic Behavior of a Tunnel Boring Machine

Fig. 7.6 Semiempirical diagram for the evaluation of clogging and liquid dispersing of soils (after Hollmann and Thewes 2013)

characterize soil 'stickiness' and laboratory experiments to assess clogging potential. A newly developed diagram for assessing clogging risks for all types of shields and a new testing scheme for evaluating sedimentary rocks in terms of clay clogging was also proposed.

Feinendegen et al. (2010) suggested a cone pull-out test to detect the adhesion/clogging propensity of a rock or soil, combined with a newly developed scheme for classifying the clogging potential. Following an extended test campaign using soils with different clay fractions and minerals, de Oliveira et al. (2018) developed a new device which adds an impulse via dropping of a 'beater' from a predetermined height. This combination could provide a more reliable assessment of the potential for clogging to occur along EPB shield tunnel drives. Furthermore, a laboratory routine to characterize the clogging and fluidity of clayey soils, including mixed soils by considering different fractions of clay was proposed by de Oliveira (2019a, b, c). Laboratory experimental work on clogging in cohesionless soils by fines have also been discussed in detail in Mehdizadeh et al. (2016, 2017).

7.3 Maintenance Challenges

7.3.1 Overview of Artificial Intelligence Techniques

The variety of geological conditions and development of clay clogging are likely to make tunnel construction more time consuming and costly. They are the main causes leading to the aforesaid mechanistic behavior. To prevent any unplanned downtimes from impeding tunnel construction, countermeasure to tackle the challenges is considered as of great necessity and must be upgraded to catch up with demands changing over time. The proliferation of tunneling parameters derived from modern TBMs presents substantial opportunity for the application of data-driven AI techniques (Jong et al. 2021) that can identify patterns in data without reference to known labels (Sheil et al. 2020). Data-driven approaches identify characteristics of the measured system by utilizing information extracted from the measured data, rather than by modeling the system response. Notwithstanding that, in most practical problems the measured data (i.e., the outputs) are not labeled. To this end, 'unsupervised' machine learning (ML) algorithms, used to infer patterns in data without reference to known outcomes, are popular. One of the main aims we set is to develop an improved understanding of existing tunneling parameters and their relationship with known geological changes throughout tunneling, in which case 'supervised' ML is the optimal technique. Some popular examples of supervised ML algorithms include multivariate adaptive regression splines (MARS), random forest (RF), and support vector machines (SVMs).

MARS is a nonparametric statistical method proposed by Friedman (1991) based on a 'divide and conquer' strategy where the training dataset is partitioned into separate piecewise linear segments (splines) of differing gradients (slope). While MARS generates a flexible model that can deal with both linear and nonlinear relationships, it is less accurate for sparse data. RF, first proposed by Breiman (2001), is a nonparametric, 'tree-based' method. Each tree learns from its predecessors and updates the residual errors successively. Trees will learn from an updated version of the residuals when they grow next in the sequence. RF stops growing when the overall mean square error (MSE) reaches a minimum value. Although RF is a highly flexible technique, it is susceptible to overfitting, particularly for small datasets. In light of these limitations, SVM is adopted herein for handling such 'classification' problem. The nonparametric nature of SVMs means that model complexity is not influenced by the number of features, and they are therefore well-conditioned for high-dimensional datasets. Further, the use of kernels allows this technique to capture complex input–output mapping.

The other aim is to tackle is the clogging while tunneling in clayey soils. Unlike the previous 'classification' problem, this is categorized as 'anomaly detection (AD)' problem. Evolutionary Polynomial Regression (EPR) has become more popular due to its more powerful ability to search a target expression than artificial neural networks (ANNs) and genetic programming (GP) (Gurocak et al. 2012; Alemdag et al. 2016; Yin et al. 2016). Usually, the global search for the best expression of the EPR equation is conducted using a genetic algorithm (GA) over the values contained in the user-defined vector of components. Notwithstanding that, for high-dimensional problems GAs require significant computational time and memory (Deep and Thakur 2007a, b). Multivariate adaptive regression splines (MARS) primarily aim to organize relationships between a set of input variables and the target dependent variable. Compared with the regression AD algorithms, 'unsupervised' clustering AD algorithms are more appropriate to process unlabeled data since the measured data (i.e., the outputs) are not labeled in most practical problems. Unsupervised clustering AD

7.3 Maintenance Challenges

algorithms are therefore employed here. A myriad of algorithms has been proposed in the literature (Zhang et al. 2020a, b); three of the more popular AD approaches are adopted, including: (1) one-class support vector machines (OCSVM), (2) isolation forest (IForest), and (3) robust covariance (Robcov). Baseline assessments are also conducted to compare their relative merits.

7.3.1.1 One-Class Support Vector Machines

Traditionally, many classification problems attempt to solve the two- or multi-class situation. The goal of the machine learning task is to discriminate between subclasses of a dataset using a 'training' portion of the data, which is not described here due to limited space. One-class support vector machine (OCSVM) denotes the case where the data comprises only 1 class and the task is to identify whether new measurements belong to that class. Figure 7.7 depicts the construction of the hyperplane by transforming the input space into a high-dimensional feature space.

Schölkopf et al. (2001) framed the OCSVM approach by considering the origin as the only member of the second class (Sheil et al. 2020). A hyperplane is constructed in the feature space to separate the dataset from the origin, using a maximal margin:

$$f(x) = (w^\wedge A \cdot x) + b \qquad (7.2)$$

where w = adjustable weight vector and b = bias. To prevent the OCSVM classifier from overfitting, slack variables ξ_i are used to create a 'soft margin' which allows some datapoints to lie within the margin. The optimal separating hyperplane can be derived by solving the below convex quadratic optimization problem (Vapnik 1995):

$$\text{Minimize} \frac{1}{2}\|w\|^2 + \frac{1}{vn} \sum_{i=1}^{n} \xi_i - \rho, i = 1, 2, \ldots, n \qquad (7.3)$$

$$\text{Subject to} (w \cdot \Phi(x_i)) \geq \rho - \xi_i, i = 1, 2, \ldots, n\xi_i \geq 0 \qquad (7.4)$$

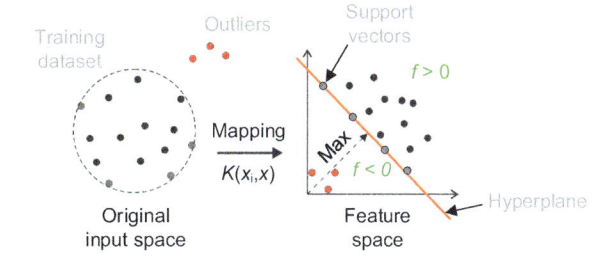

Fig. 7.7 Illustration of the construction of the hyperplane by transforming the original space into a high-dimensional feature space

where n = number of observations, ρ = margin, and ξ_i = slack variable, which is penalized in the objective function for nonzero values. The $\|w\|^2$ is an L2 regularization term to minimize overfitting, and its relative importance is determined by the parameter v. As the outcome of the decision function relies only on the dot-product of the vectors in the feature space, an explicit mapping to the feature space is deemed as of not necessity. For this reason, the 'kernel trick' is frequently employed, which allows the dot-product to be substituted using kernel functions. The decision function for a datapoint x can be written as:

$$f(x) = \text{sign} \sum_{i=1}^{N_{sv}} \alpha_i K(x_i, x) - \rho \tag{7.5}$$

where α_i = Lagrange multiplier and N_{sv} = number of support vectors. Every α_i which is > 0 is weighted in the decision function to 'support' the vector machine. Popular selections for the kernel include linear, polynomial, radial basis function (RBF), and sigmoid. RBF used here is represented as:

$$K(x_i, x) = \exp\left(-\gamma \|x_i - x_j\|^2\right) \tag{7.6}$$

where γ = kernel parameter, which is the width of the RBF and typically varies between 0 and 1, and $\|x_i - x_j\|$ = dissimilarity measure.

7.3.1.2 Isolation Forest

The IForest approach is a nonparametric method that constructs classification models in the form of a 'tree' structure (see Fig. 7.8). An isolation forest, as the name suggests, is an ensemble learning method that operates by constructing n_{tree} isolation trees and aggregating the results. The IForest approach detects anomalies by randomly partitioning the data. Partitions are created by randomly selecting a feature and then randomly producing a split value between the maximum and the minimum value of the feature. Partition creation continues until all datapoints are isolated; in most

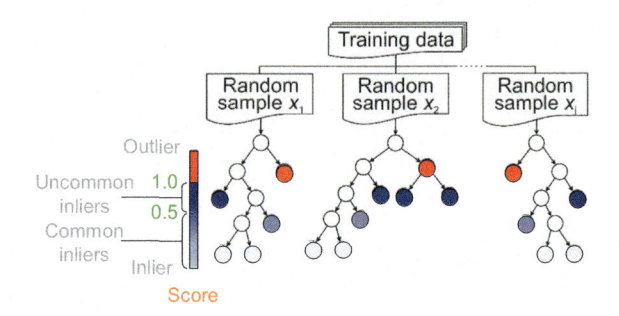

Fig. 7.8 Illustration of the construction of separate isolation trees by randomly sampling from the training dataset

cases, a limit is placed on the maximum number of partitions. Multiple training datasets are produced by sampling with replacement randomly from the original dataset, and anomalies are ultimately identified by sorting datapoints according to their corresponding path lengths (Sheil et al. 2020).

For a dataset of size n, the average, $c(n)$, of each path length, $h(x)$, is calculated as:

$$c(n) = 2H(n-1) - \left(\frac{2(n-1)}{n}\right) \quad (7.7)$$

where $H(i)$ = harmonic number ($H(i) = \ln(i) + a$ where a is Euler Mascheroni constant). The IForest anomaly score (IF) is defined as:

$$IF(x, n) = 2^{-\frac{E(h(x))}{c(n)}} \quad (7.8)$$

where $E(h(x))$ = average of $h(x)$ from a collection of isolation trees. A value of $IF > 1.0$ is categorized as anomalous (Liu et al. 2012).

7.3.1.3 Robust Covariance

The robust covariance ('Robcov') approach was first proposed by Rousseeuw (1984). The concept is to find a given proportion of 'good' observations which are not outliers and compute their empirical covariance matrix, as shown in Fig. 7.9. This empirical covariance matrix is subsequently rescaled to compensate the performed selection of observations (namely the consistency step). Having computed the Minimum Covariance Determinant (MCD) estimator, one can give weights to observations according to their Mahalanobis distance md, leading to a re-weighted estimate of the covariance matrix of the dataset (namely the reweighting step). For data following a Gaussian

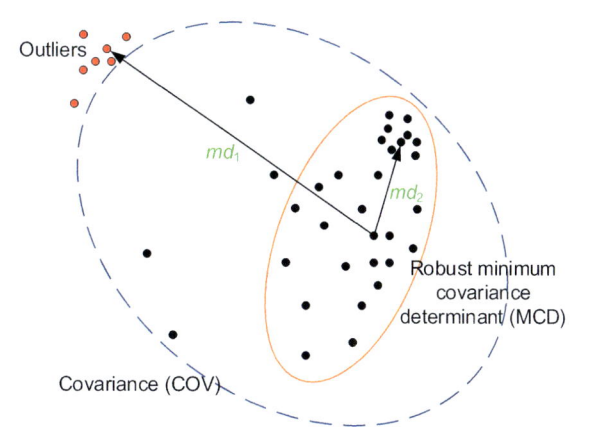

Fig. 7.9 Illustration of robust covariance approach showing the use of a robust estimator of covariance to reflect the true organization of observations using the Mahalanobis distance

distribution, the distance of an observation x_i to the mode of the distribution can be computed using its md as follows:

$$md(\mu, \Sigma)(x_i)^2 = (x_i - \mu)' \Sigma^{-1}(x_i - \mu) \quad (7.9)$$

where μ = location of the underlying Gaussian distribution and Σ = covariance of the underlying Gaussian distribution. As the usual covariance maximum likelihood estimate is very sensitive to the presence of outliers in the data, a more robust estimator of covariance is required to minimize the influence of 'erroneous' observations such that md accurately reflects the true organization of the observations. The larger the values of md, the more likely the presence of anomalous observations.

7.3.2 Implementation

7.3.2.1 Data Decomposition

Decomposition techniques, proposed by Persons (1919), isolate salient features of a dataset (e.g., trend and periodic components). Seasonal-trend decomposition using Loess smoothing (STL; Cleveland et al. 1990) is one of the most popular decomposition techniques and is also used here to partition the global series into three additive components as follows:

$$y_t = P_t + T_t + R_t \quad (7.10)$$

where P_t = periodic component, T_t = trend component, and R_t = residual component. To detect the development of clogging, two feature variables, namely the residual and trend components of the density of support slurry, the cutterwheel torque, and the jacking speed, are considered for the application of OCSVM, IForest, and Robcov. Furthermore, for the purpose of assessing the geological conditions, the same two feature variables, including the residual and trend components of the cutterwheel torque and the thrust force, are adopted for the application of SVM.

7.3.2.2 Data Scaling

The OCSVM, IForest, and Robcov algorithms were implemented using the Python module Scikit-learn (Pedregosa et al. 2011). All data were scaled using a 'min–max scaler' such that each feature varies between 0 and 1 (Masters 1993). Thus, given a set of input data $x_1, x_2, \ldots x_n$, the scaled dataset $z_1, z_2, \ldots z_n$ will be:

$$Z_i = x_i - x_{\min}/x_{\max} - x_{\min} \quad (7.11)$$

where min and max are our specified minimum and maximum values of the range to be scaled. This pre-processing maximizes the efficiency and performance of the learning process and equalizes the importance of the input dataset.

7.3.2.3 Confusion Matrix

There are four possible outcomes from the analysis, commonly summarized in a 'confusion matrix': (a) true positive (TP—system correctly identified clogging behavior), (b) false negatives (FN—system has failed to detect clogging behavior), (c) false positives (FP—system has erroneously detected clogging behavior), and (d) true negatives (TN). The four outcomes were considered here to evaluate the performance of the AD approaches using the discovery rate (DR = TP/(TP + FN)), or true positive rate, and the false alarm rate (FAR = FP/(TN + FP)), or false negative rate.

7.3.3 Tunneling in Alluvial Soils

Figure 7.10 shows the location of four drives in the soft alluvial deposits located in the Shulin district in Taipei, Taiwan. Only three (drives B, C, and D) out of the four drives are presented herein due to data completeness and measured 126 m, 75 m, and 102 m in length, respectively. The drives were located 10.5 m below ground level. The tunneling was undertaken using a 1.5-m-diameter slurry-supported shield. Given the 1.44-m-diameter trailing concrete pipes, a 30 mm overcut was created. Each pipe weighed 12.6 kN. To minimize the frictional resistance between the pipe string and surrounding soil, a highly viscous lubricant with Marsh cone viscosity of 38 min was injected into the overcut annulus.

The corresponding soil properties' profile is determined from both the field and laboratory tests, as shown in Fig. 7.11. The drives B, C, and D were excavated in gravel and sand in the main. The soils were also classified as clayey gravel and clayey sand for certain sections of the drives B, C, and D. The minor clay fraction plays a leading role in the macroscale mechanical properties of the clayey gravel or clayey sand. The phreatic surface was located at a depth of approximately 4.5 m below ordnance datum (BOD). Additional details on the project are available in Cheng et al. (2017, 2018, 2019a, b).

7.3.4 Tunneling in Loess Soil

The other tunneling project presented herein belongs to a part of Xi'an Metro line 4. There are six tunnels built using the shield tunneling method, with three cross passages, and they are responsible for joining four stations with each other, namely

Fig. 7.10 Tunneling project description: **a** location of the four drives in Taipei, Taiwan and **b** geological profile, derived from BH1 ~ 4 (after Bai et al. 2021a)

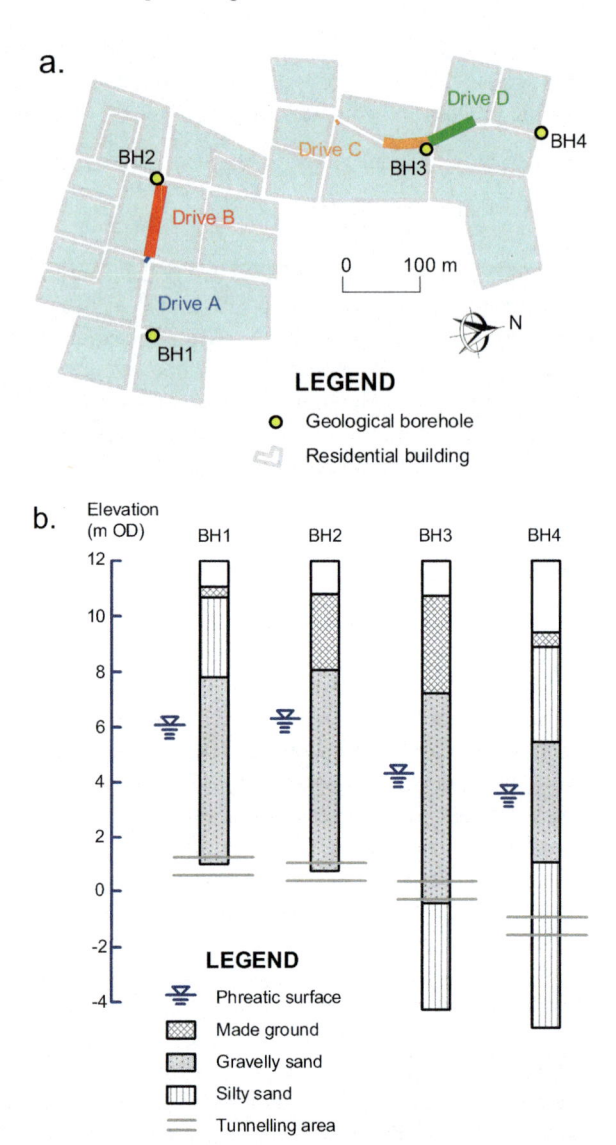

Fengcheng 9th Road Station, Fengcheng 12th Road Station, Yuanshuo Road Station, and North Railway Station. Figure 7.12 depicts the project layout.

The length of each tunnel varies from 689.6 m to 1140.1 m. This study primarily focuses on the two tunnels between the Fengcheng 12th Road Station and the Yuanshuo Road Station. Figure 7.13 shows the geological profile along the southbound tunnel. The mileage for the northbound tunnel is 689.6 m, while for the southbound tunnel it is 690.0 m. A 6.288-m-diameter EPB shield machine advanced in the water-rich sandy soils is at depths of 12–14 m. The shield machinery parameters are shown

7.3 Maintenance Challenges

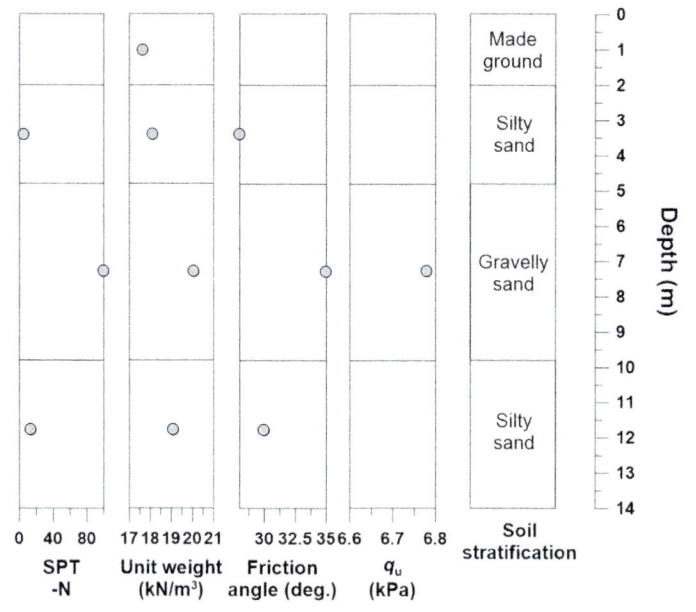

Fig. 7.11 Profile of the soil properties (after Bai et al. 2021a)

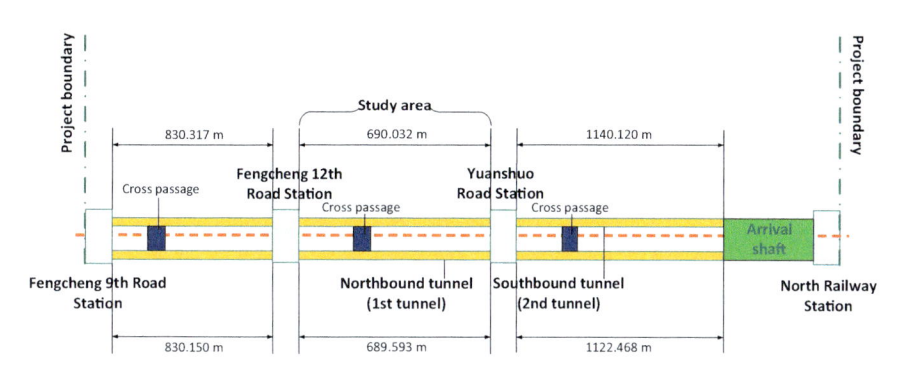

Fig. 7.12 Project layout and study area (after Bai et al. 2021b)

in Table 7.1, while the subsoil properties are shown in Table 7.2. Further, there is a great necessity of modifying the tunneling parameters while traversing underneath a culvert of G3001 expressway toward ensuring the safety of the culvert and subsequently preventing surface subsidence in excess of the designed limit. As a result, the range of cutterwheel torque changes from 4500–5500 to 3500–4500 kNm. Further, the range of advance rate is lifted from 20–35 mm/min up to 40–60 mm/min, and the face pressure changes from 40 to 60 kPa. Moreover, the greater face pressure aims to manage the development of surface subsidence. The faster the advance rate,

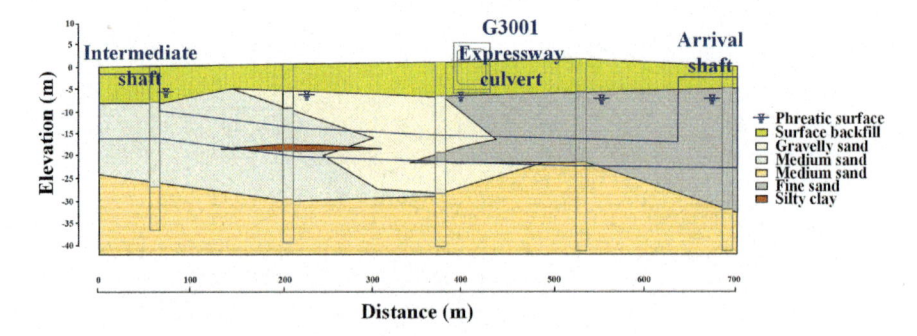

Fig. 7.13 Geological profile along the southbound tunnel (after Bai et al. 2021b)

Table 7.1 EPB shield machinery parameters

	EPB shield (Lovat)
Excavation diameter (mm)	6288
External diameter (mm)	6000
Internal diameter (mm)	5400
Shield length (incl. cutterwheel) (mm)	9070
Liner thickness (mm)	300
Liner length (mm)	1500
Cutterwheel rotational speed (rpm)	0–3.5
Total installed power (kW)	1450
Face pressure (kPa)	500
Cutterwheel opening ratio (%)	44
Maximum torque (kN·m)	6620
Maximum thrust force (kN)	36,000

Table 7.2 Subsoil properties

Soil layer	γ (kN/m³)	ω (%)	PI (%)	e (-)	G_s (-)	c' (kPa)	ϕ' (deg.)	SPT-N	E_s (kPa)
Surface backfill	11.5	20	–	1.05	–	5	8	–	–
Coarse sand	19.8	19.8	–	0.62	2.67	0	35	–	17
Medium sand	19.8	20.4	–	0.63	2.67	0	32.5	40	13
Silty clay	19.9	24.6	13.8	0.70	2.72	23	0	26	7

the smaller the disturb to surrounding environment. Tunnel lining with continuous segmental rings is usually designed against external earth pressures. In this project, a typical segmental ring is 6 m in diameter and composed of 6 segments of 0.3 m in thickness toward creating about 30 cm annular gap. The annular gap is filled with grouts using the synchronous grouting when the segmental ring has erected and left the EPB shield tail. The secondary grouting is conducted to fill cavities likely to be left after the shrinkage of injected grouts from the synchronous grouting.

The ground is composed of surface backfill, medium sand, occasional interbedded silty clay, gravelly sand, and fine sand. The moisture content ω, void ratio e, and specific gravity G_s for the coarse sand, the medium sand, and silty clay are in ranges of 19.8%–24.6%, 0.62–0.70, and 2.67–2.72, respectively. The liquid limit ω_L for the silty clay is tested as 33.6%. The permeability for the sandy soils is measured as high as 3×10^{-2} cm/s. The phreatic surface is at 6 m depth below the ground surface. Apart from that, the direct shear test results show that the friction angle for the coarse sand, medium sand, and fine sand is 35°, 32.5°, and 31.5° respectively, whereas the cohesion for the silty clay is 23 kPa according to the unconsolidated undrained test results.

7.3.5 Case Study 1: Identification of Geological Conditions

7.3.5.1 Results

Given that the classification results present henceforth correspond to the optimized hyperparameters, we therefore use the more general terminology, 'SVM classifier', in the following sections. Further, for the sake of readability we only discuss the cases that fail to correctly class the soil, i.e., 'FN' in the residual-trend torque plot and 'FP' in the residual-trend jacking force plot. The results of the SVM classifier applied to the transformed data for drive C are shown in Fig. 7.14. Figure 7.15 provides the reader with useful context in relation to the mapping of the TP, TN, FP, and FN results from the transformed feature space back to the original input space. The SVM classifier gives excellent predictions with three FNs. Here the FNs indicate that the gravels are erroneously categorized as clayey gravels. Their locations in the original space are at jacked distances of 17 m, 37 m, and 56 m, respectively. In contrast, several FPs appear in the predictions. FPs denote that the clayey gravels are erroneously classed as gravels. They appear at jacked distances of 35–36 m, 41–45 m, 52 m, and 70 m, respectively. The results of the SVM classifier applied to the transformed data for drive D are shown in Fig. 7.16. These data also mapped back to the original parameter space are shown in Fig. 7.17. The SVM classifier also provides good predictions with five FNs. The five FNs correspond to jacked distances of 22–23 m, 62 m, and 75–76 m, respectively. Further, the SVM classifier provides predictions with two FPs at jacked distances of 40 m and 95 m respectively, indicating that the majority of the clayey gravels are correctly classed.

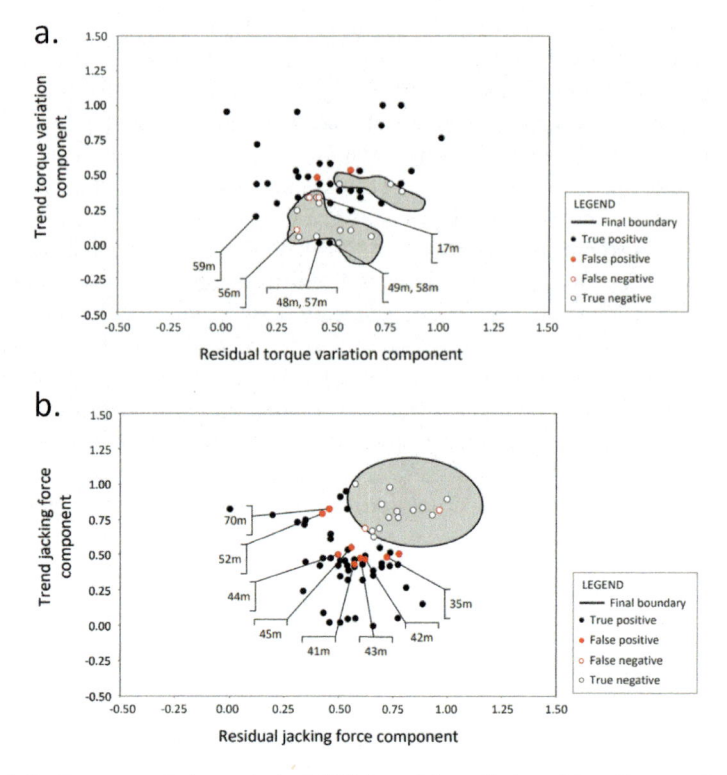

Fig. 7.14 Performance of the optimized SVM model applied to drive C mapped to high-dimensional feature space: **a** identification of gravel and **b** clayey gravel

7.3.5.2 Discussion

Most of the gravel sections confronted at drive C are successfully identified. The main cause of the three FNs is most likely due to the occasional presence of sands, reported by Cheng et al. (2017), which reduces T_t closer to the classification boundary. Furthermore, the clayey gravel is not often encountered along the tunnel alignment and the construction of the hyperplane of the SVM classifier, therefore, proves difficult. However, the gravels at jacked distances of 29 m, 48–49 m, and 57–59 m are correctly classed. The datapoints do not exactly correspond to gravels but sands with T_t below 0.33. The reason to explain why the sands can be classed as gravels is that the lack of clayey gravel datapoints leads to a difficulty in constructing the hyperplane and the SVM classifier defines the noncontinuous boundaries through bypassing the sand datapoints toward pursuing improved predictive performance.

In contrast, nine FPs present in the predictions about identification of clayey gravel at drive C. A clayey gravel can be identified by satisfying two conditions; $T_t > 0.58$ (2052 kN) and $R_t > 0.58$ (−5 kN). While the FPs at jacked distances of 35–36 and 41–45 m correspond to $R_t > 0.58$, their T_t values are below 0.58, most likely due

7.3 Maintenance Challenges

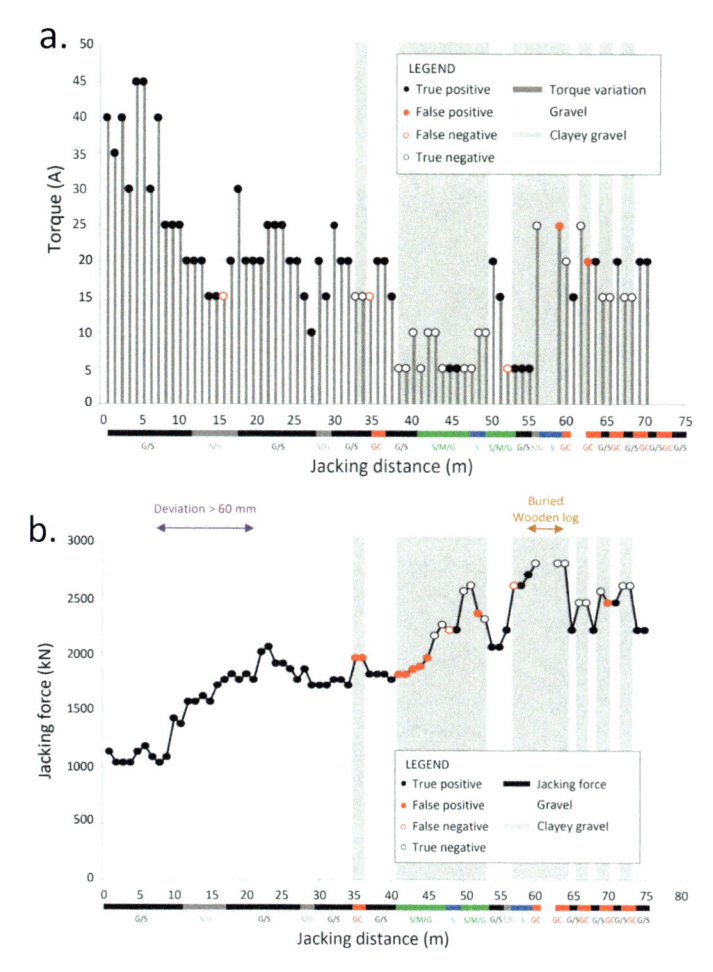

Fig. 7.15 Performance of the optimized SVM model applied to drive C remapped back to original input space: **a** identification of gravel and **b** clayey gravel

to the influence of the surrounding gravels. In spite that the FPs at jacked distances of 52 and 70 m have values of $T_t > 0.58$, the misclassifications occur due to $R_t < 0.58$, induced by traversing into the gravel from the clayey gravel. Surprisingly, the predictions appear not to be affected by pipe deviation being greater than the threshold of 60 mm between jacked distances of 8 and 21 m.

There are five FNs in the predictions regarding identification of gravel at drive D. The five FNs are formed due to their T_t equal to or below 0.33. These datapoints represent sands rather than gravels, which reduces T_t to as low as 0.12, and therefore cause these misclassifications. Notwithstanding that, it is worth noting that the sands at jacked distances of 4 and 37 m are classed as gravels, most likely because of the

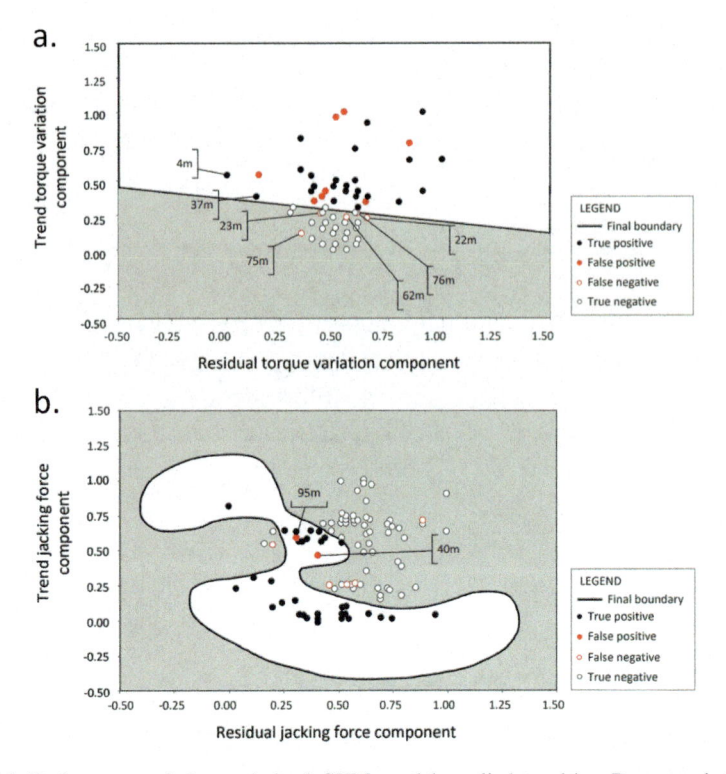

Fig. 7.16 Performance of the optimized SVM model applied to drive D mapped to high-dimensional feature space: **a** identification of gravel and **b** clayey gravel

gravels occurring at jacked distances of 3 and 36 m. The surrounding gravels elevate T_t to 0.54 and 0.39, respectively, significantly above the 0.33 threshold.

Most of the clayey gravels at drive D are correctly classed since FP = 2. A clayey gravel is herein identified as $T_t > 0.2$ (1494 kN) and $R_t > 0.58$ (−12.5 kN). The silt at a jacked distance of 40 m, while tunneling through the clayey gravel, causes the jacking force to develop a 'plateau', thereby causing a reduction in R_t. The other misclassification at a jacked distance of 95 m is formed as a result of $R_t < 0.58$, induced by jacking into the gravel from the clayey gravel, even though the corresponding value of T_t is greater than 0.2.

The discovery rate (DR) in gravels is calculated as 94.1% and 87.5% for drives C and D, respectively. This proves the use of the trend component T_t of the cutterwheel torque to identify gravels. The sands striking a buried wooden log cause misclassification of gravel at drive C; this explains the discrepancy between the hyperplane constructed for drive D (single straight line) compared to drive C (inconsistent two regions), as shown in Figs. 7.14 and 7.16. Some FNs (i.e., jacked distances of 29 m, 48–49 m, and 57–59 m at drive C and 4 m and 37 m at drive D) would have occurred if the SVM classifier does not pursue improved predictive performance by bypassing the datapoints. DR in relation to identification of clayey gravel is 59% and 96.7%

7.3 Maintenance Challenges 261

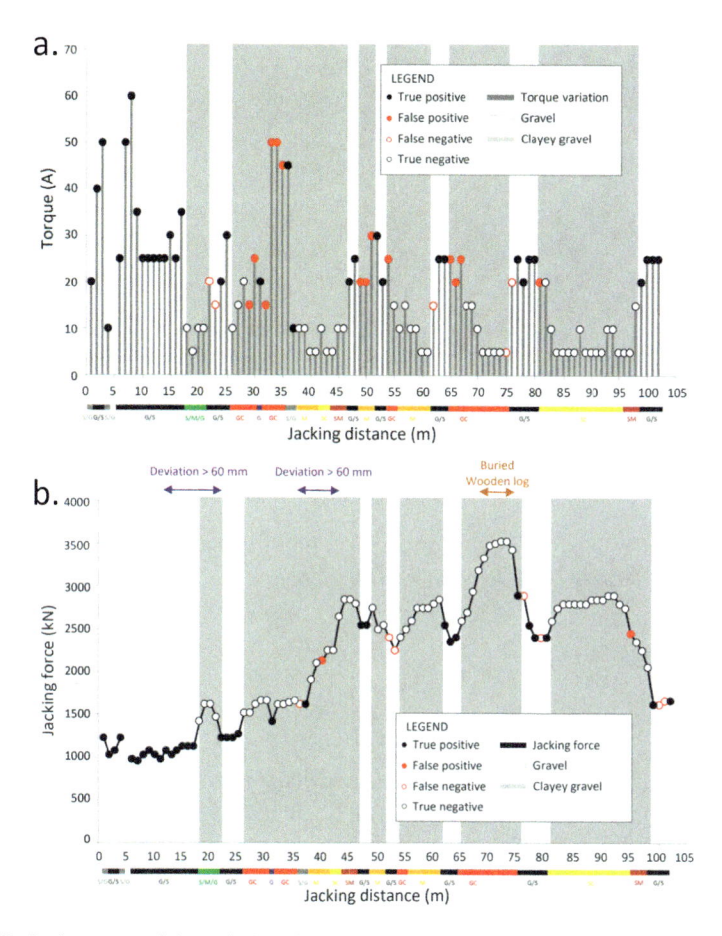

Fig. 7.17 Performance of the optimized SVM model applied to drive D remapped back to original input space: **a** identification of gravel and **b** clayey gravel

for drives C and D, respectively, indicating satisfactory performance for the identification of clayey gravels. The clayey gravel can be successfully classed by higher T_t and R_t, in accordance with the constructed hyperplanes. The lower accuracy of prediction at drive C compared to drive D could be due to the higher content of gravel which reduces T_t below the classification threshold.

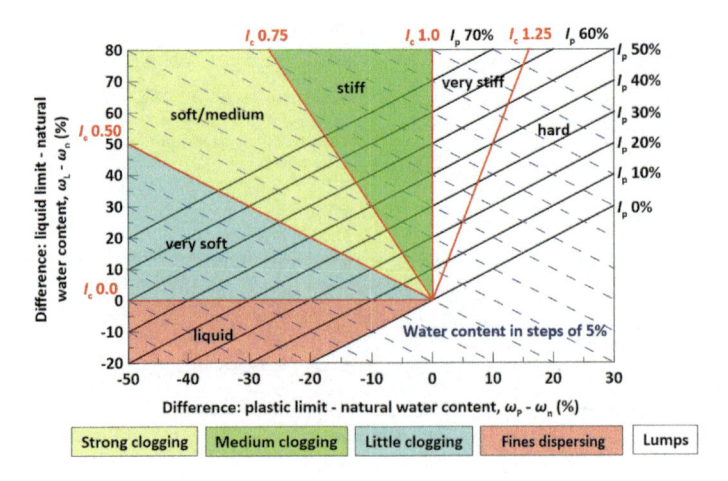

Fig. 7.18 Universal classification diagram for the assessment of anomalous clogging

7.3.6 Case Study 2: Clogging Detection

7.3.6.1 Results

Baseline Assessment Results

The use of the semi-empirical approach is considered herein as the baseline assessment of clogging potential. Avunduk and Copur (2018) reported that specimens extracted from the excavated material of an earth pressure balance (EPB) TBM, used in an Istanbul utility tunnel project, are classified as low-plasticity clay (CL) or high-plasticity clay (CH) and characterized by $I_p = 19$–31% and $I_c = 0.23$–0.91. Tokgöz (2016) analyzed a total of 264 EPB-TBM excavation performance data, when subjected to fine-grained sedimentary materials with $I_p = 40$–68% and $I_c = 0.64$–0.98. The results revealed that thrust force and cutterwheel torque both increase while jacking speed decreases with increasing I_p and I_c, and that the increase in cutterwheel torque can be explained by the higher shear strength of the clogging material. Further, one can deduce that the effect of fine-grained materials on jacking speed mainly depends upon the clay morphology which affects specific surface area (SSA) and then free water adsorption capability. For this reason, an appropriate designed soil conditioning chemical is deemed crucial toward preventing clogging and adhesion-related problems. Here, the fine-grained soil specimens from nearby geological boreholes (Cheng et al. 2017) and from those in Tamshui T1 area (Woo and Moh 1990) and their clogging potentials, as defined by Hollmann and Thewes (2013), and those reported in the literature (Tokgöz 2016; Avunduk and Copur 2018) are plotted in the universal classification diagram, as shown in Fig. 7.18. It can be observed that I_p and I_c for the clayey soils spread over the worksite and Tamshui T1 area are in the ranges of 8.3–12.9 and 0.27–0.60 respectively, substantiating all the

7.3 Maintenance Challenges

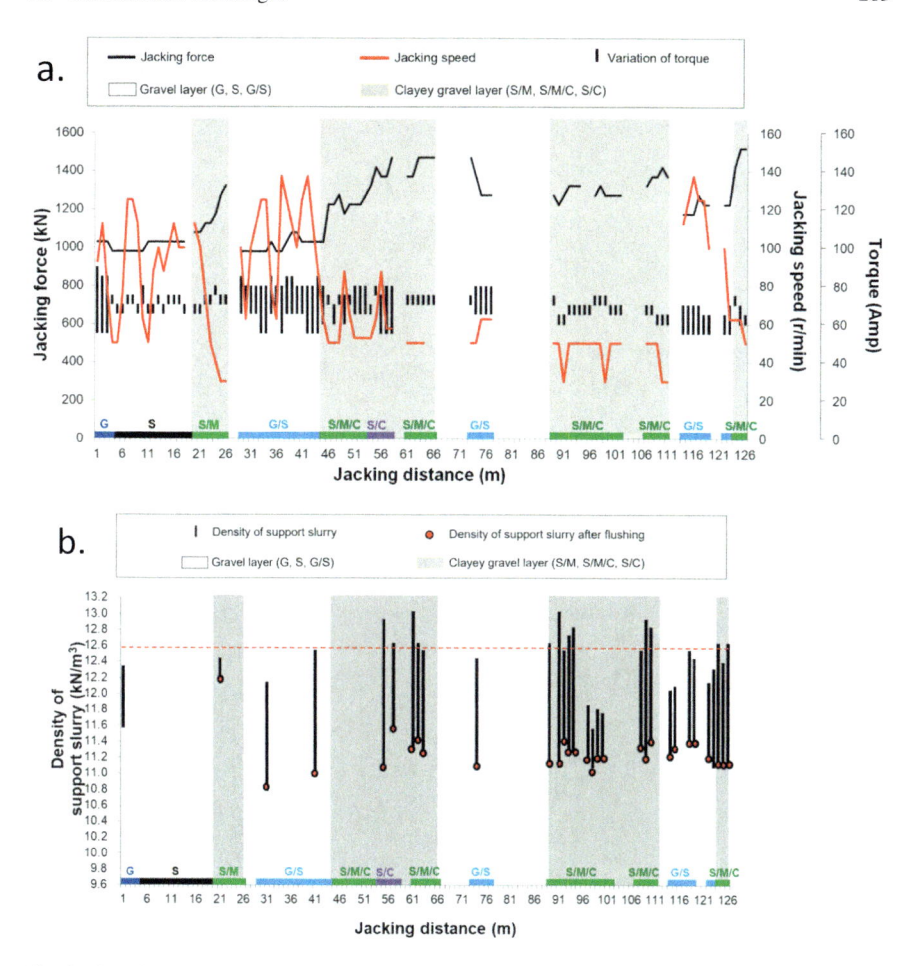

Fig. 7.19 a Tunneling activities and b variation of slurry density at drive B

specimens of clogging material with predominantly high and/or little clogging potentials. These results using the semi-empirical approach are in line with the authors' observations corresponding to jacking distances of 20–26 m at drive B and 26–47 m at drive D (see Figs. 7.19 and 7.20).

Anomaly Detection Results

The performance results of AD approaches using the transformed slurry density space are depicted in Figs. 7.21 and 7.22. At drive B, all three AD approaches provide 1 TP at 21 m jacked distance, with $R_t = 1.0$ and $T_t = 1.0$ (Fig. 7.21). There are two FPs corresponding to jacked distances of 31 and 41 m. At drive D, Robcov and IForest give one TP at 38 m jacked distance, whereas OCSVM gives one FN at 38 m jacked

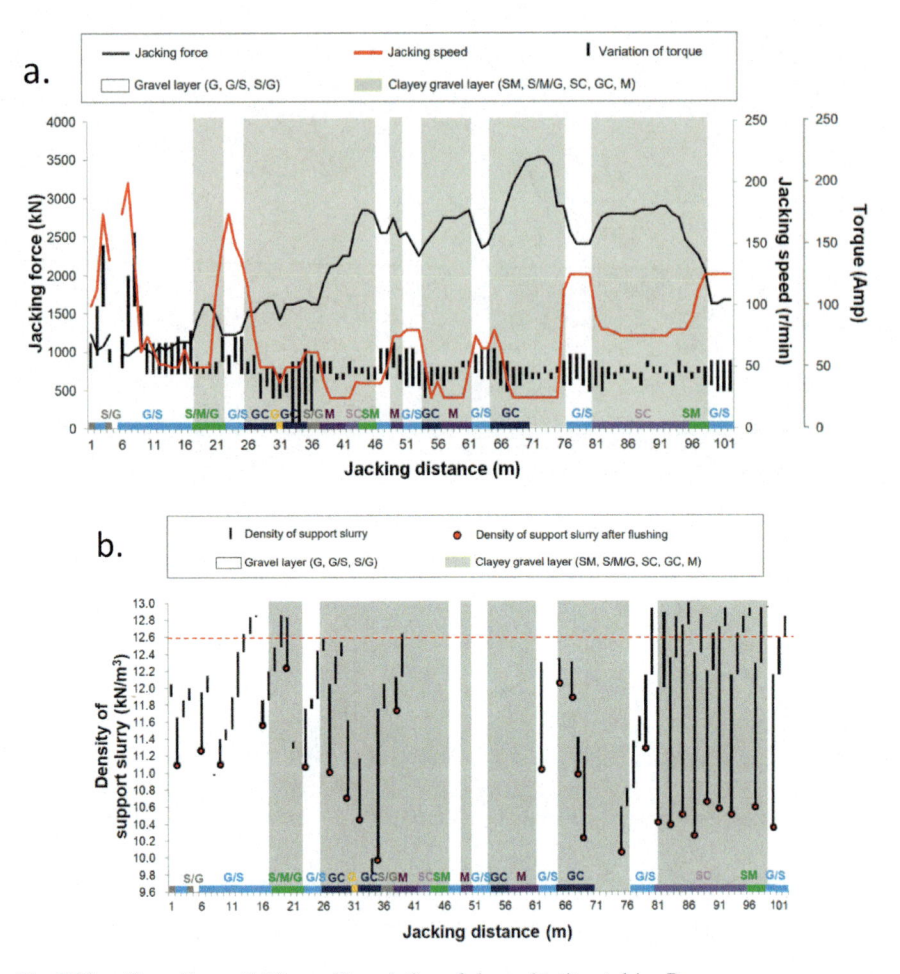

Fig. 7.20 a Tunneling activities and b variation of slurry density at drive D

distance (Fig. 7.22). There are two FPs (i.e., 65 and 79 m) distributed on the right-hand side of the feature space (large R_t) and the other FP (i.e., 75 m) distributed on the left (small R_t).

The performance results of AD approaches using the transformed maximum torque space are depicted in Figs. 7.23 and 7.24. At drive B, IForest and OCSVM produce 1 TP at 24 m jacked distance (Fig. 7.23). While Robcov gives one FN at 24 m jacked distance. There are four FPs (i.e., 110–119 m) positioned on the bottom (small T_t) in the feature space and another one FP (i.e., 10 m) distributed on the top right corner (large R_t and T_t). The other two (i.e., 54 and 90 m) sit on the left (small R_t). At drive D, OCSVM and Robcov give one TP at 38 m jacked distance. IForest produces one FN at 38 m jacked distance (Fig. 7.24). There is one FP (i.e., 89 m) distributed on the right (large R_t). The other three FPs (i.e., 19, 72, and 73 m) sit on

7.3 Maintenance Challenges

Fig. 7.21 Performance of AD approaches applied to drive B using transformed slurry density space: **a** OCSVM, **b** Robcov, and **c** IForest

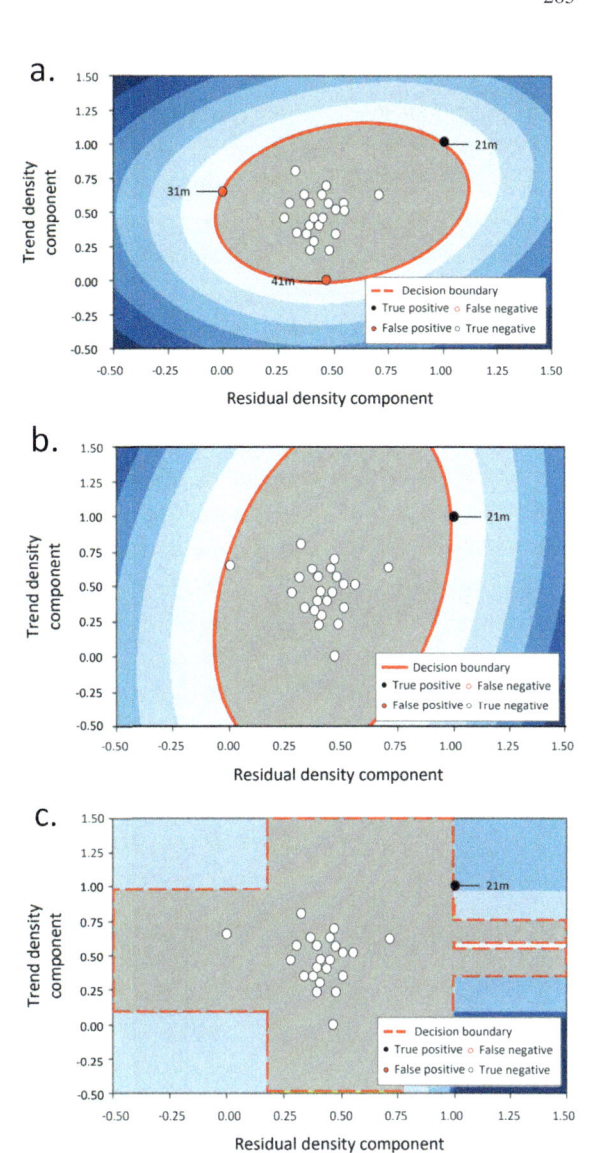

the top left corner (small R_t and large T_t) and the bottom left corner (small R_t and T_t) as well as the bottom right corner (large R_t and small T_t), respectively.

The performance results using the transformed jacking speed space are shown in Figs. 7.25 and 7.26. At drive B, IForest produces three datapoints at 20–26 m jacked distance. Robcov and OCSVM give two datapoints at 20–26 jacked distance. There are two FPs (i.e., 35 and 49 m) distributed on the top right corner (large R_t and T_t) in the feature space and another one FP (i.e., 36 m) positioned on the top left corner (small R_t and large T_t) (Fig. 7.25). The other two FPs (i.e., 6 and 11 m) sit on the

Fig. 7.22 Performance of AD approaches applied to drive D using transformed slurry density space: **a** OCSVM, **b** Robcov, and **c** IForest

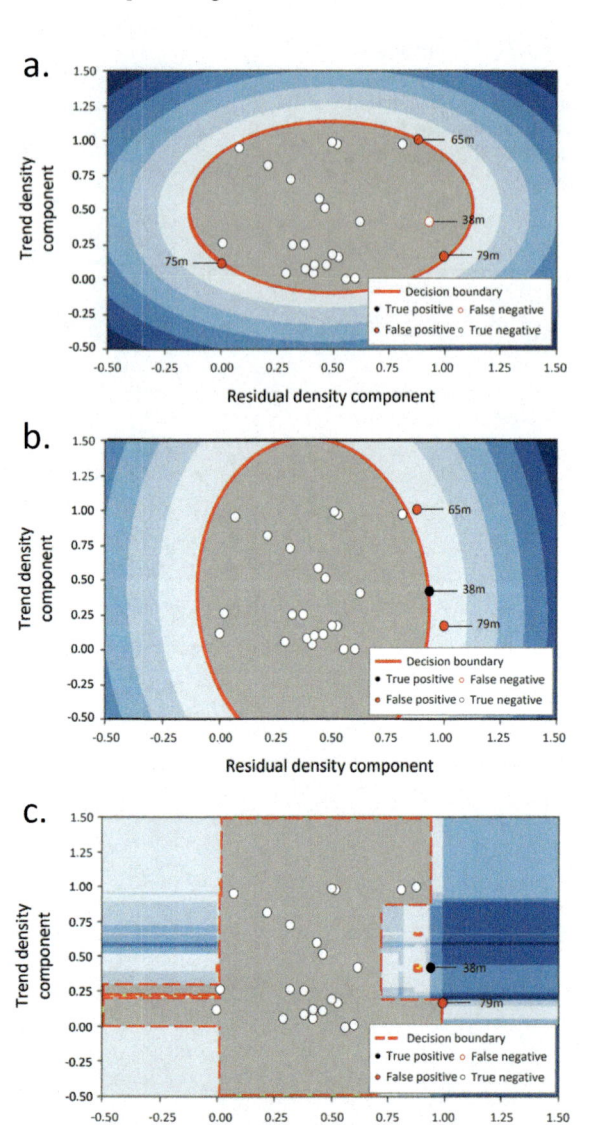

right (large R_t) and the left (small R_t), respectively. At drive D, OCSVM and Robcov produce 1 datapoint (i.e., 39 m) at 26–47 m. IForest provides one FN (i.e., 39 m) at 26–47 m. There are three FPs (i.e., 62, 81, and 82 m) distributed on the top right corner (large R_t and T_t) in the feature space and another three (i.e., 57, 67, and 68 m) sat on the bottom left corner (small R_t and T_t) (Fig. 7.26). The other two FPs (i.e., 75 and 83 m) distribute on the top left corner (small R_t and large T_t).

7.3 Maintenance Challenges 267

Fig. 7.23 Performance of AD approaches applied to drive B using transformed maximum torque space: **a** OCSVM, **b** Robcov, and **c** IForest

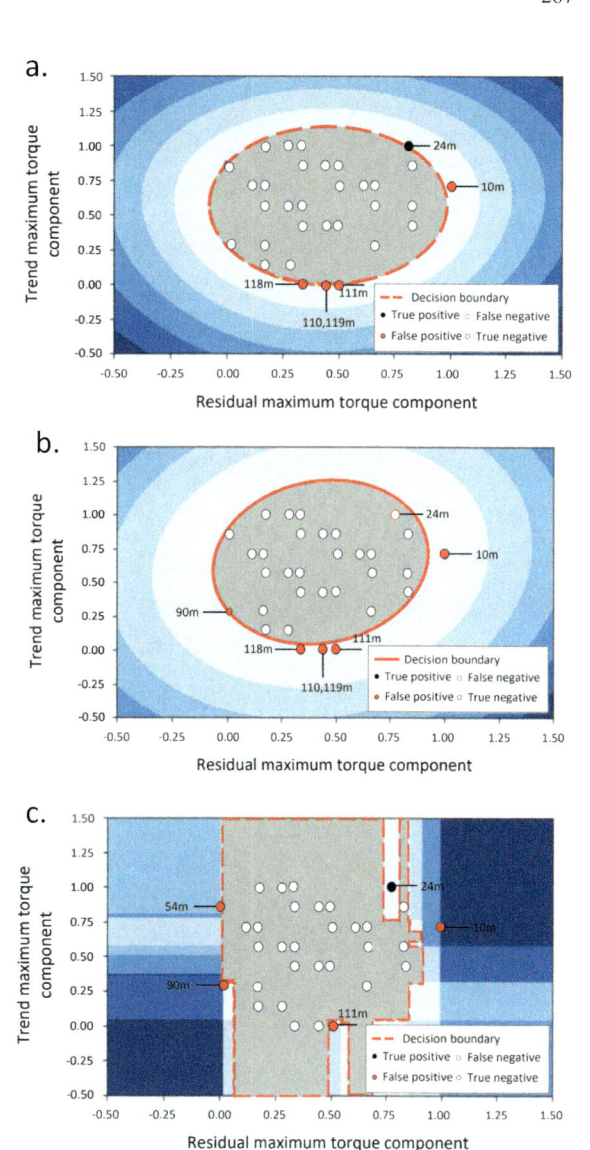

7.3.6.2 Discussion

Measurements of slurry density, cutterwheel torque and jacking speed and their difference to 'nearby' datapoints can notably affect the assessment of clogging potential. There are two types of behavior that are responsible for the formation of clogging in this work: (1) absolute values of slurry density, cutterwheel torque, and jacking speed, affect T_t and (2) the difference between 'current' measurements of slurry density, cutterwheel torque, and jacking speed from nearby datapoints affect R_t. The

Fig. 7.24 Performance of AD approaches applied to drive D using transformed maximum torque space: **a** OCSVM, **b** Robcov, and **c** IForest

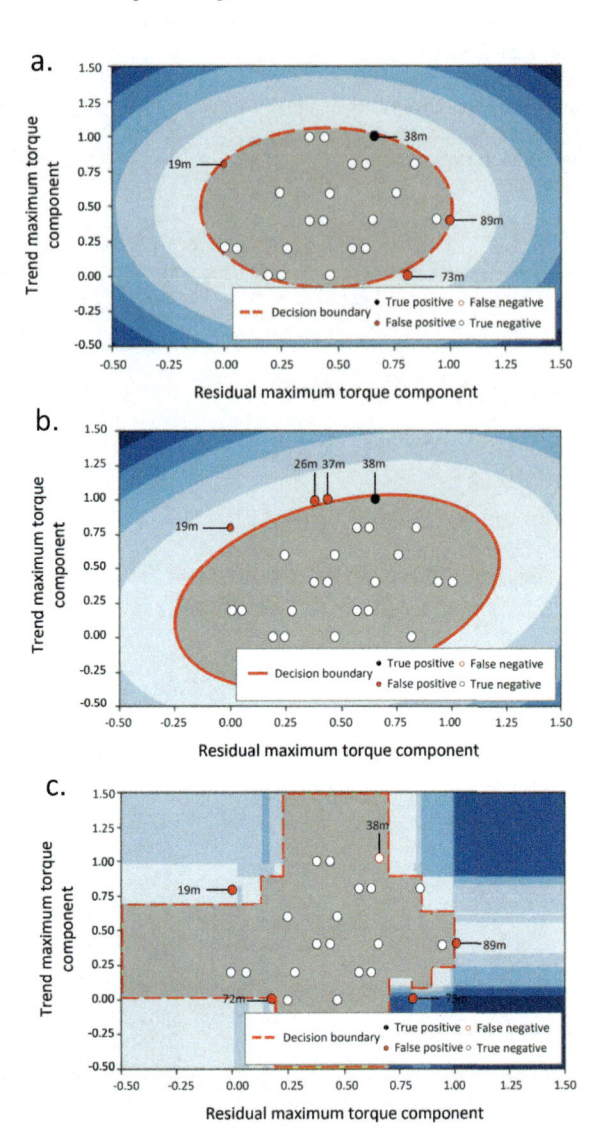

lower and upper limits of slurry density corresponded to 10.8 kN/m^3 and 12.2 kN/m^3 respectively at drive B and to 9.8 kN/m^3 and 12.8 kN/m^3 respectively at drive D. At drive B, the formation of FP at 41 m jacked distance is due to $T_t = 0$, induced by slurry density = 10.9 kN/m^3 dropping significantly below the threshold of 12.6 kN/m^3. Similarly, the FP at 65 m jacked distance of drive D is formed because of $T_t = 1.0$ and $R_t = 0.87$, induced by slurry density = 12.1 kN/m^3. These results can, therefore, be categorized as the 'type-1 behavior'.

In contrast, FP at 31 m jacked distance of drive B is formed by $R_t = 0$, induced by a decline in slurry density to 10.8 kN/m^3 from 12.2 kN/m^3 at 21 m jacked distance.

Fig. 7.25 Performance of AD approaches applied to drive B using transformed jacking speed space: **a** OCSVM, **b** Robcov, and **c** IForest

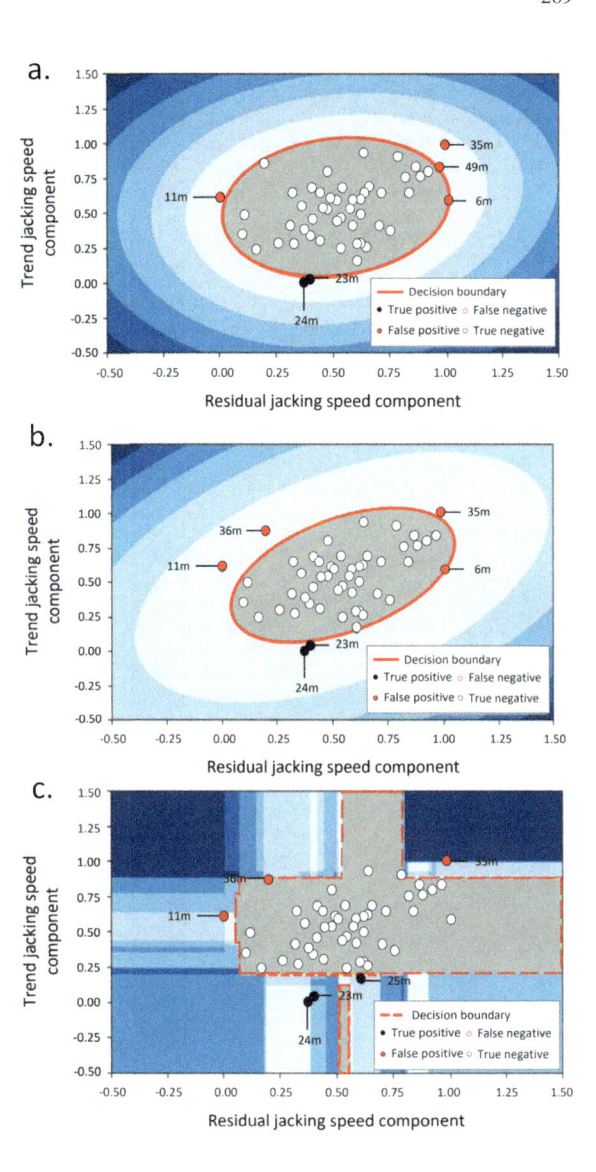

The higher slurry density from the nearby datapoint (i.e., 21 m jacked distance) increases T_t and this, in turn, reduces R_t to 0. At drive D, FP at 79 m jacked distance is developed as a result of $R_t = 1.0$, induced by an increase in slurry density to 11.3 kN/m^3 from 10.1 kN/m^3 at 75 m jacked distance. The lower slurry density from the adjoining datapoint (i.e., 75 m jacked distance) lowers T_t to about 0 and this, in turn, increases R_t to 1.0. These results can thus be categorized into 'type-2 behavior'.

The maximum of cutterwheel torque varies from 65 to 80 Amp at drive B, whereas it varies between 45 and 55 Amp at drive D. The main cause to lead to FP at 110 m

Fig. 7.26 Performance of AD approaches applied to drive D using transformed jacking speed space: **a** OCSVM, **b** Robcov, and **c** IForest

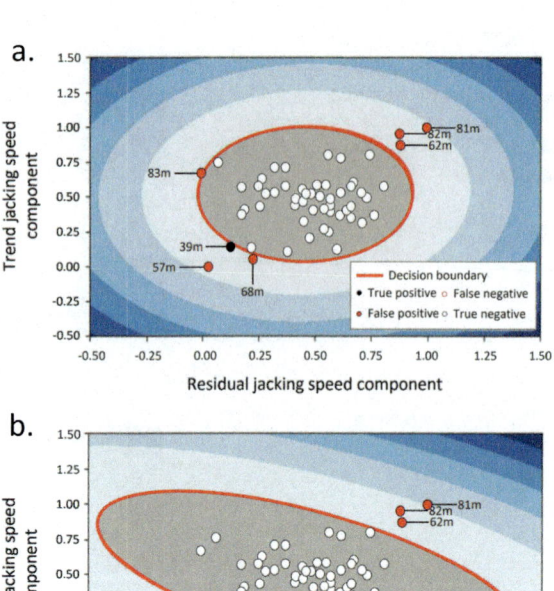

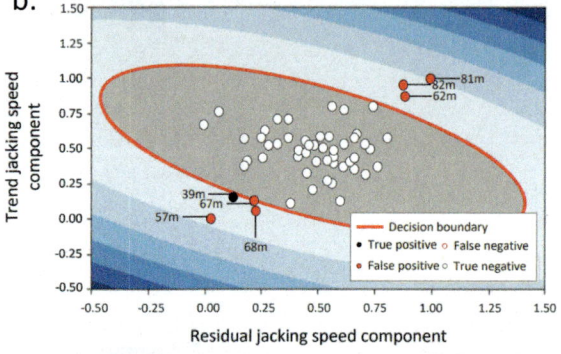

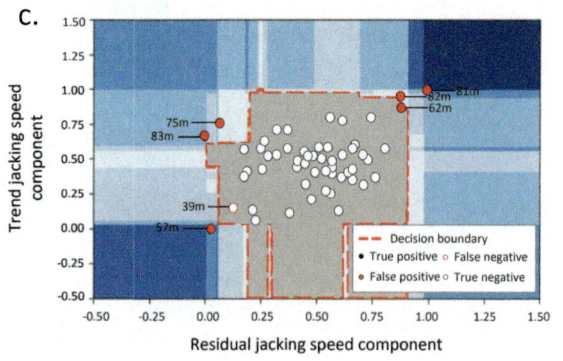

jacked distance of drive B is due to $T_t = 0$, induced by the maximum of cutterwheel torque = 65 Amp (lower limit). The maximum of cutterwheel torque from surrounding datapoints is similar, and the impact on R_t is deemed minimal. The FP at 26 m jacked distance of drive D is caused by $T_t = 1.0$ resulting from the maximum of cutterwheel torque hitting the upper limit of 55 Amp. Surrounding datapoints of similar maximum of cutterwheel torque produce a 'platform' toward leading to a negligible impact on R_t. These results can simply be classed as type-1 behavior.

7.3 Maintenance Challenges

However, the leading cause to form a FP at 10 m jacked distance of drive B is because of $R_t = 1.0$, induced by the maximum of cutterwheel torque = 80 Amp (upper limit). The maximum of cutterwheel torque = 70 Amp from the nearby datapoints, i.e., 9 and 11 m jacked distance, reduces T_t to 0.75. This, in turn, increases R_t to 1.0. A FP at 89 m jacked distance of drive D is developed when subjected to $R_t = 1.0$, induced by the maximum of cutterwheel torque = 55 Amp (upper limit). Nearby datapoints that are featured with the maximum of cutterwheel torque = 45 Amp (lower limit) increase R_t to 1.0 by lowering T_t to 0.4. The above two instances can, therefore, be classed as the type-2 behavior.

The difference in jacking speed measurements varies between −37.5–47.5 r/min and −55–65 r/min for drives B and D, respectively. A negative value of the difference in jacking speed represents a descending tendency, while a positive value indicates an ascending tendency. A FP at 35 m jacked distance of drive B ($R_t = 0.98$ and $T_t = 1.00$ is triggered by the difference in jacking speed being 47.5 r/min (upper limit). Further, the FP at 57 m jacked distance of drive D ($R_t = 0.02$ and $T_t = 0$) is formed by the difference in jacking speed reaching the lower limit of −55 r/min, which also indicates type-1 behavior. The FP at 49 m jacked distance of drive B ($R_t = 0.97$ and $T_t = 0.84$) is due to the difference in jacking speed = 37.5 r/min. The difference in jacking speed = 0 r/min from 48 m jacked distance lowers T_t to 0.84 and this, in turn, increases R_t to 0.97.

The FP at 83 m jacked distance of drive D is formed due to $R_t = 0$ and $T_t = 0.66$, induced by the difference in jacking speed = −10 r/min. The difference in jacking speed = 55 r/min from 82 m jacked distance increases T_t to 0.66, lowering R_t to 0 indicating type-2 behavior. In spite that the decision boundary established by the IForest approach presents in an irregular shape which has not been seen from the other two AD approaches and is proven rather sensitive to clogging behavior, it performs the worst among the AD approaches (see Table 7.3 for which for a given pipe jacking drive, the first, second, and third rows respectively indicate the results based upon the slurry density, torque, and jacking speed). TBM or shield machines, equipped with such an integrated AD system, can be useful in introducing countermeasures in advance of clogging during drives in clayey soils.

For clay soils with low-to-medium clogging potential, the risk of adhesion is assessed to be low, according to the practical experience of the authors. Notwithstanding that, a risk of clogged openings still remains throughout the spoil transport and disposal processes, depending upon the consistency of the material. A clay with medium-to-high clogging potential can adhere to all parts of the tunnel boring machine. Figure 7.27 shows the clay soil adhered to the cutterwheel of machine to produce clayey clogging at 20–26 m jacked distance of drive B. The field observation locates where the clayey clogging takes place and is in agreement with the clogging baseline assessment using the semi-empirical approach. Further, the seasonal-trend decomposition procedure using Loess successfully accentuates the characteristics of data in response to the development of clayey clogging toward allowing the 'slurry density' and 'maximum cutterwheel torque' as well as 'jacking speed' anomaly detectors to effectively detect the clayey clogging. The results obtained from this work

Table 7.3 Summary of discovery rate and false alarm rate

AD method	Drive	TP number	TN number	FP number	FN number	DR (%)	FAR (%)
OCSVM	B	1	26	2	0	100	7.1
		1	52	5	0	100	8.8
		1	55	4	0	100	6.8
	D	0	21	3	1	0	12.5
		1	35	3	0	100	7.9
		1	62	6	0	100	8.8
Robcov	B	1	28	0	0	100	0
		0	51	6	1	0	10.5
		1	55	4	0	100	6.8
	D	1	22	2	0	100	8.3
		1	35	3	0	100	7.9
		1	62	6	0	100	8.8
IForest	B	1	28	0	0	100	0
		1	53	4	0	100	7.0
		1	56	3	0	100	5.1
	D	1	23	1	0	100	4.2
		0	34	4	1	0	10.5
		0	62	6	1	0	8.8

give testimony of proving the reliability of the proposed approach and providing some guideposts as follows for reducing the potential of clayey clogging while tunneling in clayey soils.

It is worth noting that to use a slurry-supported shield machine in such a soil, small soil chips, narrow passages for the transport of a clay chip from the cutting face to the support slurry line, sharp angles applied to the excavation chamber, and clay agglomerations in areas prone to material settlement should be avoided. Generation of large soil chips to reduce the adhesion-prone surface of the cut lump is essential to prevent the development of clogging. Further, optimized passages, rounded angles, and turbulence (namely, flushing nozzles and agitator) in areas which are prone to material settlement should also be considered. Moreover, manipulating the ratio of the suspension flow rate to the volume of cut lump gives a benefit in preventing clay from accumulation.

Human factor is deemed as a crucial factor in anomaly detection for tunnel construction. We have also noted that the torque variation, while pipe jacking in 51–58 m distance at drive B, is more distinct than other distances of same soil layer, implying that the human factor may intervene the operation of tunnel boring machine and try to lift the machine out of the fine-grained soils by imposing 'breakout' cutter-wheel torque. The other distinct variation in torque occurs when jacking in 33–35 m distance at drive D. This is not due to the presence of the fine-grained soils but to the

7.3 Maintenance Challenges 273

Fig. 7.27 Clayey clogging with embedded gravel present while traversing through a ground consisted of the clayey gravel and the gravel at drive B: **a** before and **b** after the clay soil adhered to cutterwheel

presence of the gravel. Despite the breakout torque present in 51–58 m distance at drive B, one of the advantages for the use of artificial intelligence technologies is to accentuate patterns in the tunneling data and subsequently disaggregate time series data into feature-based sub-series. As a result, the accuracy of anomaly detection can be prevented from disturbing by, for example, human factors.

References

Alberto-Hernandez Y, Kang C, Yi Y, Bayat A (2017) Mechanical properties of clayey soil relevant for clogging potential. Int J Geotech 1–8. https://doi.org/10.1080/19386362.2017.1311086

Alemdag S, Gurocak Z, Cevik A, Cabalar AF, Gokceoglu C (2016) Modeling deformation modulus of a stratified sedimentary rock mass using neural network, fuzzy inference and genetic programming. Eng Geol 203:70–82. https://doi.org/10.1016/j.enggeo.2015.12.002

Avunduk E, Copur H (2018) Effect of clogging on EPB TBM performance: a case study in Akfirat Waste Water Tunnel, Turkey. Geotech Geol Eng 37:4789–4801. https://doi.org/10.1007/s10706-019-00938-6

Bai XD, Cheng WC, Ong DEL, Li G (2021) Evaluation of geological conditions and clogging of tunneling using machine learning. Geomech Eng 25(1):59–73. https://doi.org/10.12989/gae.2021.25.1.059

Bai XD, Cheng WC, Sheil BB, Li G (2021a) Pipe jacking clogging detection in soft alluvial deposits using machine learning algorithms. Tunn Undergr Space Technol 113:103908

Breiman L (2001) Random forests. Mach Learn 45(1):5–32. https://doi.org/10.1016/j.tust.2021.103908

Cheng WC, Ni JC, Shen JS, Huang HW (2017) Investigation into factors affecting jacking force: a case study. Proc Inst Civ Eng: Geotech Eng 170(4):322–334. https://doi.org/10.1680/jgeen.16.00117

Cheng WC, Ni JC, Arulrajah A, Huang HW (2018) A simple approach for characterising tunnel bore conditions based upon pipe-jacking data. Tunn Undergr Space Technol 71:494–504

Cheng WC, Ni JC, Huang HW, Shen JS (2019) The use of tunneling parameters and spoil characteristics to assess soil types: a case study from alluvial deposits at a pipe jacking project site. Bull Eng Geol Environ 78:2933–2942

Cheng WC, Wang L, Xue ZF, Ni JC, Rahman M, Arulrajah A (2019a) Lubrication performance of pipe jacking in soft alluvial deposits. Tunn Undergr Space Technol 91:102991

Cleveland R, Cleveland W, McRae J, Terpenning I (1990) STL: a seasonal-trend decomposition procedure based on loess. J off Stat 6(1):3–73

de Oliveira DGG, Thewes M, Diederichs MS, Langmaack L (2018) EPB tunneling through clay-sand mixed soils: Proposed methodology for clogging evaluation. Geomech Tunn 11(4):375–387

de Oliveira DGG, Diederichs M, Thewes M (2019a) EPB machine excavation of mixed soils—laboratory characterization. Geomech Tunnel J 12(4):373–385. https://doi.org/10.1002/geot.201900014

de Oliveira DGG, Thewes M, Diederichs M (2019b) Clogging and flow assessment of cohesive soils for EPB tunneling: Proposed laboratory tests for soil characterization. Tunn Undergr Space Technol 94:103110

de Oliveira DGG, Diederichs M, Thewes M (2019c) EPB excavation of cohesive mixed soils: combined methodology for clogging and flow assessment. In: ITA-WTC 2019, Tunnels and underground cities: engineering and innovation meet archaeology, Architecture and Art, Naples, Italy, pp 2008–2017

Deep K, Thakur M (2007) A new crossover operator for real coded genetic algorithms. Appl Math Comput 188(1):895–911

Deep K, Thakur M (2007) A new mutation operator for real coded genetic algorithms. Appl Math Comput 193(1):211–230

Feinendegen M, Ziegler M, Spagnoli G, Fernández-Steeger T, Stanjek H (2010) A new laboratory test to evaluate the problem of clogging in mechanical tunnel driving with EPB-shields. In: ISRM International Symposium-eurock, Lausanne, Switzerland, pp 429–432

Feinendegen M, Ziegler M, Weh M, Spagnoli G (2011) Clogging during EPB-tunneling: Occurrence, classification and new manipulation methods. In: Proceedings ITA-AITES World Tunnel Congress, Helsinki, pp 767–776

Friedman JH (1991) Multivariate adaptive regression splines. Ann Stat 19:1–67

References

Gurocak Z, Solanki P, Alemdag S, Zaman M (2012) New considerations for empirical estimation of tensile strength of rocks. Eng Geol 145–146:1–8

Hollmann FS, Thewes M (2013) Assessment method for clay clogging and disintegration of fines in mechanised tunneling. Tunn Undergr Space Technol 37:96–105. https://doi.org/10.1016/j.tust.2013.03.010

Hu W, Rostami J (2020) A new method to quantify rheology of conditioned soil for application in EPB TBM tunneling. Tunn Undergr Space Technol 96:103192. https://doi.org/10.1016/j.tust.2019.103192

Liu FT, Ting KM, Zhou ZH (2012) Isolation-based anomaly detection. ACM Trans Knowl Discov Data 6(1):1–39

Masters T (1993) Practical neural network recipes in C++. Academic Press, San Diego, CA

Mehdizadeh A, Disfani MM, Evans R, Arulrajah A, Ong DEL (2017) Mechanical consequences of suffusion on undrained behaviour of a gap-graded cohesionless soil - an experimental approach. Geotech Test J 40(6):1026–1042. https://doi.org/10.1520/GTJ20160145

Mehdizadeh A, Disfani MM, Evans R, Arulrajah A., Ong DEL (2016) Discussion of development of an internal camera-based volume determination system for triaxial testing by S. E. Salazar, A. Barnes, and R. A. Coffman. Geotech Test J 39(1):165–168. https://doi.org/10.1520/GTJ20150153

Pedregosa F, Varoquaux G, Gramfort A, Michel V, Thirion B, Grisel O, Blondel M, Prettenhofer P, Weiss R, Dubourg V, Vanderplas J (2011) Scikit-learn: machine learning in Python. J Mach Learn Res 12:2825–2830

Pelia D, Picchio A, Martinelli D, Dal Negro E (2015) Laboratory tests on soil conditioning of clayey soil. Acta Geotech 11(5):1061–1074. https://doi.org/10.1007/s11440-015-0406-8

Persons WM (1919) Indices of business conditions: an index of general business conditions. Harvard University Press, Cambridge, MA

Rousseeuw PJ (1984) Least median of squares regression. J Am Stat Assoc 79(388):871–880

Schölkopf B, Platt JC, Shawe-Taylor J, Smola AJ, Williamson RC (2001) Estimating the support of a high-dimensional distribution. Neural Comput 13(7):1443–1471

Sheil B, Suryasentana SK, Cheng WC (2020) Assessment of anomaly detection methods applied to microtunneling. J Geotech Geoenviron Eng 146(9):04020094. https://doi.org/10.1061/(ASCE)GT.1943-5606.0002326

Thewes M, Burger W (2005) Clogging of TBM drives in clay—identification and mitigation of risks. In: Proceedings ITA-AITES World Tunnel Congress, Istanbul, Turkey, pp 737–742

Thewes M, Hollmann FS (2016) Assessment of clay soils and clay-rich rock for clogging of TBMs. Tunn Undergr Space Technol 57:122–128. https://doi.org/10.1016/j.tust.2016.01.010

Thewes M (1999) Adhesion of clay soil in tunnel drives with slurry shields (In German: Adhäsion von Tonböden beim Tunnelvortrieb mit Flüssigkeitsschilden). Berichte aus Bodenmechanik und Grundbau der Bergischen Universität Wuppertal, Fachbereich Bauingenieurwesen, Bd. 21. Shaker Verlag, Aachen

Tokgöz N (2016) An assessment method for fine-grained sedimentary materials excavated by EPB TBM. Geomechanik Und Tunnelbau 9(4):326–337

Vapnik V (1995) The nature of statistical learning theory. Springer, New York

Weh M, Ziegler M, Zwick O (2009) Verklebungen bei EPB-Vortrieben in wechselndem Baugrund: Eintrittsbedingungen und Gegenmaßnahmen. Forschung + Praxis 43:185–189

Wong STY, Ong DEL, Robinson RG (2017) Behaviour of MH silts with varying plasticity indices. Geotech Res 4(2):118–135. https://doi.org/10.1680/jgere.17.00002

Woo SM, Moh ZC (1990) Geotechnical characteristics of soils in the Taipei basin. In: Proceedings 10th Southeast Asian Geotechnical Conference, Taipei, Taiwan, pp 51–65

Yin ZY, Jin YF, Huang HW, Shen SL (2016) Evolutionary polynomial regression based modelling of clay compressibility using an enhanced hybrid real-coded genetic algorithm. Eng Geol 210:158–167. https://doi.org/10.1016/j.enggeo.2016.06.016

Zhang WG, Zhang R, Wu C, Goh ATC, Lacasse S, Liu Z, Liu H (2020) State-of-the-art review of soft computing applications in underground excavations. Geosci Front 11(4):1095–1106. https://doi.org/10.1016/j.gsf.2019.12.003

Zhang WG, Li HR, Wu CZ, Li YQ, Liu ZQ, Liu HL (2020) Soft computing approach for prediction of surface settlement induced by earth pressure balance shield tunneling. Undergr Space 6(4):353–363. https://doi.org/10.1016/j.undsp.2019.12.003

Chapter 8
Decarbonizing Tunnel Design and Construction

8.1 Why 'Decarbonize' Infrastructure?

Living in the twenty-first century means designing, constructing, operating, and decommissioning infrastructure in ways that are good for planet and people. Over the last decade, research and discussion of this imperative has evolved from calls to 'green the built environment', through to calls for 'sustainable design and construction', and more recently 'low-carbon design and construction' for sustainable built environment.

As world leaders gather more frequently to determine strategies for urgent and substantial greenhouse gas emission reductions, the clear imperative is to 'decarbonize the built environment' which includes associated infrastructure, from its creation to its operation and decommissioning. The decarbonizing agenda of achieving 'low carbon' through to 'zero carbon' outcomes is central to the call for decoupling (i.e., disconnecting) environmental and social impacts from economic growth as we continue to build for increasing populations around the planet (Smith et al. 2010; Desha et al. 2005). Closely connected with this imperative is the call for 'circular economy' practices that enable economic growth (jobs) through the transition to a decarbonized way of life.

In this chapter, we introduce this context of 'decarbonizing infrastructure' to underpin the book's consideration of tunneling practices and innovation. We begin with a discussion of the critical context and language used, and then discuss some 'foundation topics' which includes:

- Tunneling processes: Introducing the differences in open-cut versus trenchless technologies, and other advances in tunnel design and construction (Sect. 8.2)
- Spoil reuse: Introducing the variety of ways that are already being considered to service other decarbonizing agendas in construction activities onsite or on other sites (Sect. 8.3)
- Circular economy in practice: Considering the practical implications for industry, to integrate these insights into the workplace (Sect. 8.4)

- Case study example: An example of how decarbonizing opportunities for tunneling and spoil can be implemented, considering the range of regulatory environments being offered (Sect. 8.4)

8.1.1 Global Sustainability Goals

Reducing the amount of carbon that we use in the design and construction of infrastructure, goods and services, is recognized internationally as a key enabler of sustainable development. In 2015, the United Nations published its Sustainable Development Goals (SDGs) for its 193 member countries to action by 2030, including 17 goals, 169 targets and 232 indicators.

In particular Goal 9 (SDG-9) directly appeals to the built environment sector to, *'build resilient infrastructure, promote inclusive and sustainable industrialization and foster innovation'* (UN 2015). Target-4 of SDG-9 states, *'By 2030, upgrade infrastructure and retrofit industries to make them sustainable, with increased resource-use efficiency and greater adoption of clean and environmentally sound technologies and industrial processes, with all countries taking action in accordance with their respective capabilities'*. The target for achieving this (Target 9.4.1) is 'carbon dioxide (CO_2) emission per unit of value added'. For tunneling, in plain-speak these SDG directives are calling for reducing the amount of greenhouse gas emissions it takes to construct and operate infrastructure—in this case tunnels.

8.1.2 Circular Economy Goals

The field of circular economy has evolved over the last two decades, with a call for project managers and clients to monitor and evaluate energy used by excavating equipment, embedded energy in the materials used for constructing the equipment being used onsite, and emissions associated with the transportation of spoil offsite (distance traveled, impact on the destination).

Such 'whole of life cycle' sustainability assessment enables accountability for actual emissions on a project, which can then be aggregated to provide sustainability performance criteria to the client and shareholders. Gijzel et al. (2019) have found that a variety of tools exist for assessing sustainability in projects ranging from checklists, to maturity models and proposed standards to impact assessments (Gijzel et al. 2019). In Australia and New Zealand, significant opportunities and support exists through the Infrastructure Sustainability Council (ISC), which works to ensure low-carbon operators are rewarded for their efforts and outcomes (ISC 2021). The organization's Infrastructure Sustainability (IS) Rating Scheme has emerged to document and incentivize decarbonizing and other sustainability efforts in a project. It assesses the sustainability performance of infrastructure at the individual assets level, for portfolios or networks, from local to regional scales (ISC 2020).

8.1.3 Life Cycle Assessment (LCA)

The life cycle assessment (LCA) addresses the environmental aspects and potential environmental impacts (e.g., use of resources and environmental consequences of releases) throughout a product's life cycle. The life cycle for a given item may range from raw material acquisition to final disposal (i.e., cradle-to-grave) through production, use, end-of-life treatment, and recycling.

An important context for LCA is its use for 'construction products' in addition to 'buildings' and 'tunnels', considering the environmental impacts throughout the infrastructure life cycle from construction to end-of-life including the operation phase and end-of-life decommissioning (Schwartzentruber 2015). It is interesting to consider that for tunnels 'end-of-life' does not really exist with a common reference of '100 years'). Instead, it might reflect the 'end of operation' that could follow decommissioning and site rehabilitation.

An example of LCA use is in the paper by Huang et al. (2013), which analyzed the life cycle impacts of a Norwegian standard road tunnel (Hammervold 2014). A standard Norwegian tunnel is a 3-km-long, 9.5-m-wide rock excavated tunnel. The construction phase is found to be the largest contributor, where the production of materials is the most dominant contributor to their included impact categories. Over the 100 years of expected service, the tunnel was calculated to produce 13 tons CO_2-eq per m of tunnel. Furthermore, the emissions from tunnel construction and operation in CO_2-eqs per m are 337 kg (lifetime 100 years) and an analysis time (40 years).

8.2 Decarbonizing Tunneling Processes

Tunneling technology is generally considered as an expensive method when compared to open-cut trenches. It is essential to assess the individual scenario and influencing factors which will have an impact on the project. As a general rule of thumb, we cannot identify the most cost-effective method solely on the technology without considering the various site parameters. This section describes a comparative study on the direct and indirect costs related to open-cut versus trenchless method. Carbon footprint and energy consumption for both cases were also reported.

8.2.1 Open-Cut Versus Trenchless Technologies

In recent years, the fast pace of development has resulted in scarcity of urban land which favors tunnel construction. Although trenchless technology has shown to have better social and economic benefits, open trenching is more commonly adopted due to the perception of being more feasible, simpler, and useful (Lu et al. 2020).

Open-cut, also known as cut and cover or trenching, had very few improvements since the early days. Mechanical excavation, wall supports for deep excavation and improved quality of pipe material are some of the refinements of open-cut technique. These improvements would have been more beneficial if the local context would have not changed. Road users and pedestrians have influenced the need for more sophisticated transport infrastructure and more utilities are now installed underground. Current urbanization rate would require more utilities such as non-potable water, pneumatic waste collection, combined heat and power pipelines to be installed underground (Hunt et al. 2011). There is an increasing attention given to the future use of underground space for living area, transportation, resource extraction, and waste storage (Parker 2004; Jefferson et al. 2006; Bobylev 2009; Evans et al. 2009; Sterling et al. 2012). Currently, ground surfaces are commonly paved or built over in urban and suburban areas. In addition, rural areas often have utility services buried under paved roads. Asset location, excavation and utility placement, maintenance, repair, and decommissioning will contribute to greater cost (Costello et al. 2007). Brundtland (1987) stated that engineers should take into consideration current and future related costs to utility placement such as indirect economic cost, environment, and society cost instead of focusing solely on the direct economic cost. Examples of indirect cost for utility placement are traffic disruption, environmental impacts, health and safety risks, possible risks of damage to paved surfaces, and adjacent infrastructure (Tighe et al. 2002). When incorporating the indirect cost, it can be argued that open-cut may not always be the preferred method for pipe installation. In London, open-cut method was found not to be suitable for electric cable installation via bored tunnels as it would result in multiple disturbance complaints due to its numerous excavation and reinstatement procedures (Hunt et al. 2014).

8.2.1.1 Cost Difference Between Open-Cut and Trenchless

In general, trenchless technology is more suitable in urban area due to its lower environmental impacts, social cost, and limitations of access. Conversely, open-cut method would be preferred for rural areas consisting of larger space with lesser restrictions which would make it less costly. Figure 8.1 shows the cost difference between open-cut and trenchless in terms of social, labor, material, and indirect cost. The social cost for open-cut method was found to be 40% of the total project cost while for trenchless method, it was only 5% of the project cost (Tabesh et al. 2016).

When comparing the life cycle cost of open-cut and trenchless methods, Najafi and Kim (2004) suggested to consider the following factors:

- Equipment and transportation
- Pipe material
- Removal of spoil
- Backfill and compaction
- Resurfacing
- Shoring and sloping of trench walls

8.2 Decarbonizing Tunneling Processes

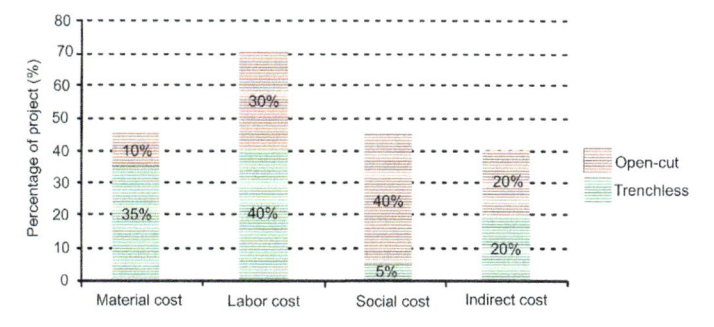

Fig. 8.1 Cost difference between open-cut and trenchless with sample breakdown (adapted from Tabesh et al. 2016)

- Detour roads
- Dewatering
- Labor

Based on the project nature and size, the cost of construction equipment would differ for each method. For the comparative life cycle assessment by Najafi and Kim (2004), trenchless was found to be more costly when compared to open-cut. The pipe material is also more costly for trenchless technique via pipe jacking where a better pipe quality is required. Transportation of tunnel boring machine (TBM) to the construction site and back would be higher compared to the open-cut equipment. Open-cut would require about 1.5 times more soil handling and removal of spoil in contrast with trenchless method. However, labor cost is generally cheaper compared to the specialized workforce required for trenchless. Najafi and Gokhale (2005) stated that resurfacing during open-cut contributes up to 70% of the total project cost which for the case of tunneling is minimal. Shoring and sloping during open-cut has shown to increase the construction cost significantly. Due to inconvenience caused to general public, existing structural and environmental damages, detour roads are required during construction which will add on to the social costs. Dewatering is an important factor which needs to be considered during open-cut construction at sites with high water table. This is not required for trenchless technique (Tabesh et al. 2016).

Sterling (2020) mentioned that for trenchless method to gain acceptance over to open-cut, the project owner needs to agree and incorporate the costs related to other parties and the cost related to environmental damage along with the direct cost of the construction. Over the past years, environmental impact and carbon emission have raised issues over construction methods and marketing for these technologies.

8.2.1.2 Carbon Footprint and Energy Consumption for Trenchless

The carbon footprint and energy consumption for open-cut and trenchless methods were analyzed by Lu et al. (2020). Various construction cases and influencing factors were taken into consideration for the analysis. Among the numerous construction methods studied, pipe installation using horizontal directional drilling (HDD) was found to reduce between 6 and 56% of the energy consumption and carbon footprint when compared to open-cut method. For installation by pipe jacking, the energy consumption and carbon footprint were 14–86% lower compared to open-cut. The outcomes were still lower for the case of shallow buried depth and short length pipelines. For pipe replacement using the pipe bursting method, a reduction of 50–82% in carbon footprint and energy consumption was recorded in contrast to open-cut method. In summary, various trenchless techniques have induced in lower energy consumption and carbon footprint relative to open-cut method. However, for short construction length or at shallow depth, the consumption of energy and carbon footprint may be higher as opposed to open-cut technique based on the trenchless method used (Lu et al. 2020).

8.2.2 Advances in Tunnel Design and Construction

Frictional force assessment and prediction are ongoing interests with aim to improve tunnel design and construction process. In today's technological era, the tunneling industry is focusing on digital analysis, simulations, and AI techniques to achieve optimal outcomes. This section shows current advancement achieved to improve the assessment of pipe frictional forces.

8.2.2.1 Pipe Friction Assessment and Prediction

Trenchless tunneling via microtunneling and pipe jacking methods comprise pipe sections or tunnel support structure which are pushed through the ground to produce a desired tunnel pipeline length. The key parameters to ensure completion of the project is the thrust required to jack through the drive and the capacity of the jacking components to withstand the thrust force. The governing factors of pipe friction assessment consist of the pipe–soil interaction, the effect of the alignment curvature (planned and unplanned curvature), the influence of the overcut region and reduced friction due to pressurized lubrication. Common practical frictional force models are developed based on simple physical models combined with prior field experiences and updated with some laboratory testing (Sterling 2020).

Popular jacking force models have been established statistically and empirically by Bennett (1998), Chapman and Ichioka (1999), Osumi (2000), Staheli (2006) and Pellet-Beaucour and Kastner (2002) to estimate jacking forces. For the assessment of frictional forces, each model takes into account its own parameters like the pipe and soil properties. However, Peerun et al. (2020) state that these jacking force models are applicable and limited to specific scenarios and ground conditions. A detailed analysis of the jacking forces with regard to various types of geology has been documented by Choo and Ong (2015), Ong and Choo (2018) and Peerun et al. (2017a, b, 2020), and

8.2 Decarbonizing Tunneling Processes

some of these established case studies are presented in Chaps. 2 and 3 of this book for further reader appreciation. Particle shape, effectiveness of lubrication and arching effect were found to significantly influence the frictional force. The vertical stress acting onto the pipe crown is used as an indication of arching effect which would significantly influence the jacking forces. From Pellet-Beaucour and Kastner (2002) model, assessment of vertical stress onto the pipe and the jacking forces requires the strength properties of the surrounding geology. These properties are friction angle and cohesion, which are commonly obtained from a laboratory direct shear test.

8.2.2.2 Shearing Stages During a Direct Shear Test

Direct shear test (DST) has been frequently used in the study of pipe jacking forces. The strength and volumetric behaviors of the tested specimen are based on particle angularity, mineralogical content and density (Peerun et al. 2016a, 2020). Hence, it is important to understand the mechanical behavior of the particles subject to shear which will be used to interpret its respective strength properties. From a conventional direct test result, the stress and volumetric behaviors are as described in Table 8.1.

To gain better insight on the stress and volumetric behaviors described in Table 8.1, the interparticle behavior during shearing is required. Li and Aydin (2010) proposed a four-stage shearing model which describes in detail how the particles interact among one another to produce the resultant shear and volumetric outcomes. Various geologies with different particle shapes and mineralogy would produce contrasting shear results. This would be advantageous for jacking force assessment of different geologies by combining the interparticle behavior and shear strength properties. The four shearing stages are:

- Stage 1: End zone deformation
- Stage 2: Particle interlocking
- Stage 3: Shear zone formation
- Stage 4: Steady shear

Figure 8.2 shows the four-stage shearing model (Li and Aydin 2010) in terms of the shear stress and volumetric plots. The first stage known as 'end zone formation' begins at the start of the shearing phase where the specimen contracts due to a change

Table 8.1 Conventional stress and volumetric behaviors for dense and loose DST specimens

Specimen	Shear stress	Volumetric behavior
Dense to medium dense	Distinct peak shear stress which then reduces to constant residual state	Initial compression followed by dilation
Loose	Increase in shear stress until constant residual state is achieved	Compression is expected until residual state (no/minimal dilation)

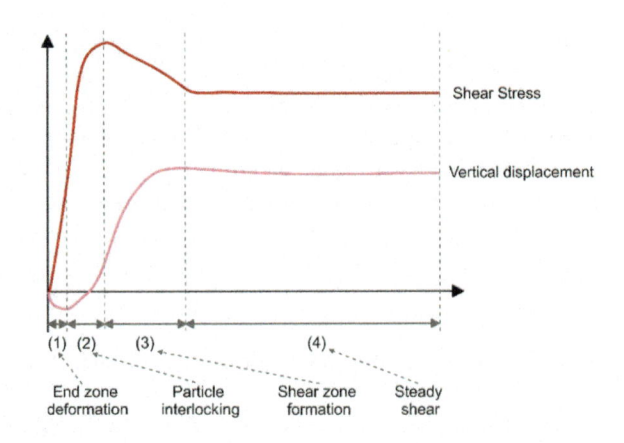

Fig. 8.2 Four-stage shearing model (adapted from Li and Aydin 2010)

in stress field acting onto the specimen. When the shear force is applied, the particles are rearranged from their initial consolidated state. From the change in stress field, the particles rearrange themselves into existing voids resulting in compression. Stage 1 ends at the lowest volumetric strain where no more compression occurs due to the start of the interlocking stage (Shimizu 1997; Peerun 2016). The second stage is particle interlocking which starts at the lowest point of the volumetric plot. During this stage, the particles in contact against each other will try to overcome interlocking. Interlocking can be defined as the force required to prevent the particles from rolling, rotating, and sliding in terms of displacement. Particle interlocking ends when the peak shear stress is achieved due to maximum interlocking activities occurred along the shear band (Peerun 2016; Peerun et al. 2019a, b). The third stage is known as 'shear zone formation' which starts from the peak shear stress until the stress stabilizes to a near constant value. The particles along the shear band would displace until a loose layer is achieved (Oda and Kunishi 1974; Fukuoka et al. 2006; Peerun 2016). At this point, the larger particles will roll or rotate while the smaller particles will end up in existing voids which will reduce the shear resistance. With large shear displacement, a constant shear band can be obtained (Li and Aydin 2010; Peerun 2016; Peerun et al. 2019a, b). 'Steady shear' is the last stage of shearing where no further compression or dilation is produced. Shearing continues with minimal change in vertical displacement which produces constant shear stress, achieving residual state. Li and Aydin (2010) state that an equilibrium in compression and dilation occurs which allows the particles to slide along the surface.

There are ongoing efforts to better understand the particle behavior described using the four-stage shearing model. The opaque nature of a typical direct shear test equipment makes it a challenge to physically and visually assess the particle activities during shearing. To improve the understanding on particles activities such as interlocking, localized compression or dilation and possible particle breakage, Peerun (2016) used a unique purpose-built transparent shear box (see Fig. 8.3) which allows clear views of the particles while they are being sheared. Sequential images were captured during the direct shear test and were processed using particle image

8.2 Decarbonizing Tunneling Processes

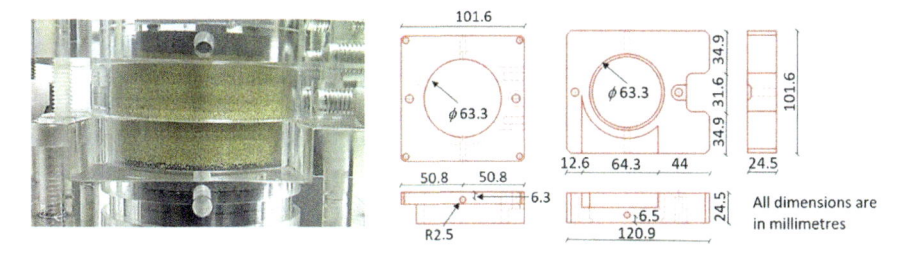

Fig. 8.3 Transparent shear box and AutoCAD drawing with dimension

velocimetry (PIV) technology. The PIV technology generates vector plots of the particle displacements during shear, which visually compliments the DST results.

8.2.2.3 Particle Image Velocimetry

Particle image velocimetry (PIV) makes use of sequential images to analyze patches of texture within the image to produce displacement vectors. PIV technology was developed by Adrian (1991) to study velocity of fluids. GeoPIV makes use of PIV principles to analyze particle movements in geotechnical test applications. GeoPIV was developed by White and Take (2002) as a MATLAB module which makes use of command lines to analyze a batch of sequential images and producing displacement vector plots. The predefined input parameters of GeoPIV are (i) patch size, (ii) patch spacing, (iii) searchzone pixel, (iv) frame rate and (v) leapfrog.

These parameters need to be defined by the user, and the values would vary based on the application and image acquisition technique. To ensure reliable vector plot is obtained from the GeoPIV analysis, calibration and validation exercises are essential. In order to ensure that GeoPIV conducts adequate analysis without any external influence, Peerun et al. (2016b) completed two validation exercises prior to working on the calibration and actual GeoPIV analysis. The first validation consisted of analyzing two identical images where the vector plot should not record any displacement. The second validation makes use of digitally inserting a dot in the first image and moving it to a known direction in the second image. The vector plot was able to produce the correct displacement of the single dot with no other interference. Figure 8.4 shows the interface of GeoPIV.

Once the GeoPIV process has been validated, the input parameters need to be calibrated to ensure reliable vector plots are obtained (Peerun et al. 2016b, c). A parametric study can be done to define the optimum value of each parameter. Each parameter is varied individually to understand its influence on the vector plot. The most suitable value is selected based on a clear and detailed vector plot with no wild vectors.

Patch size refers to a grid of test patches (grids) drawn on the first image. The patch size dimension is in pixels. The test patch is analyzed between the sequential images to produce the particle displacement (White et al. 2003). White and Take

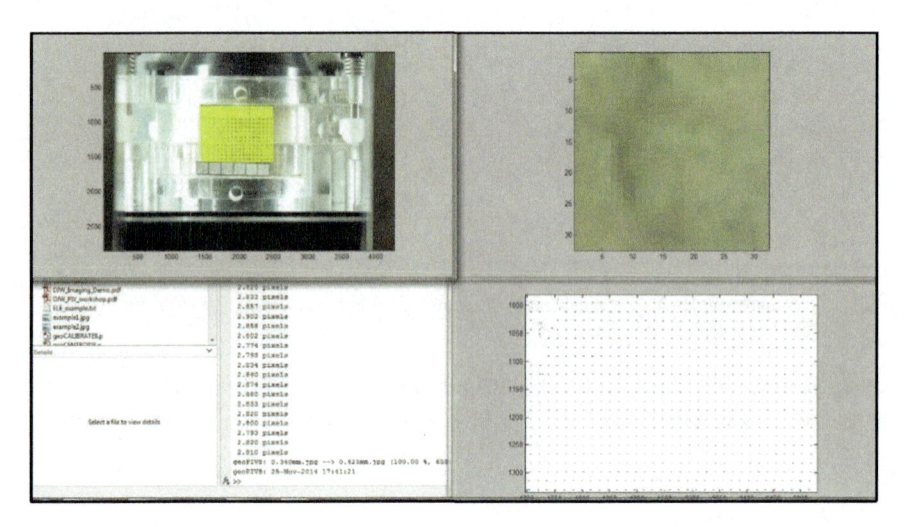

Fig. 8.4 GeoPIV interface

(2002) stated that the image resolution and patch size would determine the precision of the vectors. Figure 8.5 shows the calibration results for patch size parameter using three patch sizes, namely 16 × 16, 32 × 32, and 64 × 64 pixels. A patch size of 16

Fig. 8.5 Calibration results for patch size parameter using 16 × 16, 32 × 32, and 64 × 64 pixel sizes

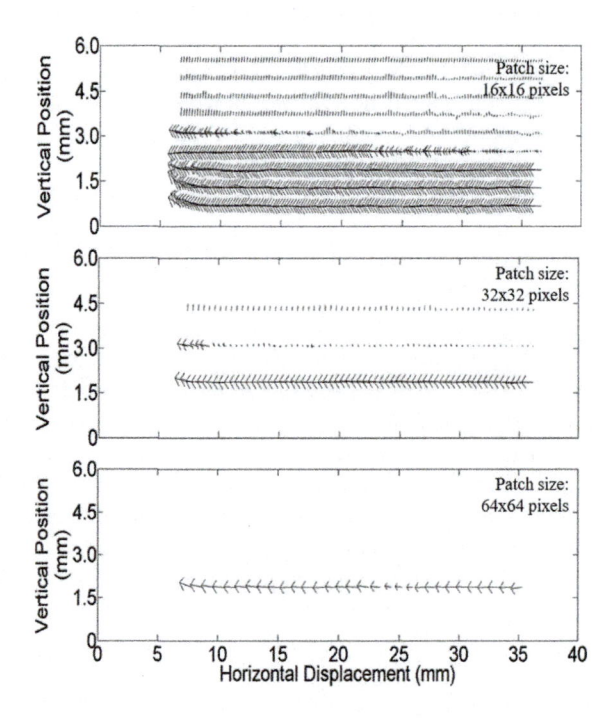

8.2 Decarbonizing Tunneling Processes

Fig. 8.6 Image acquisition setup for GeoPIV during direct shear test

× 16 pixels was found to be most suitable where more details can be obtained as opposed to the other patch sizes.

Patch spacing is defined as the overlapping area between two successive test patches during the analysis. The recommended spacing is half of the patch size where each half of the patches will be covered twice (Berghe 2012; Peerun 2016). Kelly (2014) also stated that the vector precision can be maximized when adopting an overlapping area of 50%. The defined spacing should be able to capture the particle displacement between the test patches of two successive images.

Searchzone pixel is the maximum zone extended from the patch size, where GeoPIV will analyze a particle displacement within two successive images. The user must ensure that the searchzone pixel adopted is larger than the maximum displacement recorded within two successive images (Grognet 2011).

Frame rate is the time interval used to capture the sequential images. Based on the experiment, a suitable frame rate should be adopted to avoid large particle displacement between images. Larger frame rates would produce wild vectors (erroneous vectors) due to the particles being out of range from GeoPIV analysis. Smaller frame rates would produce more accurate and precise vectors but the computational time would increase significantly. White and Take (2002) stated that larger frame rates can be used as long the number of wild vectors is admissible and does not change the real data.

The leapfrog value defines how frequent the reference image should be updated to minimize precision error. However, the use of leapfrog is optional and for most cases a leapfrog value of one is suitable, as long there is no presence of wild vectors.

Peerun (2016) used a modified direct shear test with a transparent shear box to capture sequential images during the test. The images were remotely captured using a computer software connected to the camera to avoid any disturbance or shakiness. A total of 175 images were captured for a shearing displacement of 15 mm. To eliminate shadows and improve visibility of the test specimen, dim lights were installed near to the shear box. Figure 8.6 shows the image acquisition setup for GeoPIV analysis during direct shear test.

Peerun (2016) conducted direct shear tests on sand and used GeoPIV to study the particle behavior during shearing. Figure 8.7 shows a typical vector plot which demonstrate particle movements according to the specific four-stage shearing model as discussed in Sect. 8.2.2. The vector plot (Fig. 8.7) was obtained for sand subjected to a normal stress of 500 kPa and at shearing stage 3. The vector arrows depict the particle displacement direction, and its magnitude is proportional to the length of the arrow. There are distinct movements in terms of vectors where the upper and lower halves of the shear box can be recognized. Larger displacement is obtained at the bottom of the shear box which shows the shear displacement to the left direction while relatively smaller displacements are recorded at the top half of the shear box. The vector plot can capture localized particle activities such as interlocking (localized dilation) or possible breakages (localized compression). Figure 8.8 shows typical vector patterns describing (a) particle interlocking, (b) possible particle breakage and (c) no vertical displacement (Peerun 2016).

When integrating the GeoPIV vector plots with the direct shear test results, distinct particle behavior can be related to the various shearing stages (Peerun 2016). Localized particle interlocking would result in sample dilation, which can be confirmed

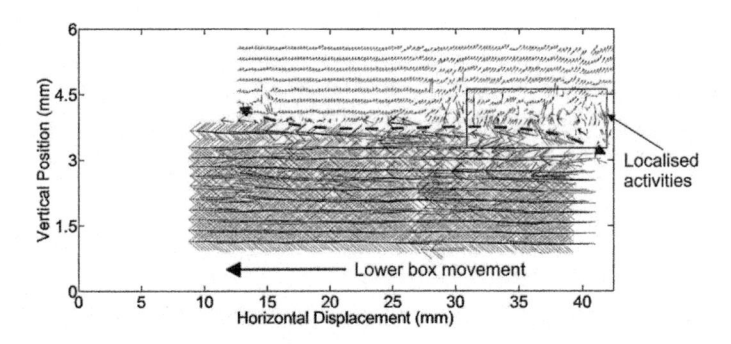

Fig. 8.7 Typical vector plot of DST obtained from GeoPIV analysis (adapted from Peerun 2016)

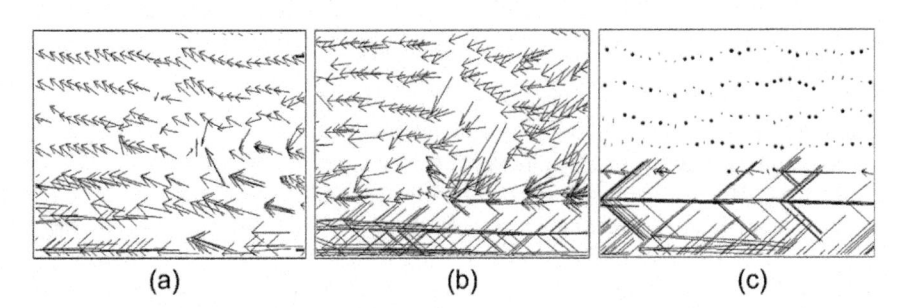

Fig. 8.8 Typical vector patterns describing **a** particle interlocking, **b** possible particle breakage and **c** no vertical displacement (adapted from Peerun 2016)

8.2 Decarbonizing Tunneling Processes

by the GeoPIV vector plots. Conversely, localized compression shown in the vector plot is an indication of possible particle breakage which will be reflected as specimen compression at that specific shearing stage. Figure 8.9 shows a combined DST and GeoPIV results with clear distinction of each shearing stage (Peerun 2016).

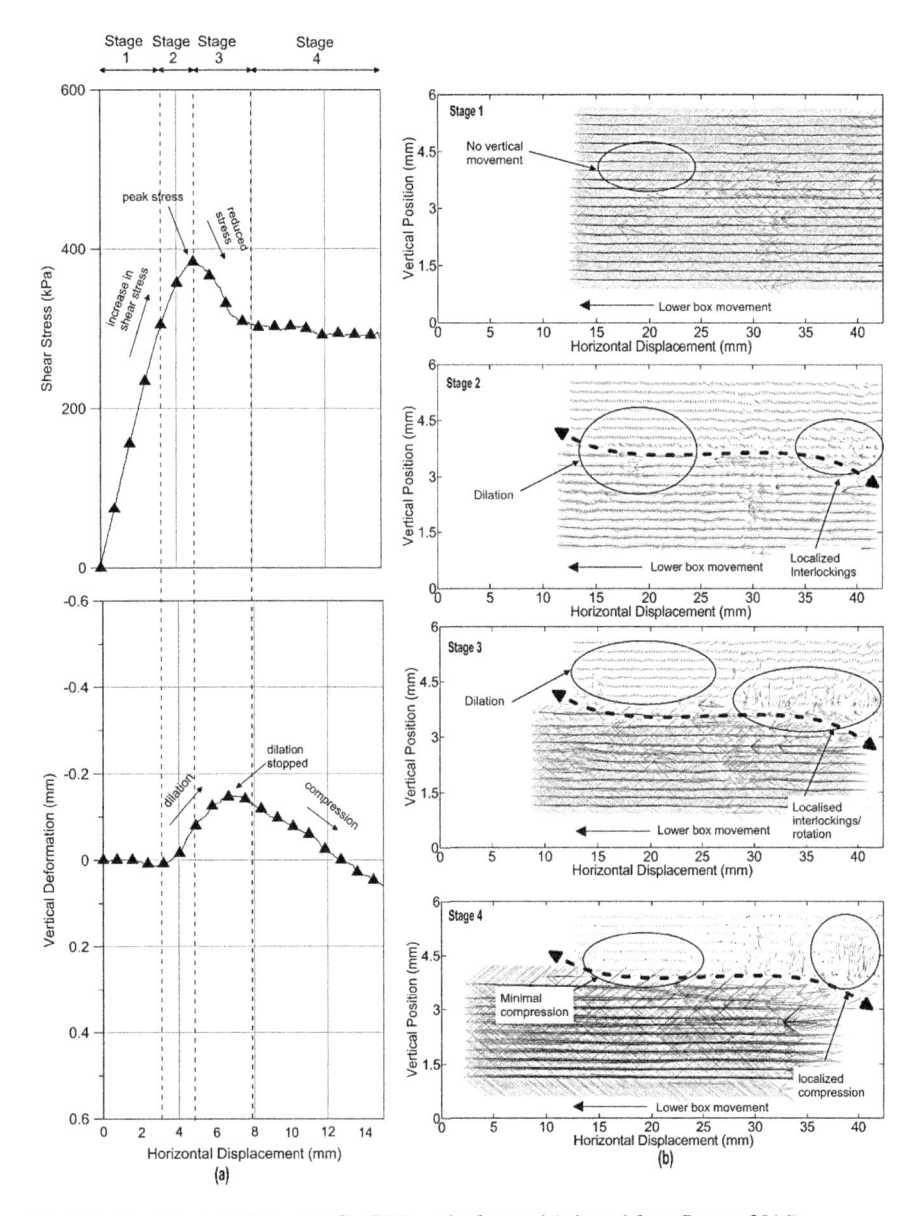

Fig. 8.9 Combined **a** DST and **b** GeoPIV results for sand (adapted from Peerun 2016)

During Stage 1 shearing, no vertical movement was recorded as the vectors showed horizontal movements to the left. This describes the initial stage of shearing where the volumetric plot recorded horizontal displacements with no vertical movements. In Stage 2, shear stress increased until a peak state was reached due to particle interlocking. This statement is confirmed in Stage 2 vector plot where localized interlocking and dilation were observed. Stage 3 experienced a reduction in shear stress as described by Li and Aydin (2010). The vector plot for Stage 3 shows reduced dilation with localized interlocking and rotation of particles. At this stage, particles were rearranging themselves where the larger particles would experience interlocking and smaller particles would fill the voids. This activity reduced the shear stress which was confirmed in Stage 3 of the stress plot. Lastly, Stage 4 experienced minimal compression where a distinct localized compression was observed in the vector plot. While steady shear was almost achieved, the specimen experienced compression throughout Stage 4 as witnessed by the volumetric and vector plots.

Peerun (2016) and Peerun et al. (2020) described the important contributions of GeoPIV to the interpretation of direct shear test results. With the aid of the GeoPIV generated vector plots, Peerun (2016) concluded the following:

- Particle breakage will contribute to the reduction of peak shear stress
- Particle interlocking will produce higher apparent cohesion
- At higher normal stress, particle interlocking will contribute to higher apparent cohesion with a reduction in friction angle value
- Apparent cohesion was found to reduce the vertical stress, and this is analogous to the stress acting on the pipe during pipe jacking

8.2.2.4 Discrete Element Modeling (DEM)

With today's digital advancement, discrete element modeling (DEM) is referred as one of the essential tools used to analyze geotechnical problems. Complex engineering scenarios such as tunneling, deep excavation, and pavement loading can be simulated using DEM. Interface shearing can be simulated with DEM to produce strength properties such as friction angle and cohesion, which can then be used as an input for the assessment of pipe jacking frictional forces. To ensure that realistic failure patterns and properties are obtained from the model, realistic input parameters are vital (Peerun et al. 2019a, b). The conventional approach to DEM simulation makes use of spheres or clump of spheres as representation of the complex particle shape and users often utilizes generalized input parameters and curve fitting exercises to achieve this, which may not be ideal if fundamental particulate behavior was to be better understood. However, with current advances in technology, there are better ways to improve the DEM simulation. Realistic input parameters and realistic particle shapes can be incorporated in DEM with the aid of X-ray scanning and 3D-printed synthetic particles.

Current advancement in DEM has showed that the use of spheres to simulate particle interactions is outdated. This is simply due to the fact that granular particles are far from having a perfect spherical shape. Particle shape was found to be an influencing factor which governs the interparticulate behavior and the strength properties such as friction angle and cohesion (Peerun et al. 2019a). The recommended framework for advanced DEM consists of extracting realistic particle shapes of the granular particles by means of Micro-CT scan. The scanned images can then be converted into a 3D surface mesh format called Standard Tessellation Language (STL) (Peerun et al. 2019a). A novel and advanced DEM software called 'Rocky DEM' has shown great modeling updates in terms of DEM capabilities to solve complex engineering problems. Using the scanned STL file, realistic particle shapes can be imported into the DEM simulation without the use of standard, less accurate spheres, and clumps of spheres. Advanced options such as simulation of particle breakage and coupled finite element modeling (FEM)-DEM, and interactions with mechanical components can be achieved. For large-scale complex DEM simulation consisting large number of particles, greater computational power such as workstation with enhanced random access memory (RAM) and graphic processing unit (GPU) would be highly beneficial. Therefore, it is suggested to explore the possibility of customized high-end workstation or cloud computing services for this purpose. The authors have previously utilized a high-performance RAM-intensified virtual instance provided by QRISCloud, a cloud computing service supported by the Queensland government and Australian Research Data Commons (ARDC).

To improve the generalized input parameters while avoiding curve fitting exercises, an appropriate calibration and validation exercise would be required to ensure that the outcomes of the advanced DEM simulation is accurate. Peerun et al. (2019c, 2021) recommended to synthetically reproduce the scanned particles by means of advanced 3D printing, a process called selective laser sintering. The STL file obtained from Micro-CT scans were used to print the synthetic particles using PA2200 nylon material. The printed particles were then used as specimen for a laboratory direct shear test. The 3D printing material has known and established engineering properties which can be incorporated in the DEM simulation. The laboratory test using 3D printed synthetic particles and DEM simulations can be used to calibrate and validate the input parameters for a more enhanced DEM simulation. This advanced DEM technique would contribute to a more reliable and accurate assessment of jacking forces in microtunneling works.

Figures 8.10, 8.11, 8.12 and 8.13 show an overview of the improved methodology for an advanced DEM technique making use of Micro-CT scan and 3D printing. Sandstone aggregates were scanned, and layers of X-ray images were obtained (Fig. 8.10). Each particle was individually extracted, where the layers of X-ray images were reconstructed in three-dimension (Fig. 8.11) and then exported in STL format (Fig. 8.12). The STL file was then used to 3D print the synthetic particle using PA2200 nylon material (Fig. 8.13).

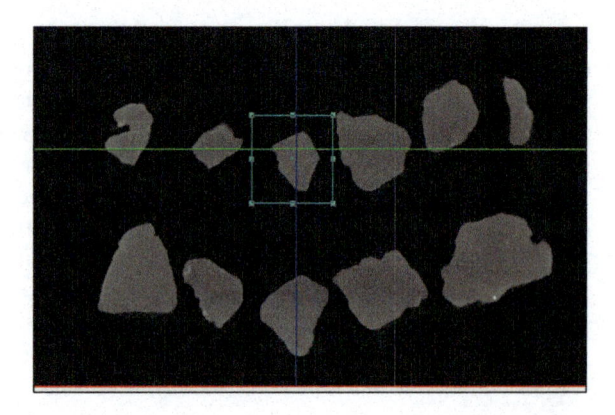

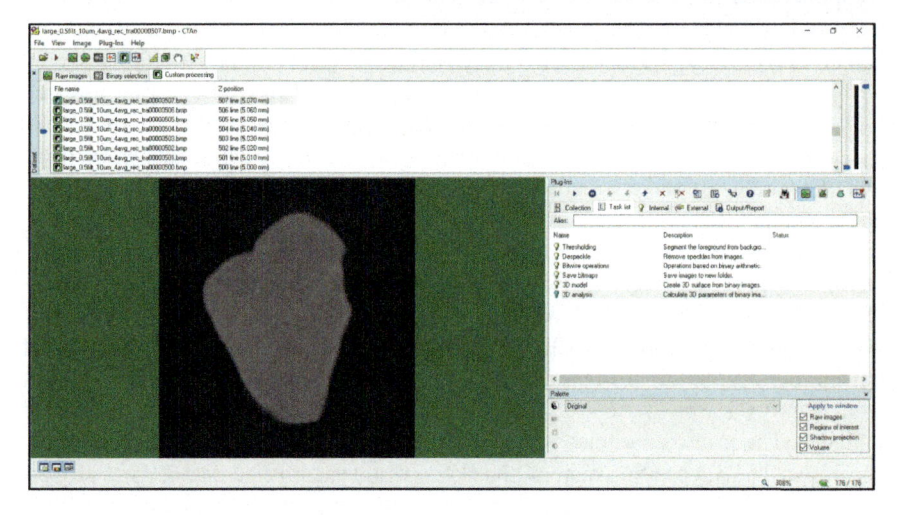

Fig. 8.10 Micro-CT scanned images of sandstone aggregates

Fig. 8.11 Reconstructed layers of X-ray images in 3D

8.2.2.5 Immersive Analysis and Visualization During Tunnel Design

During conceptual planning, detail design, operation, and maintenance phases of a project, several engineering aspects are required to be modeled and interlinked with other components. These components could be from geotechnical, structural, mechanical, and architectural requirements, which may need to be modeled together for better understanding of the overall project. There is always a challenge when working with different software platforms where the data cannot be incorporated for additional/further analysis. Bentley Systems is an engineering software solution provider which ensures interoperability among the whole range of software packages. Some of the key areas where Bentley Systems has contributed to include transportation, mining, utility network, rail, and transit, water and wastewater, bridges,

8.2 Decarbonizing Tunneling Processes

Fig. 8.12 Exporting scanned particle shape in STL format

Fig. 8.13 Synthetic particle produced by means of 3D printing

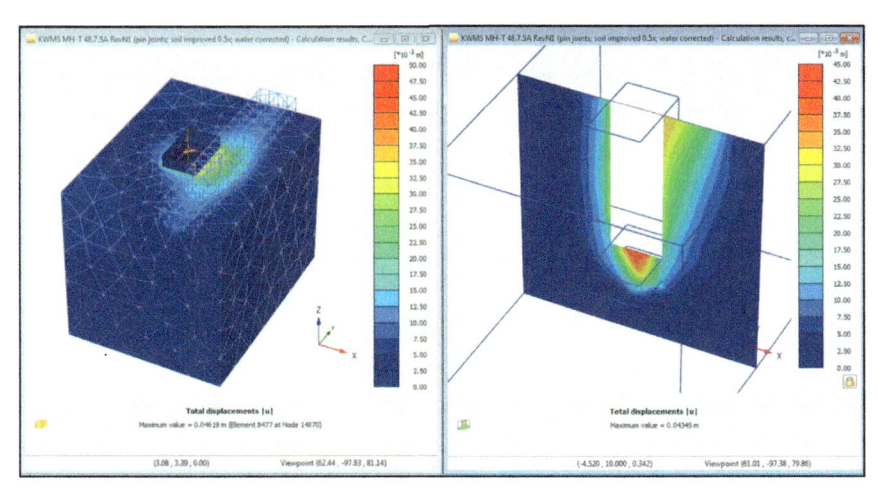

Fig. 8.14 Geotechnical software interface solution PLAXIS by Bentley systems

Fig. 8.15 Visualization software interface solution LumenRT by Bentley systems

processing plants and offshore structures. Figure 8.14 shows the geotechnical software interface using PLAXIS, and Fig. 8.15 shows the visualization software interface solution LumenRT provided by Bentley Systems. For the tunneling sector, the following solutions by Bentley Systems can be adopted:

- Geotechnical analysis: PLAXIS 2D and 3D
- Road design: OpenRoads ConceptStation/Designer
- Rail design with tunnels: OpenRail ConceptStation/Designer
- Project management: Synchro Pro
- Augmented Reality/Visualization: LumenRT

Conceptual design can now be drafted within a few hours while incorporating GIS data. Cost estimate and material quantities can be obtained based on customized local rates. Parametric studies can be conducted to identify the best design option for the project where the model can be easily modified. Design templates and building information modeling (BIM) components based on relevant standards can be incorporated in the analysis. Synchro Pro provides an immersive project management tool in four-dimensions where the 3D model can be visualized in virtual reality and the timeline can be changed to suit various tasks of the projects.

LumenRT is a visualization tool which allows non-technical audience to appreciate the final product by bringing it to life using realistic high-definition 3D animations. Many industries have now included the immersive analysis and visualization components during the design and planning stages of their projects. Benefits for adopting these tools include optimizing limited resources and reducing potential construction risks.

8.2.2.6 Artificial Intelligence Techniques

Apart from the use of digital simulations such as DEM to improve our understanding on the design process of microtunneling works, as discussed in Chap. 7, artificial intelligence (AI) techniques have recently emerged to be a potential solution to supplement existing pipe jacking design methods in the industry. In the recent years, AI techniques have been proven to be capable of solving many engineering problems that are difficult to be solved using conventional methods. Applications of AI techniques in geotechnical engineering can be found in many aspects such as site characterization, tunneling, deep excavations, ground improvement, foundations, and liquefaction. By learning the inputs and outputs presented to the AI models, useful interpretation can be obtained to improve our understanding on various geotechnical problems. With the vast applications of AI in geotechnical engineering, it is believed that AI has a great potential to be used to further understand the behavior of underground construction due to pipe jacking. This can be achieved by developing design models that are robust and data-driven, which can process complex relationships between several variables and produce meaningful output for interpretation.

Due to improved computational capabilities and increasing use of Information Technology (IT) in data collection over the years, massive amount of data such as settlement, lateral displacement, and ground water fluctuations can be collected on site during and after construction. The availability of such data can be utilized to improve the design and construction process of pipe jacking projects. Imagine if you have collected instrumentation monitoring data for a pipe jacking project for several months or years and the data are readily available, why not use this opportunity to better understand the project conditions by using AI techniques to process the data? Not only that AI techniques are powerful in processing complex datasets, but they can also predict the desired design parameters efficiently based on the data provided, which may reflect actual ground conditions more accurately.

In this section, the successful AI application in microtunneling is further reviewed to demonstrate the advantages of employing AI techniques in studying problems related to pipe jacking design. Then, some challenges that may be faced when applying AI techniques and recent development of AI are discussed to provide the readers with insights into common issues and possible countermeasures that can be considered to tackles those issues. Lastly, potential opportunities for using other AI techniques in pipe jacking works for future applications are discussed.

Review of AI Applications: Prediction of Frictional Jacking Forces

The prediction of jacking forces is an important component during the design of pipe jacking works. Sheil (2021) presented a probabilistic observational approach to predict the jacking forces during microtunneling using Gaussian Process (GP) regression based on the monitoring data collected. The proposed method was applied to two recent microtunneling projects in the UK, and the results were compared with current design methods used in the industry. The study showed that GP regression could predict the jacking forces accurately and thus highlighted the opportunity of complementing existing pipe jacking design methods using data-driven probabilistic technique.

Sheil (2021) stated that GP regression is a stochastic method that performs Bayesian inference about functions in a nonparametric way. The use of covariance functions in GP regression allows prior assumptions about the parameters to be incorporated into the model, thus enabling the parameters of interest to be learned during the process, instead of generating only a mapping between data input and output. The proposed GP regression method was developed using Python. Field observations comprising of tunneling and lubrication data from two pipe jacking drives (termed as "drive A" and "drive B") in Blackpool, UK were considered in the application of the proposed probabilistic approach. Data from drive A was used for model training, whereas drive B was used for model validation.

Feature selection process was first carried out to identify the most discriminant parameters in the microtunnel boring machine (mTBM) data for the estimation of total jacking force. As a result, mTBM records of time and jacked distance were selected as the input parameters, whereas total jacking force was considered as the model output. The GP model was trained to develop a relationship between jacked distance, time, and total jacking force autonomously to provide the output probabilistically. The root mean squared error (RMSE) was then calculated to measure the error between the mean GP prediction and monitored data. Finally, the prediction accuracy was assessed by averaging over the test sets. After training the model using data from drive A, the model was validated using drive B. The model predictions obtained from drive B were then compared with several design guidelines commonly used in the industry, including the design guidance provided by Pipe Jacking Association (PJA 1995), Staheli (2006), and Concrete Pipe Association of Australasia (CPAA 2013).

Sheil (2021) noted that the proposed probabilistic approach was able to provide improved predictions of jacking forces as compared to existing design methods used in the industry, which are unable to capture site-specific changes during construction and over-predict the jacking forces significantly. By integrating observational approach within a Bayesian framework in the prediction of jacking forces, the uncertainties associated with the design parameters can be quantified and the model is flexible in defining a probabilistic design profile based on the desired risk level. Therefore, this study presents the potential of applying data-driven techniques to supplement existing pipe jacking design methods in the industry.

Challenges and Recent Development of AI

Although AI techniques have been shown to be powerful in processing complex datasets, there are also challenges that may be faced when adopting AI techniques. Some common challenges one may encounter when applying AI methods to tackle these challenges/limitations are briefly discussed in the section hereinafter. Further

8.2 Decarbonizing Tunneling Processes

Fig. 8.16 Some challenges that may be faced when applying AI techniques

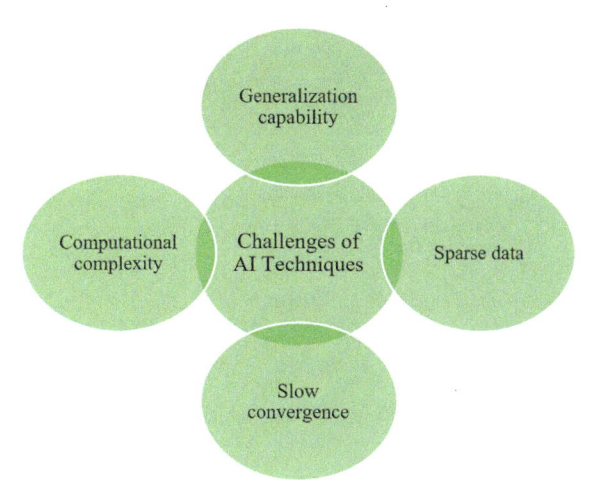

details of the applications of various AI techniques in underground geotechnical problems can be found in Jong et al. (2021).

Challenges

The main challenge when applying AI techniques is the generalization capability of the techniques. Since these techniques are data-driven, the performance of the techniques significantly depends on the data provided during the training process. The predictive models developed using data-driven approaches are generally applicable to problems with similar geological conditions; however, it may be challenging to apply them to different project conditions if the new datasets are not within the range of datasets used for model training. This challenge has also been mentioned by Sheil et al. (2020) and Sheil (2021) in their studies in which the proposed approaches using different machine learning techniques are limited to specific tunnel geometries, ground conditions, and construction techniques. In this case, data that includes more complex project conditions are required to be trained so that the models can adapt to the new data and better generalize the conditions (Goh et al. 2018; Zhang et al. 2018, 2019). When more data can be collected from the field, the 'noise' in the data may increase due to the increasing number of parameters in the observations. Sparse data may also be collected from project sites with varying geological conditions. Slow convergence is another challenge when employing AI techniques due to the tendency of the developed model being trapped in local minimum while searching for global solutions (Mottahedi et al. 2018; Armaghani et al. 2020). Computational complexity may also present a challenge during model training for some techniques. It is found that higher computational capability may be required when training some AI models to process more complex datasets (Kohestani et al. 2015). Figure 8.16 shows the summary of some challenges that may be experienced when applying different AI techniques in the study of geotechnical problems.

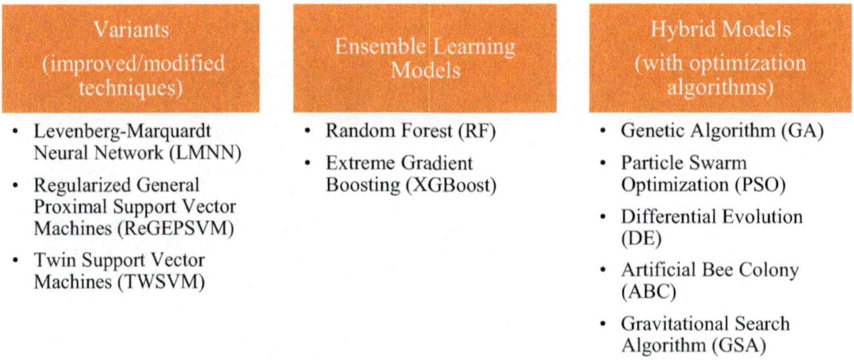

Variants (improved/modified techniques)	Ensemble Learning Models	Hybrid Models (with optimization algorithms)
• Levenberg-Marquardt Neural Network (LMNN) • Regularized General Proximal Support Vector Machines (ReGEPSVM) • Twin Support Vector Machines (TWSVM)	• Random Forest (RF) • Extreme Gradient Boosting (XGBoost)	• Genetic Algorithm (GA) • Particle Swarm Optimization (PSO) • Differential Evolution (DE) • Artificial Bee Colony (ABC) • Gravitational Search Algorithm (GSA)

Fig. 8.17 Development of existing AI techniques in the recent years

Recent Development

To overcome the challenges faced in the AI applications, existing techniques have been improved or modified to boost their performances. The improved techniques are also termed as "variants" of the techniques. Sometimes, two or more AI techniques can be combined to form hybrid models that can tackle limitations of existing techniques with better model performance. Figure 8.17 shows some of the development that have been made to existing techniques in the recent years.

Many improvements or modifications have been made to existing AI techniques to enhance their efficiencies. For example, Shahri (2016) integrated Levenberg–Marquardt algorithm into the artificial neural network (ANN) model to speed up the training process. Two variants of support vector machines (SVM) that are developed, known as Regularized Generalized Proximal SVM (ReGEPSVM) and Twin SVM (TWSVM), are found to be able to reduce the model runtime by half and improve the capability of the model to generalize to other datasets when compared with other regression methods (Moayedi and Hayati 2019).

Different techniques can also be combined to form a single model by using ensemble learning methods that integrate a few algorithms in one model to improve model performance. Ensemble models can effectively overcome the limitations of existing AI techniques while providing excellent performance. Random forest (RF) is one of the ensemble models that provides strong data mining capability and high prediction accuracy by combining a large set of randomly trained decision trees (Zhou et al. 2019). Extreme Gradient Boosting (XGBoost) is another ensemble learning method that has been successfully developed by Chen and Guestrin (2016). This method can improve model accuracy by integrating decision trees and capable of handling sparse data.

Apart from ensemble models, hybrid models can be formed by combining AI techniques with optimization algorithms such as genetic algorithm (GA) and particle swarm optimization (PSO) to optimize the performance of certain techniques. PSO can achieve an efficient global search process and prevent the developed model from being trapped at local minima (Ismail et al. 2013) while GA can reduce the

8.2 Decarbonizing Tunneling Processes

runtime for parameter tuning (Elbaz et al. 2019). Other evolutionary algorithms such as differential evolution (DE), artificial bee colony (ABC) and gravitational search algorithm (GSA) can improve the generalization capability of developed models and increase the processing speed and model accuracy (Fattahi and Babanouri 2017).

Recommendations for Future Research

Based on the recent trend in applying machine learning techniques to study different problems in pipe jacking, more opportunities for employing AI have been identified and proposed for future applications. Sterling (2020) stated that the assessment of frictional jacking forces remains an area of strong interest in research as it is one of the most crucial aspects in pipe jacking design that has overlapping concerns from designers, contractors, and equipment manufacturers. There are also limited considerations given for pipe jacking drives traversing rocks Choo and Ong (2017). Therefore, it is believed that there is an opportunity to use AI techniques to assess the jacking forces in such geological conditions. Particularly, it is envisaged that a Bayesian probabilistic framework can be potentially used to predict the jacking forces for pipe jacking in highly weathered and highly fractured geology.

Bayesian method is a data-driven technique that is developed based on Bayes' theorem (Bayes and Price 1763). Bayesian inference enables uncertainties of parameters to be quantified using probability (Murphy 2012). Bayesian method considers probability as the evolution of the degree of belief in parameters of interest, before observing the data, and after studying the data in the analysis (Jin et al. 2018). The probability distribution of parameters of interest, θ given the observed data, D can be described as $p(\theta|D)$, which is known as posterior distribution of the parameters. It can be calculated using Eq. 8.1. The prior distribution represents the knowledge on the parameters prior to the observation of data. The likelihood function represents how likely the observed data are given the parameters. The model evidence is the normalizing constant for computing the posterior distribution of parameters. The details of a Bayesian model can be found in Houlsby and Houlsby (2013).

where

$$p(\theta|D) = \frac{p(\theta)p(D|\theta)}{p(D)} \qquad (8.1)$$

$p(\theta|D)$ = the posterior distribution of parameters.
$p(\theta)$ = the prior distribution of parameters.
$p(D|\theta)$ = the likelihood function.
$p(D)$ = the model evidence.

Bayesian method has also been widely used to update soil parameters based on instrumentation monitoring data. Miro et al. (2015) proposed a probabilistic back-analysis to evaluate the effects of uncertainties of subsoil parameters on tunneling-induced ground movements. The uncertainties of soil parameters were quantified and updated using Bayesian method to increase the reliability of model output. Jin et al. (2018) applied sequential Bayesian inference to estimate the most probable soil parameters and predict the wall deformations for subsequent excavation stages accurately. Bayesian method is commonly used to back-analyze soil parameters because it is a probabilistic method that considers soil parameters as random variables, as opposed to other AI techniques that produce deterministic models (Ching et al. 2010; Juang et al. 2013). Besides, Bayesian method can incorporate prior information such as case histories and field observations into the model (Wang and Cao 2013). This allows the model parameters to be updated as the project progresses by integrating previous output as prior into the model because monitoring data are usually collected at different construction stages. The parameter uncertainty can also be reduced through Bayesian updating process, increasing the model accuracy in making predictions. Hence, Bayesian method is found to be superior in back-analysis and updating of soil parameters.

The application of Bayesian method to predict jacking forces can be achieved by integrating existing knowledge such as published reports, case histories, and engineering judgment with field observations, to estimate the parameters of interest probabilistically. Therefore, it is in the authors' opinion that Bayesian method can be potentially applied to microtunneling in challenging geological formations, to estimate the jacking forces for pipe jacking traversing highly weathered and highly fractured rocks.

Apart from predicting jacking forces during pipe jacking, it is believed that Bayesian approach can be further applied to study the ground conditions during pipe jacking. A variant of Bayesian method, known as Bayesian network, may be able to be used for classifying the soil types of surrounding ground during pipe jacking. Bayesian network is a probabilistic graphical model that illustrates a set of uncertain variables and expresses their dependency by using a directed acyclic graph (DAG) (Khalaj et al. 2020). The DAG consists of: (i) nodes, which represent a set of variables, and (ii) directed edges, which link two variables with conditional dependencies. The joint probability distribution can then be calculated based on the conditional dependencies as shown in Eq. 8.2.

$$P(a_1, a_2, \ldots, a_n) = \prod_{i=1}^{n} P(a_i | Pa(a_i)) \qquad (8.2)$$

where $Pa(a_i)$ is the parent set of variable a_i. For example, Fig. 8.18 illustrates the schematic of a Bayesian network model that contains five variables (or nodes), $a_1, a_2, \ldots, a_5$. In this instance, node a_4 is considered as a child of nodes a_2 and a_3; whereas nodes a_2 and a_3 are considered as the parents of node a_4. Each variable is associated with a conditional probability distribution given its parents in the model. The probability distributions for all the variables can be written as: $P(a_1)$, $P(a_2|a_1)$, $P(a_3|a_1, a_2)$, $P(a_4|a_2, a_3)$, and $P(a_5|a_4)$, respectively. The model then computes the joint probability distribution for all the variables based on the conditional dependencies and chain rule (Khalaj et al. 2020), as shown in Eq. 8.3:

Fig. 8.18 Schematic of a Bayesian network model

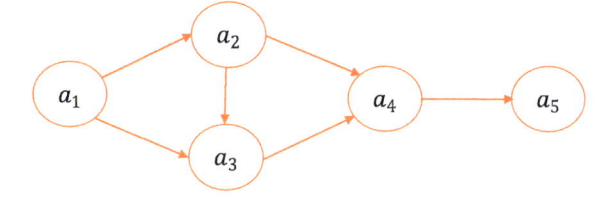

$$P(a_1, a_2, a_3, a_4, a_5) = P(a_1)P(a_2|a_1)P(a_3|a_1, a_2)P(a_4|a_2, a_3)P(a_5|a_4) \quad (8.3)$$

Based on the monitoring data collected, it is postulated that a Bayesian network model can be used to identify the relationships between parameters that may affect the ground conditions during pipe jacking (i.e., to establish conditional dependencies between the variables), and subsequently classify the soil types of surrounding ground by calculating the probability of each soil type based on the given data.

8.3 Decarbonizing Through Spoil Reuse

Tunneling spoil provides an opportunity for construction projects to reduce their overall carbon footprint. For example, tunneling spoil can be potentially used as pipe beddings, backfill materials, and recycled aggregates in light-traffic and unbounded road pavements. These examples are focused on the local council requirements in Brisbane, Australia. For other parts of the world, it is recommended to consult the relevant council guidelines and standards.

8.3.1 Bedding Material for Pipe Installation

Procedures and standard requirements for installing and maintaining utility services have been documented by Brisbane City Council (BCC) in the report specification S145. For the installation of UPVC and polyethylene pipe systems, the bedding material must consist of clean river sand. Recycled concrete or clean crushed rock of nominal size ranging from 5 to 7 mm can also be used but should be free of clay, salt, or organic matter. A sieving methodology can be adopted to assess the suitability of the tunneling rock spoil (sieved to 5–7 mm) to conform to the specification of the bedding materials. Table 8.2 shows the grading requirements for the bedding material. The bedding material can be used for pipe underlay, side support, and overlay with a minimum of 75 mm thick compacted underlay and a minimum cover of 150 mm over the pipe (Brisbane Infrastructure Division 2016).

Table 8.2 Grading requirement for bedding material (Brisbane Infrastructure Division 2016)

Australian standard sieve (mm)	Percentage passing by weight	
	Bedding sand	5–7 mm screenings
9.50	–	100
6.70	–	85–100
4.75	100	30–85
2.36	80–100	0–30
1.18	–	0–5
0.075	0–15	–

8.3.2 Backfill Material for Trenches

The backfill material used in trenches must not hold water nor obstruct existing drainage paths through the soil. A simple field test can be conducted to determine the permeability characteristics of the material as described in Sect. 7.3 of the Specification S145 for the installation and maintenance of utility services (Brisbane Infrastructure Division 2016). The acceptable backfill materials are 'Class 3 material', granular fill, sand, and stabilized sand.

The Class 3 material should fulfill the requirements from Specification S300 Quarry Products except that the plasticity index must not be less than 4%. For granular fill, a 75-mm maximum-size crushed rock, non-plastic open-graded material or crusher run recycled concrete can be used (Brisbane Infrastructure Division 2016). The tunneling rock spoil would stand a very good chance to satisfy the granular fill category by incorporating a sieving process to segregate particles larger than 75 mm. The granular fill material should be compacted in uniform layers. Each layer is to be rolled until no permanent visible lowering of the surface occurs. The minimum thickness of uncompacted layers is 150 mm. The maximum thickness of uncompacted layers is determined from Table 8.3.

Table 8.3 Maximum thickness of uncompacted layers (Brisbane Infrastructure Division 2016)

Static module weight or vibrating drum equivalent (tons)	Uncompacted layer maximum thickness (mm)	
	Voids not filled	Voids filled by adding finer materials
5	400	300
10	600	400
15	900	600
20	1200	800

8.3.3 Road Pavement

Construction and demolition wastes such as concrete, brick, and glass can be used as alternatives to natural aggregates and sand in road bases. Up to 8000 tons of construction and demolition waste has been shifted from landfill per kilometer of road in Queensland, Australia (Queensland Government 2021). The Department of Transport and Main Roads (TMR) has provided a summary of recycled materials, which are currently permitted for various applications according to TMR Specifications (see Table 8.4). Materials from natural, quarried, and recycled sources, or a combination of these can be used for the construction of unbound granular pavements (Department of Transport and Main Roads 2020a). This statement shows that the tunneling spoil is qualified to be used as recycled materials in unbound pavements. However, the spoil will have to meet detail specifications and requirements from TMR prior for use. Additionally, the tunneling spoil can also be used as base and subbase course materials for roads subject to light traffic. The specification S150 by BCC describes the requirements for the qualified materials to be used for light-traffic road construction.

Table 8.4 Overview of recycled material uses and relevant specifications (Department of Transport and Main Roads 2020a)

Application	Recycled material							
	Crushed concrete	Crushed brick	Crushed glass	RAP	Crumb rubber	Fly ash and slag	In situ material	Recycled plastic
Unbound pavement	P	P	P	P	–	–	–	–
Stabilization	P	P	P	P	–	P	P	–
Sprayed sealing	–	–	–	–	P	–	–	R
Asphalt	–	–	P	P	R/D	P	P	R
Concrete	R	–	R	–	–	P	–	–
Concrete pavements	–	–	–	–	–	P	–	–
Earthworks, drainage and backfill	P, R	P, R	P, R	P, R	–	–	P	–

P = Permitted within limits/uses, R = Research underway, D = Demonstration projects underway

8.4 Decarbonizing in Practice: From Compliance to Best Practice

We can see from the discussion of advances in tunnel design and construction, that tunneling processes are rapidly innovating to provide more energy efficient and cost-effective methods for extracting and transporting spoil. As tendering metrics shift from costing the number of tons transported to valuing the efficiencies achieved, the market-benefit of innovations such as the VAC Group's machinery will become clearer. While there are no regulatory requirements yet for complying with carbon (emission) metrics during the construction process, the international pressures discussed in the introduction signal this as a potential reality in future.

The regulatory mechanisms have progressed further when considering the use of tunneling spoil within the project and/or elsewhere in other projects, with specific requirements and standards from national, state/regional and local authorities. In addition, industry practice may also be influenced by best practice requirements through rating schemes and performance standards. These could relate to considerations covered in this chapter, including the type of tunneling equipment used, the type of processes used to tunnel (e.g., consumed carbon in operating the equipment and transporting spoil, the energy source—solar-powered electric or other), and the type of management to be used for extracted spoil.

In the following sections, examples are provided in terms of legal requirements, standards, and best practice, including some worked examples of considerations when 'decarbonizing tunneling activities' in projects, focusing on spoil use in Sects. 8.4.1 and 8.4.2, then both tunneling operations and spoil reuse in Sect. 8.4.3.

8.4.1 State Government Requirements

Each state government in Australia manages its own standards. For example, in Queensland, the Department of Transport and Main Roads uses the Technical Specification MRTS05 to set regulatory requirements for the supply and construction of unbound granular pavements. Material from natural, quarried, and recycled sources, or a combination of these, can all be used. Various specifications set by TMR for the use of recycled aggregates are listed in Table 8.5. Tunneling spoil should be assessed and classified accordingly.

As a worked example, consider the process for evaluating 'Brisbane Tuff' as spoil (from the ongoing Cross River Rail project at time of writing) for use in road construction. This would definitely reduce the amount of virgin materials needed for the road project and thus provides a useful end-point for the excavated tunneling spoil.

Using Table 8.5, we can see that MRTS05 governs the specifications for 'unbound pavements'. Brisbane Tuff can be classified as acid igneous rock or intermediate igneous rock according to clause 2.0 of MRTS05 (Department of Transport and

8.4 Decarbonizing in Practice: From Compliance to Best Practice

Table 8.5 TMR specifications for use of recycled aggregates (Department of Transport and Main Roads 2020a)

Application	Specification	
Unbound pavement materials	Unbound pavements	MRTS05
Stabilized pavements	In situ stabilized pavements using cement or cementitious blends	MRTS07B
	In situ stabilized pavements using foamed bitumen	MRTS07C
	Plant-mixed heavily bound (cemented) pavements	MRTS08
	Plant-mixed pavement layers stabilized using foamed bitumen	MRTS09
	Plant-mixed lightly bound pavements	MRTS10
Unbound pavements and asphalt	Recycled glass aggregate	MRTS36
Asphalt	Asphalt pavements	MRTS30
	Aggregates for asphalt	MRTS101

Main Roads 2020b). Material used for unbound pavement can be classified as High Standard Granular (HSG) (Type 1), Standard Material (Type 2 and 3), and Non-standard Material (Type4). All pavement materials incorporated into the finished pavement shall be free from sticks, organic matter, clay lumps, and other deleterious material or contamination.

In managing the use of the spoil, it would be normal to visually inspect the spoil material during and after placement to ensure compliance with clause 7.5 of MRTS05. In this example, from studying the soil and rock profiles suitable for testing, we will assume that the tunneling rock spoil is to be classified as HSG. The requirements for HSG are that it must be assessed for its particle size distribution.

Table 8.6 shows the property requirements for coarse component when using Type 1 (HSG) materials, and Table 8.7 describes the properties required for the fines component of Type 1 (HSG) material.

Table 8.8 shows the particle size distribution requirement for Type 1 (HSG). The minimum and maximum percentage passing limits designate the allowable zone of target grading required from any pavement layer sampling during construction. Additionally, for Type 1 (HSG) material, a ratio of 0.30–0.55 should be adopted for fines. Prior to pavement construction using Type 1 (HSG) material, a trial pavement of more than 1000 m^2 is required. Once approved by the local authority, the trial pavement using the Type 1 (HSG) material can then be integrated into the permanent works (Department of Transport and Main Roads 2020b).

Table 8.6 Type 1 (HSG) material properties for coarse component (Department of Transport and Main Roads 2020b)

Properties	Material group*			
	Acid igneous	Basic igneous	Intermediate igneous and metamorphic	Sedimentary duricrust
Wet strength* (kN)	≥130	≥150	≥140	≥130
Wet/Dry strength variation* (%)	≤40	≤30**	≤35	≤40
Degradation factor	≥40	≥50	≥45	-
Flakiness index (%)	≥35	≤35	≤35	≤35
Water absorption (%)	≤2.0	≤2.0	≤2.0	≤2.0

*To group materials based on major granular component except where two or more components consist of 30% each of the total material
** If wet strength is ≥60 kN, wet/dry strength variation requirement can be omitted

Table 8.7 Fines component properties—Type 1 (HSG) (Department of Transport and Main Roads 2020b)

Properties	Values
Liquid limit	≤25
Linear shrinkage (%)	1.5–3.5

Table 8.8 Grading—Type 1 (HSG) (Department of Transport and Main Roads 2020b)

Sieve size (mm)	Percentage passing by weight		
	Target	Minimum	Maximum
37.5	100	100	100
26.5	100	100	100
19.0	100	95	100
13.2	85	78	92
9.5	73	63	83
4.75	54	44	64
2.36	39	30	48
0.425	18	14	22
0.075	9	7	11

8.4.2 Case Study: The Cross River Rail Project

The Queensland government has committed $5.4 billion to improve the transport network in Brisbane, Australia. The Cross River Rail (CRR) project consists of a 10.2 km rail line and 5.9 km twin tunnels which will traverse under the Brisbane River and the CBD. The new infrastructure of CRR will consist of four underground stations

8.4 Decarbonizing in Practice: From Compliance to Best Practice

and upgrades to eight above-ground stations (Cross River Rail Delivery Authority 2020). The initial phase of construction which includes demolition and removal of existing infrastructure within the project zone has been completed at the time of writing. From the demolition sites, 20,500 tons of concrete, 1094 tons of ferrous metals, and 275 tons of aluminum and copper were recycled from the Woolloongabba station, a nine-story building (Cross River Rail Delivery Authority 2020). Along with these demolition wastes, 2.1 million m^3 of tunneling spoil (National Building Program 2010) will need to be moved from the tunneling sites, and this will create challenges in terms of logistics and landfill capacity.

Crushed concrete, bricks, and glass can be used as recycled aggregates replacing sand or natural/quarried aggregates. The local council has shown commitment for the use of these recycled materials by incorporating their relevant classification systems in the design specifications. Similarly, the large amount of tunneling spoil should be considered and assessed as another viable form of recycled materials to be used in engineering works such as in pipe beddings or road pavements. Some of the benefits of possibly using the tunneling spoil are:

- Potential cost-effectiveness in using approved recycled materials / tunneling spoil
- Reduction in the use of non-renewable/virgin mined resources
- Reduction in the amount of wastes sent to landfill
- Reduction in greenhouse gas emissions
- Equivalent or improved performance when compared to traditional materials when used for adequate application

The CRR alignment was designed to traverse mainly in rocks. Geotechnical investigations have been conducted at different locations such as near to the stations and the proposed river crossing. Along with existing geotechnical records and with the new data acquired, a geotechnical model was made for the CRR corridor (National Building Program 2010). From the geotechnical model, it is possible that majority of the 2.1 million m^3 of tunneling spoil from the CRR tunnels would be rocks originated from Brisbane Tuff, Aspley Formation, and Neranleigh-Fernvale Beds.

Brisbane Tuff is a volcanic sediment which consists of rock fragments, pumice, and lava that were welded together during volcanic explosion to form layers of tuff. Brisbane Tuff is used as building stone, kerbing, and road-making materials (Windsor and Districts Historical Society Inc 2020). Schuh (2007) categorized tuff into four structural types based on the rock mass classification. The four types of tuff with its definition and estimated unconfined compressive strength (UCS) have been shown in Table 8.9. Aspley formation consists of alluvial sediments which are formed on irregular surface erosion of the tuff. The Aspley formation is a conglomerate of coarse sandstone and siltstone interbedded together (McQueen et al. 2019). The Neranleigh-Fernvale bed consists of primarily fine to medium grained greywacke and very fine-grained argillite (Tugun Bypass Alliance 2004). The carboniferous-age rocks are mainly found under the northern part of the site.

Based on the mineralogical content and estimated strength of the Brisbane Tuff, there is great potential in exploring the feasibility of reusing the tunneling spoil as recycled aggregates as bedding material for pipe laying and as sub base material for

Table 8.9 Brisbane Tuff classification with average unconfined compressive strength (Schuh 2007)

Rock mass unit	Definition	UCS (MPa)
Tuff 1	Fresh to slightly weather, rough joints, narrow sheared zones, high to very high strength	~100
Tuff 2	Narrow shear zones, rough opened stained joints, slightly weathered to fresh, high to very high strength	~80
Tuff 3	Narrow shear zones, rough open joints, distinctly to slightly weathered, medium to high strength	~30
Tuff 4	Narrow shear zones, rough open stained joints, extremely to distinct weathered, very low to medium strength	~3

light traffic roads. The potential to reduce 2.1 million m^3 of spoil will significantly decarbonize the Cross River Rail project.

8.4.3 Local Government Requirements

For works within local government or city council, there is often a local specification that also refers to extracts or whole parts of the relevant state government specifications. Another avenue of possibly using the 'Brisbane Tuff' tunneling spoil could be as base and sub base course for roads subject to light traffic and managed by a local government authority.

Considering this potential use of the spoil within the locality of Brisbane City Council (BCC, Southeast Queensland), the reference specifications for civil engineering work are S150 Roadworks. The specification by BCC refers to numerous Australian Standards when defining the requirements for roadworks. Specification S150 states that in order to use Class 1 material base course for roads subject to light traffic, the sub-base course should be Class 2 material (or Class 1 alternatively) for a minimum 100-mm-thick top layer. A Class 3 material can be used for the subsequent sub-base courses to achieve the design pavement thickness (Brisbane City Council 2018). The material properties and classifications are defined in Specification S300—Quarry Products and are an extract from TMR Specification MRTS 05 for unbound pavements as described in the previous section.

8.4.4 Rating Scheme Incentives

Within the design and construction industry, there are rapidly emerging best practice schemes for decarbonizing the industry, in response to the international and national drivers discussed earlier. Depending on the preference of the Client—which could be from the public or private sector—the call for tender and subsequent contract documentation may be explicit in requiring companies that tender for the works to be

8.4 Decarbonizing in Practice: From Compliance to Best Practice

certified in particular rating schemes. The call for tenders may also require that certain components of design and construction comply with rating scheme requirements. The Infrastructure Sustainability Council of Australia (ISCA) is an example of an organization driving the decarbonizing agenda (ISCA n.d.).

8.5 Conclusions

In this chapter, we have presented precedents and opportunities for decarbonizing infrastructure as an essential step forward in the global pursuit of sustainable development that meets the needs of the present while not compromising the ability of future generations to meet their own needs.

Within the infrastructure sector, there are significant opportunities to decarbonize tunneling practices in excavation activities and spoil reuse. Through examples and case study, we have discussed the regulatory requirements and incentives being offered to overcome challenges in mainstreaming improved practices.

These topics covered in this chapter raise significant questions for decision-makers and practitioners, regarding innovation opportunities not being taken up more readily. We intend that the discussion points, examples, and case studies presented hereinbefore will provide a robust foundation from which we learn to embrace the decarbonizing agenda in tunneling design and construction.

References

Adrian RJ (1991) Particle imaging techniques for experimental fluid mechanics. J Fluid Mech 23:261–304

Armaghani DJ, Asteris PG, Fatemi SA et al (2020) On the use of neuro-swarm system to forecast the pile settlement. Appl Sci 10:1904. https://doi.org/10.3390/app10061904

Bayes T, Price R (1763) An essay towards solving a problem in the doctrine of chances by the Late Rev. Mr. Bayes. Philos Trans 53:370–418

Bennett RD (1998) Jacking forces and ground deformations associated with microtunneling. PhD thesis, University of Illinois

Berghe TV (2012) Image processing for a LSPIV application on a river. MSc. thesis, Universiteit Gent

Bobylev N (2009) Mainstreaming sustainable development into a city's Master plan: a case of urban under-ground space use. Land Use Policy 26(4):1128–1137

Brisbane Infrastructure Division (2016) Reference specifications for civil engineering work S145 installation and maintenance of utility services. Revision 2.0, May 2016

Brisbane City Council (2018) References specifications for civil engineering work, S300 Quarry Products. Revision 4.0 – November 2018.

Brundtland GH (1987) Our common future. Oxford University Press, Oxford

Chapman DN, Ichioka Y (1999) Prediction of jacking forces for microtunnelling operations. Tunn Undergr Space Technol 14:31–41. https://doi.org/10.1016/S0886-7798(99)00019-X

Chen T, Guestrin C (2016) XGBoost: a scalable tree boosting system. In: Proceedings of the ACM SIGKDD international conference on knowledge discovery and data mining, pp 785–794

Ching J, Phoon KK, Chen YC (2010) Reducing shear strength uncertainties in clays by multivariate correlations. Can Geotech J 47:16–33. https://doi.org/10.1139/T09-074

Choo CS, Ong DEL (2015) Evaluation of pipe-jacking forces based on direct shear testing of reconstituted tunneling rock spoils. J Geotech Geoenviron Eng 141(10). https://doi.org/10.1061/(ASCE)GT.1943-5606.0001348

Choo CS, Ong DEL (2017) Impact of highly weathered geology on pipe-jacking forces. Geotech Res 4:94–106. https://doi.org/10.1680/jgere.16.00022

Costello SB, Chapman DN, Rogers CDF (2007) Underground asset location and condition assessment technologies. Tunn Undergr Space Technol 22(6):524–542https://doi.org/10.1016/j.tust.2007.06.001

CPAA (2013) Jacking design guidelines. Concrete Pipe Association of Australasia

Cross River Rail Delivery Authority (2020) Rail route, cross river rail route and stations. https://crossriverrail.qld.gov.au/about/rail-route/

Department of Transport and Main Roads (2020a) Technical Note TN193—use of recycled materials in road construction, The State of Queensland

Department of Transport and Main Roads (2020b) Technical specification—MRTS05 unbound pavements, The State of Queensland

Desha C, Birkeland J, Pears A (2005) Greening the built environment. In: Smith M, Hargroves K (eds) The natural advantage of nations. Earthscan Publications, United Kingdom, pp 346–370

Elbaz K, Shen SL, Zhou A et al (2019) Optimization of EPB shield performance with adaptive neuro-fuzzy inference system and genetic algorithm. Appl Sci 9:780. https://doi.org/10.3390/app9040780

Evans D, Stephenson M, Shawn R (2009) The present and future use of 'land' below ground. Land Use Policy 26:302–316

Fattahi H, Babanouri N (2017) Applying optimized support vector regression models for prediction of tunnel boring machine performance. Geotech Geol Eng 35:2205–2217. https://doi.org/10.1007/s10706-017-0238-4

Fukuoka H, Sassa K, Wang G et al (2006) Observation of shear zone development in ring-shear apparatus with a transparent shear box. Landslides 3(3):239–251. https://doi.org/10.1007/s10346-006-0043-2

Gijzel D, Bosch-Rekveldt M, Schraven D, Hertogh M (2019) Integrating Sustainability into Major Infrastructure Projects: Four Perspectives on Sustainable Tunnel Development. Sustainability 12(1) 6-10.3390/su12010006

Goh ATC, Zhang W, Zhang Y et al (2018) Determination of earth pressure balance tunnel-related maximum surface settlement: a multivariate adaptive regression splines approach. Bull Eng Geol Environ 77:489–500. https://doi.org/10.1007/s10064-016-0937-8

Grognet M (2011) The boundary conditions in direct simple shear tests, development for peat testing at low vertical stress. MSc. thesis, Delft University of Technology

Hammervold J (2014) Towards greener infrastructure. Philosophiae doctor. Norwegian University of Science and Technology

Houlsby NMT, Houlsby GT (2013) Statistical fitting of undrained strength data. Geotechnique 63:1253–1263. https://doi.org/10.1680/geot.13.P.007

Huang L, Bohne RA, Bruland A et al. (2013) Life cycle assessment of norwegian standard road tunnel. In: The 6th international conference on life cycle management, Gothenburg

Hunt DVL, Jefferson I, Rogers CDF (2011) Assessing the sustainability of underground space usage—a toolkit for testing possible urban futures. J Mt Sci 8(2):212–222

Hunt DVL, Nash D, Rogers CDF (2014) Sustainable utility placement via Multi-Utility Tunnels. Tunn Undergr Space Technol 39:15–26. https://doi.org/10.1016/j.tust.2012.02.001

ISC (2020) IS v2.1 Technical manual. Infrastructure Sustainability Council, Melbourne. Available at: https://www.iscouncil.org/is-v2-1/

ISC (2021) Impact report 2021. Infrastructure Sustainability Council, Melbourne. Retrieved from: https://www.iscouncil.org/wp-content/uploads/2021/10/IMPACTS-REPORT-October-21-2021-Web.pdf

Ismail A, Jeng DS, Zhang LL (2013) An optimised product-unit neural network with a novel PSO-BP hybrid training algorithm: applications to load-deformation analysis of axially loaded piles. Eng Appl Artif Intell 26:2305–2314. https://doi.org/10.1016/j.engappai.2013.04.007

Jefferson I, Rogers CDF, Hunt DVL (2006) Achieving Sustainable Underground Construction in Birmingham Eastside? The 10th International Congress of the IAEG, Nottingham, UK, Sept 2006

Jin Y, Biscontin G, Gardoni P (2018) A Bayesian definition of 'most probable' parameters. Geotech Res 5:130–142. https://doi.org/10.1680/jgere.18.00027

Jong SC, Ong DEL, Oh E (2021) State-of-the-art review of geotechnical-driven artificial intelligence techniques in underground soil-structure interaction. Tunn Undergr Space Technol 113. https://doi.org/10.1016/j.tust.2021.103946

Juang CH, Luo Z, Atamturktur S, Huang H (2013) Bayesian updating of soil parameters for braced excavations using field observations. J Geotech Geoenviron Eng 139:395–406. https://doi.org/10.1061/(ASCE)GT.1943-5606.0000782

Kelly P (2014) Soil structure interaction and group mechanics of vibrated stone column foundations. Ph.D. thesis, University of Sheffield

Khalaj S, BahooToroody F, Mahdi Abaei M, et al (2020) A methodology for uncertainty analysis of landslides triggered by an earthquake. Comput Geotech 117:103262. https://doi.org/10.1016/j.compgeo.2019.103262

Kohestani VR, Hassanlourad M, Ardakani A (2015) Evaluation of liquefaction potential based on CPT data using random forest. Nat Hazards 79:1079–1089. https://doi.org/10.1007/s11069-015-1893-5

Li YR, Aydin A (2010) Behavior of rounded granular materials in direct shear: mechanisms and quantification of fluctuations. Eng Geol 115:96–104. https://doi.org/10.1016/j.enggeo.2010.06.008

Lu H, Matthews J, Iseley T (2020) How does trenchless technology make pipeline construction greener? A comprehensive carbon footprint and energy consumption analysis. J Clean Prod 261. https://doi.org/10.1016/j.jclepro.2020.121215

McQueen LB, Purwodihardjo A, Barrett SV (2019) Rock mechanics for design of Brisbane tunnels and implications of recent thinking in relation to rock mass strength. J Rock Mech Geotech Eng 11:676–683. https://doi.org/10.1016/j.jrmge.2019.02.001

Miro S, König M, Hartmann D, Schanz T (2015) A probabilistic analysis of subsoil parameters uncertainty impacts on tunnel-induced ground movements with a back-analysis study. Comput Geotech 68:38–53. https://doi.org/10.1016/j.compgeo.2015.03.012

Moayedi H, Hayati S (2019) Artificial intelligence design charts for predicting friction capacity of driven pile in clay. Neural Comput Appl 31:7429–7445. https://doi.org/10.1007/s00521-018-3555-5

Mottahedi A, Sereshki F, Ataei M (2018) Overbreak prediction in underground excavations using hybrid ANFIS-PSO model. Tunn Undergr Space Technol 80:1–9. https://doi.org/10.1016/j.tust.2018.05.023

Murphy KP (2012) Machine learning: a probabilistic perspective. MIT Press

Najafi M, Kim OK (2004) Life-cycle-cost comparison of trenchless and conventional open-cut pipeline construction projects. In: Proceedings of ASCE international pipelines conference, San Diego, CA, July 2004

Najafi M, Gokhale S (2005) Trenchless technology, pipeline and utility design, construction, and renewal. McGraw-Hill, New York.

National Building Program (2010) Reference design overview. CrossRiverRail. https://www.cabinet.qld.gov.au/documents/2010/oct/cross%20river%20rail/Attachments/pdf_cross_river_rail_reference_design_overview[1].pdf

Oda M, Kunishi J (1974) Microscopic deformation mechanism of granular material in simple shear. Jpn Soc Soil Mech Found 14(4):25–38

Ong DEL, Choo CS (2018) Assessment of non-linear rock strength parameters for the estimation of pipe-jacking forces. Part 1. Direct shear testing and backanalysis. Eng Geol 244:159–172. https://doi.org/10.1016/j.enggeo.2018.07.013

Osumi T (2000) Calculating jacking forces for pipe jacking methods. No-Dig Int Res 40–42

Parker HW (2004) Underground space: good for sustainable development, & vice versa. In: Proceedings, WTC, ITA, Singapore, May 2004

Peerun MI (2016) Behaviour of reconstituted sand-sized tunnelling rock spoils during shearing using GeoPIV technology for the assessment of soil arching effect during pipe-jacking works. MEng thesis, Swinburne University of Technology

Peerun MI, Ong DEL, Desha C, Oh E, Choo CS (2021) Advances in the study of micromechanical behaviour for granular materials using Micro-CT scanner and 3D printing. In: 16th international conference of the international association for computer methods and advances in geomechanics (16th IACMAG), 5–8 May 2021, Torino, Italy, Politecnico Di Torino

Peerun MI, Ong DEL, Choo CS et al (2020) Effect of interparticle behavior on the development of soil arching in soil-structure interaction. Tunn Undergr Space Technol 106:103610. https://doi.org/10.1016/j.tust.2020.103610

Peerun MI, Ong DEL, Choo CS (2019a) Interpretation of geomaterial behavior during shearing aided by PIV technology. J Mater Civ Eng (ASCE) 31(9). https://doi.org/10.1061/(ASCE)MT.1943-5533.0002834

Peerun MI, Ong DEL, Desha C et al (2019c) Recent advancements in fundamental studies of particulate interaction and mechanical behaviour using 3-D printed synthetic particles. In: 1st Malaysian Geotechnical Society (MGS) and Geotechnical Society of Singapore (GeoSS) Conference 2019, Petaling Jaya, Malaysia, 24–26 June 2019

Peerun MI, Ong DEL, Desha C et al (2019b) Influences of geological characteristics on the construction of tunnels. World Engineers Convention 2019, 20–22 November 2019, Engineers Australia

Peerun MI, Ong DEL, Choo CS et al (2017a) Novel methods in estimating pipe-jacking forces in highly fractured rocks. Indian Society for Trenchless Technology, NoDIG India, vol XIII no. 1, pp 16–26

Peerun MI, Ong DEL, Choo CS (2017b) Development of laboratory based jacking mechanism considering soil-pipe interaction. In: Southeast Asian conference and exhibition in tunnelling and underground space 2017 (SEACETUS2017), Subang Jaya, Malaysia 18–19 Apr 2017

Peerun MI, Ong DEL, Choo CS (2016a) Effect of particle shapes on shear strength during direct shear testing using GeoPIV technology. In: The Institution of Engineers Malaysia, Proceedings of the 19th SEAGC and 2nd AGSSEAC, 31 May–3 June 2016, Subang Jaya, Malaysia

Peerun MI, Ong DEL, Choo CS (2016b) Calibration and parametric studies using GeoPIV technology to track particle movements in a transparent shear box. In: The Institution of Engineers Malaysia, Proceedings of the young geotechnical engineers conference 2016, 30 May 2016, Selangor, Malaysia

Peerun MI, Ong DEL, Choo CS (2016c) Behaviour of reconstituted sand-sized particles in direct shear tests using PIV technology. In: Japanese Geotechnical Society, Proceedings of the 15th Asian regional conference on soil mechanics and geotechnical engineering, Japanese Geotechnical Society, Tokyo, Japan, pp 354–359

Pellet-Beaucour AL, Kastner R (2002) Experimental and analytical study of friction forces during microtunneling operations. Tunn Undergr Space Tech 17:83–97. https://doi.org/10.1016/S0886-7798(01)00044-X

PJA (1995) Guide to best practice for the installation of pipe jacks and microtunnels. Pipe Jacking Association, London

Queensland Government (2021) Building sustainable roads. Department of Transport and Main Roads, accessed on 05/11/2021. https://www.tmr.qld.gov.au/Buildingsustainableroads

Schuh SA (2007) Estimating rippability of Brisbane Tuff using quantitative and qualitative characteristics and a modified approach to an existing rippability rating method. Faculty of Engineering and Surveying, University of Southern Queensland

Schwartzentruber LD (2015) LCA (Life Cycle Assessment) applied to the construction of tunnel, ITA WTC 2015 Congress and 41st General Assembly, Dubrovnik, Croatia

References

Shahri AA (2016) An optimized artificial neural network structure to predict clay sensitivity in a high landslide prone area using piezocone penetration test (CPTu) data: a case study in Southwest of Sweden. Geotech Geol Eng 34:745–758. https://doi.org/10.1007/s10706-016-9976-y

Sheil B (2021) Prediction of microtunnelling jacking forces using a probabilistic observational approach. Tunn Undergr Sp Technol 109:103749. https://doi.org/10.1016/j.tust.2020.103749

Sheil B, Suryasentana SK, Cheng WC (2020) Assessment of anomaly detection methods applied to microtunneling. J Geotech Geoenviron Eng 146(9):04020094

Shimizu M (1997) Strain fields in direct shear box tests on a metal-rods model of granular soils. In: Asaoka, Adachi, Oka. Deformation and Progressive Failure in Geomechanics, pp 151–156

Staheli K (2006) Jacking force prediction: an interface friction approach based on pipe surface roughness. PhD thesis, School of Civil and Environmental Engineering, Georgia Institute of Technology

Sterling RL (2020) Developments and research directions in pipe jacking and microtunneling, Undergr Space 5:1–19. https://doi.org/10.1016/j.undsp.2018.09.001

Sterling RL, Admiraal H, Bobylev N et al (2012) Sustainability issues for underground space in urban areas. Urban design and planning. In: Proceedings of the institution of civil engineers, special issue on urban development and sustainability

Tabesh A, Najafi M, Korky SJ et al (2016) Comparison of trenchless and open-cut methods for construction of an underground freight transportation (UFT) system. North American Society for Trenchless Technology (NASTT) NASTT's 2016 No-Dig Show, Dallas, Texas

Tighe S, Knight M, Papoutsis D et al (2002) User cost savings in eliminating pavement excavations through employing trenchless technologies. Can J Civ Eng 29:751–761. https://doi.org/10.1139/l02-071

Tugun Bypass Alliance (2004) Tugun bypass environmental impact statement. Technical paper number 4 geotechnical assessment, Queensland Department of Main Roads–South Coast Hinterland District

UN (2015) Transforming our world: the 2030 Agenda for Sustainable Development A/RES/70/1. Retrieved from: https://www.refworld.org/docid/57b6e3e44.html

Wang Y, Cao Z (2013) Probabilistic characterization of Young's modulus of soil using equivalent samples. Eng Geol 159:106–118. https://doi.org/10.1016/j.enggeo.2013.03.017

White DJ, Take WA (2002) GeoPIV: Particle Image Velocimetry (PIV) software for use in geotechnical testing. Eng. Dept. Cambridge University

White DJ, Take WA, Bolton MD (2003) Soil deformation measurement using particle image velocimetry (PIV) and photogrammetry. Géotechnique 53(7):619–631

Windsor and Districts Historical Society Inc (2020) Brisbane Tuff. https://windsorhistorical.org.au/brisbane-tuff/

Zhang W, Zhang R, Goh ATC (2018) Multivariate adaptive regression splines approach to estimate lateral wall deflection profiles caused by braced excavations in clays. Geotech Geol Eng 36:1349–1363. https://doi.org/10.1007/s10706-017-0397-3

Zhang W, Zhang R, Wang W et al (2019) A multivariate adaptive regression splines model for determining horizontal wall deflection envelope for braced excavations in clays. Tunn Undergr Space Technol 84:461–471. https://doi.org/10.1016/j.tust.2018.11.046

Zhou Y, Li S, Zhou C, Luo H (2019) Intelligent approach based on random forest for safety risk prediction of deep foundation pit in subway stations. J Comput Civ Eng 33:1–14. https://doi.org/10.1061/(ASCE)CP.1943-5487.0000796

Printed in Great Britain
by Amazon